General Theory of Relativity

General Theory of Relativity

S. P. Puri

Former Professor
Panjab University, Chandigarh

Pearson

Assistant Editor–Acquisitions: Jigyasa Bhatia
Editor–Production: Nikhil Rakshit

ISBN: 978-81-317-9568-2

First Impression

Published by Dorling Kindersley (India) Pvt. Ltd., lincencees of Pearson Education in South Asia.

Head Office: Pearson Education, 7th Floor, Knowledge Boulevard, A-8 (A), Sector-62, Noida, India
Registered Office: 11 Community Centre, Panchsheel Park, New Delhi, 110017, India

Compositor: PageTech Publishing Services
Printed in India at Repro India Limited

Contents

Preface

This book, 'General Theory of Relativity', is intended as a textbook for the Physics and Mathematics students at the B.Sc. (Hons.) and M.Sc. level of our national universities. The theory of relativity developed in two stages—Special Theory of Relativity and the General Theory of Relativity. Whereas the special theory dispensed with the concepts of absolute space and time of Newtonian mechanics and adapted the concept of inertial frame to the basic postulate of the constancy of the velocity of light; the general theory required that the laws of physics must be same for all observers, accelerated or not and thereby extended it to non-inertial frames. Alternatively stated, general theory of relativity is a relativistic theory of gravitation in which gravitation emerges as a property of the space-time structure, described by the metric tensor $g_{\mu\nu}$. It unifies special relativity, Newton's law of universal gravitation and the insight that gravitational acceleration can be described by the curvature of the space-time, and as such, it is the geometrical theory of gravitation. The curvature of space-time is caused by the mass-energy and momentum of the matter in the space-time.

At the time of Einstein's death in 1955, there was no evidence of general relativity in action. However, the scenario changed drastically in 1960, when new discoveries of neutron stars, black holes etc. were made by astronomers. Due to the massive nature of these objects, these put a severe dent in the shape of space-time.

A brief review of special relativity is provided to begin with, especialy of those concepts, required to understand the general theory.

A detailed study of Tensor Analysis is covered including parallel displacement and covariant differentiation, affine connection and its relation to metric tensor, Riemannian–Christoffel curvature tensor, Ricci tensor, Einstein tensor etc., as a foreground for the treatment of general relativity.

The modern versions of equivalence principle: the weak, Einsteinian and the strong and the covariance principle are introduced, before stating the principles of general relativity. Next, Einstein's field equations are derived from the general principles as well as the variational technique. Newton's law of gravitation is shown to follow as an approximation from the Geodesic equation and Poisson's equation as an approximation of the field equations, thus affirming our faith in the correctness of field equations.

The application of Einstein's law of gravitation to empty space (Schwarzschild solution) and how it leads to the existence of black holes, are discussed next. Alongwith the three classical tests, are discussed the modern tests of general relativity in great detail. Furthermore, the solution of Einstein's equations is sought for non-empty space and Schwarzschild's exterior and interior solutions are studied in considerable detail. Next, are treated the conservation laws in curved space and the pseudo-energy tensor for the gravitational field.

The gravitational waves are treated by linearizing the field equations through successive approximations. Furthermore, we consider the emission of gravitational radiation by a rotating source and the effect of radiation on a test particle. Beginning in 1974, Hulse, Taylor and others have studied the behaviour of binary pulsars and provided an indirect evidence for the existence of gravitational waves.

The subject of black holes is interesting, but complicated and has been introduced at an introductory level. The method of extracting energy from the black holes, *i.e.*, Penrose process, has been discussed alongwith the evidence for the existence of black holes.

Cosmology has been treated in some detail, starting with the Robertson–Walker line element and Friedmann equations alongwith their consequences. The standard model of the early universe, nucleosynthesis and CBR have been presented in support of Big Bang.

Lastly, astrophysics is presented, beginning with Tolman–Oppenheimer–Volkoff equation, polytropic stars, the Lane–Emden equation and Chandrashekhar mass limit. The theory of the formation of white dwarfs, neutron stars and black holes constitutes the finale of the treatment.

I express my indebtedness to the authors of standard works on relativity, which had influenced me greatly. The writing of this book leaves me obligated to my former students at the Universities of

Roorkee, Roorkee and Panjab, Chandigarh, whose criticism contributed to my own understanding and subsequent improvement in the presentation of the subject matter. Special mention must be made of my colleagues, Late Prof S.N. Gupta at Roorkee and Prof Vishwamittar at Chandigarh for their readiness to discuss some of the problems, which arose when the book was being written. I thankfully acknowledge their help in the venture.

Lastly, my special thanks are due to my wife Ramesh, whose deep interest and constant encouragement served as a beacon of hope throughout.

Any suggestions or comments to improve the subject matter will always be most welcome.

—S.P. Puri

Historical Perspective

The romance of creative achievement can be better expressed by reciting the story itself, as how it finally evolved historically. General theory of relativity is a relativistic theory of gravitation and for comprehending its background, we have to delve to find how it developed. The heliocentric model of the solar system introduced by Copernicus in the sixteenth century, was the starting point for considerations of gravitation. Basing on Kepler's laws of planetary motion and Galileo's observations on the falling bodies, Newton presented his theory of gravitation in the *Principia* in 1687. According to his law of gravitation, the force of attraction between two bodies of masses m_1 and m_2, separated by a distance r is given by

$$F = \frac{Gm_1m_2}{r^2}$$

where G is the constant of gravitation. This theory is invariant under Galilean transformation. The force of gravity is tacitly assumed to propagate with infinite velocity and consequently is an action at a distance theory. It can be visualized by imagining that all masses are removed from the universe except one mass under consideration. If another mass is introduced, each will influence the other instantly. Through the remarkable contributions of Euler, Lagrange, Hamilton and Jacobi, Newton's laws of motion were formulated into general laws through variational calculus and in the process invoking new analytic quantities, such as potential, related to force but remote from everyday experience. Poisson, introducing the concept of gravitational potential, deduced Poisson's equation, which unlike Newton's laws, could be solved under rather general conditions. Although Newton's theory was greatly successful in explaining the planetary motion, but for one lingering limitation as how each of the two bodies knew the other was present. The concept of the action at a distance as envisaged in the mechanistic theory of gravitation, was replaced by the concept of field in the formulation of electrodynamics by Maxwell, based on the work of Faraday, Maxwell and Hertz. Maxwell noted that there is a paradox caused by the attraction of like bodies. The energy of the medium must decrease by the presence of the bodies and he remarked,

> *As I am unable to understand in what way a medium can possess such properties, I cannot go further in the direction in searching for the cause of gravitation.*

In 1900, Lorentz suggested that gravitation may propagate with the velocity of light. Poincaré in his July 1905 paper, submitted days before Einstein's special relativity paper, stated that all forces should transform according to Lorentz transformations. He also noted that Newton's law is not valid and proposed gravitational waves travelling at the velocity of light.

In 1907, two years after submitting his paper on special relativity, while preparing a review of special relativity, Einstein suddenly realized how Newton's gravitation would have to be modified so as to be compatible with special relativity. A body in free fall experiences no gravitational field and behaves like an inertial system. Einstein described his thought as the *happiest thought of his life*. He then stated the equivalence principle as follows:

> *We shall therefore assume the complete physical equivalence of a gravitational field and the corresponding acceleration of the reference frame. This assumption extends the principle of relativity to the case of uniformly accelerated motion of the reference frame.*

Till 1911, Einstein published nothing further on gravitation. He realized that the bending of sunlight in the gravitational field, was a consequence of equivalence principle. In addition, he thought of the gravitational redshift of light, since light leaving a massive body, will get redshifted by the energy loss of escaping the gravitational field.

Einstein published further papers on gravitation in 1912. He concluded that Lorentz transformations will not hold in the general case. Further that the field equations for gravitation will be nonlinear and the equivalence principle will hold only locally. In addition, he realized that if all accelerated systems are equivalent, the Euclidean geometry cannot hold in all of them.

Einstein realized that the foundations of geometry have physical significance, in analogy with Gauss's theory of surfaces which he had studied as a student. His friend Grossman made Einstein aware of the contributions of

Riemann, Ricci and Levi–Civita in tensor calculus. Jointly they published a paper in 1913, using the tensor calculus of Ricci and Levi–Civita. The Riemann–Christoffel and Ricci tensors were the basic elements in the future development of general theory. The gravitation was described for the first time by the metric tensor. Planck paid a visit to Einstein in 1913 and when Einstein told Planck the state of theory of gravitation, Planck remarked:

As an older friend, I must advise you against it for in the first place you will not succeed, and even if you succeed, no one will believe you.

Planck was wrong but partly since when Einstein did succeed in formulating his theory of gravitation, but it was not readily accepted.

Einstein wrote a paper in October, 1914 and nearly half of it is on tensor analysis and differential geometry. Levi-Civita pointed out the technical error in Einstein's paper. Einstein was delighted in exchanging ideas with Levi-Civita whom he found more receptive to his ideas on gravitation.

Hilbert at Gottingin and Einstein had noticed the flaws in Einstein's October, 1914 paper and exchange of correspondence between them took place in November, 1915. Both of them came up with the final version of the gravitational field equations within a few days of each other.

Einstein, since 1911, had realized that observational evidences were important for the successful confirmation of his theory. The advance of the perihelion of planet Mercury was accurately known to be 43″ per century by 1882. On applying his theory, Einstein was able to account for the advance of the perihelion of planet Mercury by 43″ per century. This set aside the earlier ad hoc assumptions, such as Venus being 10% heavier, the existence of another planet in the Mercury orbit, or the Sun being more oblate than observed.

On re-analysis, Einstein discovered that the bending of light by a massive object was out by a factor of 2 in his 1911 work, and should be 1.74″. Two British expeditions confirmed Einstein's predictions by obtaining 1.98″ ± 0.30″ and 1.60″±0.30″. Einstein became a world celebrity overnight, and the news of the success of his theory of gravitation appeared in all the leading newspapers of the world. On being asked by his secretary, 'how he would have felt, if his theory was not confirmed by the observations of Sir Arthur Eddington,' he remarked:

I would have felt sorry for the dear Lord.

It was on 25 November, 1915 that Einstein submitted his paper giving the correct field equations. The calculations of the bending of light and the advance of Mercury's perihelion, were the same as calculated by him a week earlier. A paper was submitted by Hilbert five days before Einstein submitted his 25 Nov paper; also had the correct field equations for gravitation. Hilbert made use of variational calculus to gravitation and his paper contains some important contribution to relativity which are not found in Einstein's paper.

Carl Schwarzschild in 1916 found a mathematical solution to the field equations, which corresponds to the gravitational field of a massive compact object. Although at that time, it was purely a theoretical work but the present day work in astronomy such as neutron stars, pulsars and black holes, relies entirely on Schwarzschild's work.

It may be pointed out that Einstein arrived at the final version of general theory on a slow road to success and many errors on the way. In December 1915, he commented on himself:

That fellow Einstein suits his convenience. Every year, he retracts what he wrote the year before.

Today, the general theory has a central role in many areas, cosmology, the Big Bang theory etc. In addition, a program of precision tests was started in 1959 that the various predictions of general relativity, be tested to a higher degree of accuracy in the weak gravitational field limit, severely limiting possible deviations from the theory. Beginning in 1974, Hulse, Taylor and others have studied the behaviour of binary pulsars, experiencing much stronger fields than found in our solar system.

On the largest spatial scales, such as galactic and cosmological scales, general relativity has not been subjected to precision tests. Some have interpreted observations supporting the presence of dark matter and dark energy as a failure of general relativity at large distances, small accelerations or small curvatures. The very strong gravitational fields that must be present due to black holes, especially those super massive black holes which are thought to power quasars and less active galactic nuclei, belong to a field of intense active research.

1 A Brief Review of Special Relativity

1.1 INTRODUCTION

It is in terms of particles and fields that all natural phenomena are accounted for. The behaviour of bulk matter under terrestrial conditions, is governed by non-relativistic classical mechanics, which is an adequate formulation for a large range of phenomena in which the speed of light is taken as infinite and the value of Planck's constant is assumed to be zero. It is based on the assumption of instantaneous propagation of the interactions and embodies the concepts of absolute space and absolute time. However, when the velocities involved become comparable to the velocity of light, certain modifications in measures of length and time become essential, in order to explain the experimentally observed facts. Special relativity provides a formulation that is applicable at all speeds upto and including the speed of light. In addition, there is a conceptual jump between the relativistic and pre-relativistic mechanics, as the space and time intervals are no longer independent of the reference frames. These turn out to be greatly different in reference frames moving with large relative velocities with respect to each other. The concept of simultaneity is not an absolute one and has to be qualified with additional conditions. Some of these features regarding the kinematical and dynamical consequences of special relativity, will be investigated in detail. The special theory is the correct formulation since its predictions are borne out by experimental facts.

As the book is primarily about Einstein's general theory of relativity, the discussion of special relativity is only introductory in nature and is a brief overview of concepts that promote the understanding of general relativity.

1.2 SOME COMMON DEFINITIONS

As some terms, such as particle, reference frame, clock and event occur repeatedly, it is worthwhile to comment on these basic notions, at the very outset.

1.2.1 Particle

A particle is a system that is localized at a geometrical point. It is characterized by its mass (that determines its response to the impressed force) and its charge (determining its interaction with the electric charges in the universe). There are a host of elementary particles, such as electrons, protons, neutrons, and pions, which in addition to mass and charge, have additional attributes, such as spin (intrinsic angular momentum) and magnetic dipole moment. Some of these particles are unstable and have a half-life. During its lifetime, the unstable particle is a system specified by constant values of these attributes.

1.2.2 Rigid Body

A rigid body is defined as a body if the distance between any two particles of which remains unaltered under conditions of rest or motion. The distance can be measured by a measuring rod that itself is a rigid body. This implies that the definition of a rigid body is circular and one is to accept the rigid body as a basic concept. Its existence is essential to our idea of the reference frame, which is involved to measure the motion.

1.2.3 Reference Frame

The position of a body can be specified only with respect to another body. A reference frame is the space determined by a rigid body that is considered as a base. The extension of the rigid body is visualized as far as desired, by a lattice of measuring rods. A point in space is located by the three-coordinates (x, y, z) with respect to the origin of the reference system.

The state of rest of a body is specified with reference to frame when it occupies the same position, whereas that of state of motion implies a change in the position.

However, when considering relativistic motion, one has to include the timekeeping in addition to position-measuring in the reference system. The time-measuring is achieved by imagining identical clocks at the lattice points, distributed throughout the surrounding space.

1.2.4 Clock

An entity like a pendulum or an alternating electromagnetic field that repeats itself regularly, is taken as clock. In relativity, one has to synchronize the clocks according to a specified formula by a master clock and then distribute these clocks throughout the space. However, there is no such requirement of clock synchronization in Newtonian mechanics, since time is considered absolute and flowing uniformly for all bodies in the universe. The absolute nature of time in Newtonian mechanics is a direct consequence of the assumption of instantaneous action at a distance. When the signal velocity is finite, time is not absolute as in electrodynamics. Each reference frame has its own array of clocks in relativity.

1.2.5 Event

An event is any happening in space–time. It could be the collision of two particles, the emission of a flash of light or a particle just passing by. It is denoted by the space coordinates of the point of its location as well as the time of its occurrence. Thus, it is specified by four coordinates (x, y, z, t). Evidently, the transformations that relate an event as observed in two reference frames, involve time and space coordinates.

1.2.6 Inertial Reference Frames

Absolute space is an imagined framework in which bodies move and is always similar and immovable without any relation to anything external. Since experimentally, we can only observe relative motion, absolute space has no physical significance. One can locate one body with respect to another, say the earth in relation to the sun, the sun in relation to the stars, the stars with respect to the globular clusters and so on. As there is no absolute space, there is no absolute frame of reference.

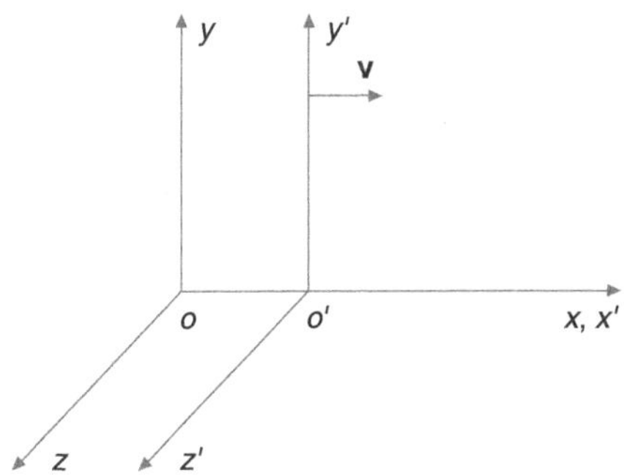

Figure 1.1 *The primed system is moving with velocity v relative to the unprimed system.*

Two frames of reference are inertial frames of reference with respect to each other when these are either at rest or in uniform motion with respect to each other. In these frames, Newton's first law of motion holds good.

Another criterion for defining inertial frames, is the one according to which, the equation of motion of a body has the simplest form, so that it is free from certain additive terms, which arise from the rotation of the frame or translational acceleration. These terms are called inertial forces. The fixed star frame is an example of an inertial frame.

A local inertial frame is defined as a reference frame in which a body, shielded from external influences, has zero acceleration. This can be achieved if the local inertial frame is in free fall in the existing gravitational fields, the gravitational field due to the material content of the frame, being itself zero. Furthermore, the frame has no spin with reference to the fixed stars. An orbiting satellite without spin and bereft of drag, will simulate the local inertial frame. Inside such a satellite, the physical laws will hold as if it is in a laboratory on earth.

1.2.7 Galilean Transformations

In special relativity, we study how a physical phenomenon appears to different observers that are in relative motion with constant velocity with respect to each other. Before the advent of special relativity, such an analysis was accomplished by *Galilean transformations*. These transformations connect the coordinates of the inertial frames that are in relative motion with uniform velocity, with respect to each other.

Consider two inertial frames S and S' and assume these are in *standard configuration*, Fig.1.1. This means that frame S' is moving in x-direction at constant velocity v relative to frame S. The axes y and z are parallel for both the frames. Let the origin of the primed axes be $(vt, 0, 0)$, which in the primed axes implies that the motion is in the OX or $O'X'$ direction and at $t = 0$, the origins O and O coincide.

It is intended to seek the relationship between the sets of coordinates (x, y, z, t) and (x', y', z', t'). All physicists before 1900 asserted that

$$\begin{aligned} x' &= x - vt \\ y' &= y \\ z' &= z \\ t' &= t \end{aligned} \tag{1.1}$$

These equations are known as Galilean transformation equations. The inverse Galilean transformations are obtained by changing primed into unprimed quantities and reversing the sign of relative velocity.

Thus,

$$\begin{aligned} x &= x' + vt \\ y &= y' \\ z &= z' \\ t &= t' \end{aligned} \tag{1.2}$$

The equation of motion of a particle of mass m subject to the force $\mathbf{F} = (F_x, F_y, F_z)$ are given by Newton's law as

$$\begin{aligned} m\frac{d^2x}{dt^2} &= F_x \\ m\frac{d^2y}{dt^2} &= F_y \\ m\frac{d^2z}{dt^2} &= F_z \end{aligned} \tag{1.3}$$

On transforming these equations under the Eqs. (1.1), we get

$$\begin{aligned} m\frac{d^2x'}{dt^2} &= F_x \\ m\frac{d^2y'}{dt^2} &= F_y \\ m\frac{d^2z'}{dt^2} &= F_z \end{aligned} \tag{1.4}$$

It is obvious that the equations of motion are invariant (preserve their form) in going from one system to another that is moving with constant velocity with respect to the former. In this manner, one can introduce arbitrarily many inertial frames of reference, moving uniformly with respect to one another. However, no one of these inertial frames is unique or pre-eminent. Neumann called this set of inertial frames, 'The Body Alpha'. Thus, the concept of absolute space has been replaced by 'The Body Alpha', but that of time is left intact.

Furthermore, the transformations (1.1), predict that the velocity of light should be different in the frames S and S' or in other words, the Newtonian law of addition of velocities, predicts that the velocity of light is different in different directions and depends on the relative uniform motion of the observer, the medium and the source.

Imagine a source of light at the origin O of the unprimed system and c, the velocity of a point on the wave surface emitted by the source, then

$$\dot{\mathbf{r}} = c\mathbf{n} \tag{1.5}$$

where $\mathbf{n}$ is a unit vector along r.

Transforming Eq. (1.5), to the primed system S', we get

$$\dot{\mathbf{r}}' = c\mathbf{n} - \mathbf{v} \tag{1.6}$$

Thus, the velocity in the primed system is no longer c and as it depends on direction, the waves are not spherical.

Classical mechanics tacitly assumes the action at a distance principle, according to which the interactions propagate with instantaneous or infinite velocity. The forces exerted on a particle by others at a particular time depend entirely on the positions of the particles at this instant. But the velocities with which we have to deal are usually so small that the velocity of interaction or *signal velocity* when taken infinite, does not affect the accuracy of the results.

1.3 THE CONSTANCY OF VELOCITY OF LIGHT

The origin of the special theory of relativity lies in a set of paradoxes arising out of the study of electromagnetic phenomena. James Clerk Maxwell, a Scottish physicist, published a set of equations, called Maxwell's equations. These equations summarize in differential form, the following basic laws of electromagnetism (in Gaussian units).

Faraday's law
$$\nabla \times \mathbf{E} = -\frac{1}{c}\frac{\partial \mathbf{B}}{\partial t} \tag{1.7a}$$

Ampere's law for time-dependent fields
$$\nabla \times \mathbf{H} = \frac{1}{c}\left[\frac{\partial \mathbf{D}}{\partial t} + 4\pi \mathbf{J}\right] \tag{1.7b}$$

Absence of free magnetic poles
$$\nabla \cdot \mathbf{B} = 0 \tag{1.7c}$$

Guass's law
$$\nabla \cdot \mathbf{E} = 4\pi\rho \tag{1.7d}$$

where $\mathbf{E}, \mathbf{H}$ are electric and magnetic field intensities, $\mathbf{D}, \mathbf{B}$ the electric displacement and magnetic induction vectors; ρ, the charge density and $\mathbf{J}$, the vector current density. In free space, we have

$$\mathbf{D} = \mathbf{E}$$
$$\mathbf{H} = \mathbf{B}$$

and in the absence of charges and currents, Eqs. (1.7a)–(1.7d), become

$$\nabla \times \mathbf{E} = -\frac{1}{c}\frac{\partial \mathbf{B}}{\partial t} \tag{1.8a}$$
$$\nabla \times \mathbf{B} = \frac{1}{c}\frac{\partial \mathbf{E}}{\partial t} \tag{1.8b}$$
$$\nabla \cdot \mathbf{B} = 0 \tag{1.8c}$$
$$\nabla \cdot \mathbf{E} = 0 \tag{1.8d}$$

Now
$$\begin{aligned}\nabla \times \nabla \times \mathbf{B} &= \nabla(\nabla \cdot \mathbf{B}) - \nabla^2 \mathbf{B} \\ &= \frac{1}{c}\frac{\partial}{\partial t}(\nabla \times \mathbf{E}) \\ &= -\frac{1}{c^2}\frac{\partial^2 \mathbf{B}}{\partial t^2}\end{aligned} \tag{1.9}$$

so that the vector **B** satisfies the wave equation

$$\nabla^2 \mathbf{B} - \frac{1}{c^2}\frac{\partial^2 \mathbf{B}}{\partial t^2} = 0 \tag{1.10}$$

Analogously, we obtain for **E**

$$\nabla^2 \mathbf{E} - \frac{1}{c^2}\frac{\partial^2 \mathbf{E}}{\partial t^2} = 0 \tag{1.11}$$

It follows that the velocity of electromagnetic waves or any disturbance in the electromagnetic fields, will propagate with velocity c, which is the ratio of electromagnetic to the electrostatic unit of electrical charge. The experimental value of c is 2.99792×10^8 m/s, which agrees with the velocity of light in vacuo. The agreement establishes that light is electromagnetic radiation.

The above analysis leads to the conclusion that the velocity of light is always the same, irrespective of the motion of the observer. The result was accepted by Einstein at its face value and made one of the postulates of the special relativity.

The negative results of Michelson–Morley type of experiments along with those about expected effects of relative velocity between two inertial frames of reference on wave propagation, led to the inference that Maxwell's equations hold in all frames of reference with the universal value of the velocity of light. Poincaré, a famous French mathematician, argued that the motion referred to the universal frame of reference cannot be observed and stated the principle of relativity, '*The laws of physics are the same for two observers in relative motion with constant velocity*'.

The combination of Poincaré principle of relativity with the finiteness of the velocity of interaction or signal velocity, was advanced by Einstein in 1905 as Einstein principle of relativity.

1.4 THE POSTULATES OF SPECIAL RELATIVITY

The theory of special relativity is based on the following two postulates:

1. It is impossible to trace any distinction between any two inertial frames that are in uniform relative motion, to each other, by any physical measurement. Alternately, all the laws of physics are the same in all inertial frames, so that no inertial frame is pre-eminent or unique and all inertial frames are equivalent.
2. The speed of light in vacuum is invariant, *i.e.*, light always propagates in vacuum with a definite velocity c, which is independent of motion of the observer.

1.5 INTERVAL

An event is any happening in space–time, such as a collision between two particles, emission of a flash of light or a particle just passing by. It is characterized by the spatial location where it occurs (*i.e.*, the coordinates x, y, z) and the time of its occurrence. In an imaginary four-dimensional space called *Minkowski space* or *Minkowski world*, an event is represented by a point called a world point, the three space coordinates x, y, z are laid off along the three axes and the time t along the fourth imaginary axis. With time, a world point moves in the four-dimensional space along a straight line called a world line.

Consider two events represented by world points (x_1, y_1, z_1, t_1) and (x_2, y_2, z_2, t_2) like the emission of a laser pulse at t_1 and its arrival at a point at t_2. The space and time coordinates of these events are related according to the relation

$$c^2(t_2 - t_1)^2 - (x_2 - x_1)^2 - (y_2 - y_1)^2 - (z_2 - z_1)^2 = 0 \tag{1.11}$$

The interval between two points in the four-dimensional space, out of analogy with distance (interval) between the two points in the three-dimensional space is defined by

$$s_{12}^2 = c^2(t_2 - t_1)^2 - (x_2 - x_1)^2 - (y_2 - y_1)^2 - (z_2 - z_1)^2 \tag{1.12}$$

The infinitesimal interval between two infinitely close events, is defined by ds and its square is given by

$$ds^2 = c^2dt^2 - dx^2 - dy^2 - dz^2 \tag{1.13}$$

The interval between two events like the emission of a light signal and its arrival at another point is zero, *i.e.*,

$$\Delta s^2 = c^2t^2 - \Delta l^2 \tag{1.14}$$

Equation (1.11) is valid in any inertial frame, since the velocity of light is constant. If it is zero in one inertial frame *S*, it will be zero in another inertial frame S', that is moving with uniform velocity w.r.t. *S*. Thus, the interval Δs between two events in *S*, must be related to the interval between the same events in S' by the relation

$$\Delta s = a\Delta s' \tag{1.15}$$

According to the principle of equivalence of all inertial frames, we get

$$\Delta s' = a\Delta s \tag{1.16}$$

Substituting Eq. (1.16) in Eq. (1.15), one gets

$$\begin{aligned} \Delta s &= a^2 \Delta s \\ \Rightarrow a^2 &= 1 \\ \Rightarrow a &= \pm 1 \end{aligned} \tag{1.17}$$

But the sign of interval between two events must be same, and therefore, choosing the positive sign, we get

$$\Delta s = \Delta s' \tag{1.18}$$

Thus, the interval between two events is invariant and the invariance is a consequence of the principle of equivalence of all inertial frames.

Therefore,

$$\begin{aligned} \Delta s^2 &= c^2\Delta t^2 - \Delta l^2 \\ &= c^2\Delta t'^2 - \Delta l'^2 \end{aligned} \tag{1.19}$$

Assuming that $\Delta s^2 > 0$, *i.e.*, the interval is real, one can find an inertial frame S' in which $\Delta l'$ will equal zero. Both the events will occur at the same point and the time interval between them in S' will be

$$\Delta t' = \frac{\Delta s}{c} \tag{1.20}$$

Such events are time-like.

Further, assuming that, $\Delta s^2 < 0$, *i.e.*, the interval is imaginary, one can find a reference frame S' in which, $\Delta t' = 0$, *i.e.*, the events occur simultaneously. The interval between them in S' is

$$\Delta l' = i\Delta s \tag{1.21}$$

The imaginary events are called space-like.

The events occurring with the same particle, can be separated only by a time-like interval and this can be easily seen as follows. Any material particle cannot travel at a velocity greater than the velocity of light c, so the distance Δl travelled by the particle in time Δt will be less than, $c\Delta t$, *i.e.*, $\Delta l \le c\Delta t$, implying that $\Delta s^2 > 0$.

A space-like interval separates events, which are acausal (have no causal relation), since if $\Delta s^2 < 0$, then $\Delta l > c\Delta t$. Thus, no signal from one point will be able to influence the other point and there is no causal relationship with each other.

Consider a particle moving with a uniform velocity v with respect to frame S and let the time interval between two events that occur with this particle be dt. We choose another frame S' with respect to which the particle is at rest and if the time interval between the events is dt', it is given by

$$dt' - \frac{ds}{c} \tag{1.22}$$

Proper time is defined as the time measured by a clock moving with the particle, so

$$d\tau' = \frac{ds}{c} \tag{1.23}$$

As ds is invariant and c is a universal constant, $d\tau'$ is also invariant. Expressing ds in space and time coordinates, we get

$$\begin{aligned} d\tau' &= \frac{ds}{c} = \frac{\sqrt{c^2dt^2 - dl^2}}{c} \\ &= dt\sqrt{1 - v^2/c^2} \end{aligned} \tag{1.24}$$

As is evident from Eq. (1.24), the proper time of a particle is always less than the lab time.

1.6 LORENTZ TRANSFORMATION

With a view to putting time and space on different footings, we choose to work in pseudo-Euclidean space, where a world point has coordinates x^0, x^1, x^2, x^3 defined as

$$x^0 = ct, \quad x^1 = x, \quad x^2 = y, \quad x^3 = z \tag{1.25}$$

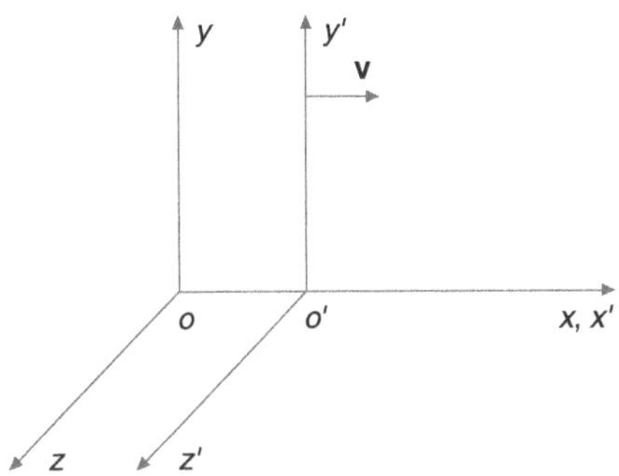

Figure 1.2 *Inertial frame S′ is moving with uniform velocity v, with respect to S and the axes OX and O′X′ coinciding at t = t′ = 0.*

The square of the amplitude of the position vector is

$$(x^0)^2 - (x^1)^2 - (x^2)^2 - (x^3)^2 \tag{1.26}$$

and the metric of the pseudo-Euclidean space is given by the tensor

$$(g^{\mu\nu}) = (g_{\mu\nu}) = \begin{pmatrix} 1 & 0 & 0 & 0 \\ 0 & -1 & 0 & 0 \\ 0 & 0 & -1 & 0 \\ 0 & 0 & 0 & -1 \end{pmatrix} \tag{1.27}$$

The square of the four-position vector is

$$\sum_{\mu=0}^{3} x^\mu x_\mu = x^0 x_0 + x^1 x_1 + x^2 x_2 + x^3 x_3 \tag{1.28}$$

The invariance of the interval in four-space, implies that the Lorentz transformations are orthogonal transformations in Minkowski space. Hence, the transformation of the components of a four-position vector is given by the relation

$$x'^\mu = \sum_{\nu=0}^{3} \alpha_\nu^\mu x^\nu \tag{1.29}$$

Consider two inertial frames S and S' in relative motion, with the axes OX and $O'X'$ parallel along the direction of relative velocity **v**, the origins O and O' coinciding at $t = t' = 0$, Fig 1.2.

The inverse transformation is given as

$$x^\mu = \sum_{\nu=0}^{3} \overline{\alpha_\nu^\mu} x'^\nu \tag{1.30}$$

where $\overline{\alpha_\nu^\mu}$ are the coefficients of the inverse transformation.

The invariance of the square of the four-position vector, requires that

$$\sum_{\mu,\nu=0}^{3} g_{\mu\nu} x^\mu x^\nu = \sum_{\mu,\nu=0}^{3} g_{\mu\nu} x'^\mu x'^\nu \tag{1.31}$$

Substituting the values of x'^{μ} from Eq. (1.29) into Eq. (1.31), one gets

$$\sum_{\mu,\nu=0}^{3} g_{\mu\nu} x^{\mu} x^{\nu} = \sum_{\mu,\nu=0}^{3} g_{\mu\nu} \sum_{\rho=0}^{3} \alpha^{\mu}_{\rho} x^{\rho} \sum_{\sigma=0}^{3} \alpha^{\nu}_{\sigma} x^{\sigma} \tag{1.32}$$

Exchanging on the r.h.s. the dummy indices μ and ρ, also the indices ν and σ, we get

$$\sum_{\mu,\nu=0}^{3} g_{\mu\nu} x^{\mu} x^{\nu} = \sum_{\mu,\nu=0}^{3} x^{\mu} x^{\nu} \sum_{\rho,\nu=0}^{3} g_{\rho\sigma} \alpha^{\rho}_{\mu} \alpha^{\sigma}_{\nu} \tag{1.33}$$

The condition of orthogonality, to be satisfied by the coefficients of Eq. (1.33), is

$$\sum_{\rho,\nu=0}^{3} g_{\rho\sigma} \alpha^{\rho}_{\mu} \alpha^{\sigma}_{\nu} = g_{\mu\nu} \tag{1.34}$$

As $g_{\rho\sigma}$ is a diagonal matrix, it will differ from zero only when $\rho = \sigma$, the Eq. (1.34) in the simplified form, becomes

$$\sum_{\rho,\nu=0}^{3} g_{\rho\rho} \alpha^{\rho}_{\mu} \alpha^{\rho}_{\nu} = g_{\mu\nu} \qquad (\mu,\nu = 0,1,2,3) \tag{1.35}$$

For $\mu = \nu = 0$, the above equation becomes

$$\left(\alpha^{0}_{0}\right)^{2} - \left(\alpha^{1}_{0}\right)^{2} - \left(\alpha^{2}_{0}\right)^{2} - \left(\alpha^{3}_{0}\right)^{2} = 1 \tag{1.36}$$

For $\mu = 1$ and $\nu = 2$, Eq. (1.35) becomes

$$\alpha^{0}_{1}\alpha^{0}_{2} - \alpha^{1}_{1}\alpha^{1}_{2} - \alpha^{2}_{1}\alpha^{2}_{2} - \alpha^{3}_{1}\alpha^{3}_{2} = 0 \tag{1.37}$$

Similarly, one can establish relationships between the coefficients of the inverse transformation. Next we find the relation between the coefficients of the direct and inverse transformations. Applying the transformations (1.29) and (1.30) consecutively, one obtains

$$x^{\mu} = \sum_{\rho=0}^{3} \overline{\alpha^{\mu}_{\rho}} x'^{\rho} = \sum_{\rho=0}^{3} \overline{\alpha^{\mu}_{\rho}} \sum_{\nu=0}^{3} \alpha^{\rho}_{\nu} x^{\nu} = \sum_{\nu=0}^{3} x^{\nu} \sum_{\rho=0}^{3} \overline{\alpha^{\mu}_{\rho}} \alpha^{\rho}_{\nu}$$

Writing x^{μ} as $\sum_{\nu} \delta^{\mu}_{\nu} x^{\nu}$, we get

$$\sum_{\rho=0}^{3} \overline{\alpha^{\mu}_{\rho}} \alpha^{\rho}_{\nu} = \delta^{\mu}_{\nu} \tag{1.38}$$

which leads to

$$\overline{\alpha^{0}_{0}} = \alpha^{0}_{0}, \quad \overline{\alpha^{0}_{i}} = -\alpha^{i}_{0}, \quad \overline{\alpha^{i}_{0}} = -\alpha^{0}_{i}, \quad \overline{\alpha^{i}_{k}} = \alpha^{k}_{i} \quad (i,k = 1,2,3) \tag{1.39}$$

Axes OY, OZ being parallel to axes $O'Y'$, $O'Z'$, $x^2 = y$ and $x^3 = z$, remain unchanged. As a result, the matrix of transformation, may be written as

$$\left(\alpha_\nu^\mu\right) = \begin{pmatrix} \alpha_0^0 & \alpha_1^0 & 0 & 0 \\ \alpha_0^1 & \alpha_1^1 & 0 & 0 \\ 0 & 0 & 1 & 0 \\ 0 & 0 & 0 & 1 \end{pmatrix} \tag{1.40}$$

In the limiting process, when the relative velocity $v \to 0$, the systems S and S' coincide and the transformation matrix reduces to the identity transformation

$$\left(\alpha_\nu^\mu\right)_{v\to 0} = \begin{pmatrix} 1 & 1 & 0 & 0 \\ 1 & 1 & 0 & 0 \\ 0 & 0 & 1 & 0 \\ 0 & 0 & 0 & 1 \end{pmatrix} \tag{1.41}$$

Therefore,

$$\begin{aligned} \underset{v\to 0}{Lt}\, \alpha_0^0 &= \underset{v\to 0}{Lt}\, \alpha_1^1 = 1 \\ \underset{v\to 0}{Lt}\, \alpha_1^0 &= \underset{v\to 0}{Lt}\, \alpha_0^1 = 0 \end{aligned} \tag{1.42}$$

Next, we show that the coefficients α_ν^μ differing from 0 and 1 are not independent. Eq. (1.36) for $\mu = \nu = 0$, in the light of Eq. (1.42), is

$$\left(\alpha_0^0\right)^2 - \left(\alpha_0^1\right)^2 = 1 \tag{1.43}$$

A similar relation is obtained for the coefficients of inverse transformation

$$\left(\overline{\alpha_0^0}\right)^2 - \left(\overline{\alpha_0^1}\right)^2 = 1 \tag{1.44}$$

which in the light of Eq. (1.39) becomes

$$\left(\alpha_0^0\right)^2 - \left(\alpha_1^0\right)^2 = 1 \tag{1.45}$$

The condition of orthogonality, Eq. (1.35) for $\mu = 0$, $\nu = 1$, becomes

$$\alpha_0^0 \alpha_1^0 - \alpha_0^1 \alpha_1^1 = 0 \tag{1.46}$$

Comparing Eqs. (1.43) and (1.45), we get

$$\left(\alpha_0^1\right)^2 = \left(\alpha_1^0\right)^2$$
$$\Rightarrow \alpha_0^1 = \pm\alpha_1^0 \tag{1.47}$$

Comparing Eqs. (1.47) and (1.46), yields the following alternatives

$$\alpha_0^0 = \alpha_1^1 \quad \text{if} \quad \alpha_0^1 = \alpha_1^0 \tag{1.48}$$

or

$$\alpha_0^0 = -\alpha_1^1 \quad \text{if} \quad \alpha_0^1 = -\alpha_1^0 \tag{1.49}$$

It is the first alternative that is compatible with Eq. (1.41). Thus, we assume that

$$\alpha_0^0 = \alpha_1^1 \quad \text{and} \quad \alpha_1^0 = \alpha_0^1 \tag{1.50}$$

Therefore, the transformation matrix becomes

$$\left(\alpha_\nu^\mu\right) = \begin{pmatrix} \alpha_0^0 & \alpha_1^0 & 0 & 0 \\ \alpha_1^0 & \alpha_0^0 & 0 & 0 \\ 0 & 0 & 1 & 0 \\ 0 & 0 & 0 & 1 \end{pmatrix} \tag{1.51}$$

It may be remarked that the coefficients α_0^0 and α_1^0 will depend only on the relative velocity **v**.

From Eq. (1.29), we get for x'^1

$$x'^1 = \alpha_1^0 x^0 + \alpha_0^0 x^1$$

Expressing it in terms of *x* and *t*, one gets

$$x' = \alpha_1^0 ct + \alpha_0^0 x \tag{1.52}$$

Next writing the expression for the origin of the primed system O', for which $x' = 0$ and $x = vt$ and so Eq. (1.52) yields

$$0 = \alpha_1^0 ct + \alpha_0^0 vt$$
$$\Rightarrow \alpha_1^0 = -\alpha_0^0 v/c \tag{1.53}$$

The Eq. (1.45) in view of Eq. (1.53) becomes

$$\left(\alpha_0^0\right)^2 - \left(\alpha_0^0\right)^2 (v/c)^2 = 1$$

$$\Rightarrow \alpha_0^0 = \frac{1}{\sqrt{1 - v^2/c^2}} \tag{1.54}$$

Thus,

$$\alpha_1^0 = -\frac{v/c}{\sqrt{1 - v^2/c^2}} \tag{1.55}$$

Finally, the transformation matrix for special Lorentz transformation, Eq. (1.46) is

$$\left(\alpha^{\mu}_{\nu}\right)=\begin{pmatrix} \dfrac{1}{\sqrt{1-v^2/c^2}} & -\dfrac{v/c}{\sqrt{1-v^2/c^2}} & 0 & 0 \\ -\dfrac{v/c}{\sqrt{1-v^2/c^2}} & \dfrac{1}{\sqrt{1-v^2/c^2}} & 0 & 0 \\ 0 & 0 & 1 & 0 \\ 0 & 0 & 0 & 1 \end{pmatrix} \tag{1.56}$$

The components of a position four-vector transform as

$$\begin{aligned} x'^0 &= \frac{x^0-\beta x^1}{\sqrt{1-\beta^2}} \\ x'^1 &= \frac{-\beta x^0+x^1}{\sqrt{1-\beta^2}} \\ x'^2 &= x^2 \\ x'^3 &= x^3 \end{aligned} \tag{1.57}$$

These in terms of x, y, z, t become

$$\begin{aligned} t' &= \frac{t-vx/c^2}{\sqrt{1-\beta^2}} \\ x' &= \frac{x-vt}{\sqrt{1-\beta^2}} \\ y' &= y \\ z' &= z \end{aligned} \tag{1.58}$$

The inverse transformations are obtained by replacing $\beta \to -\beta$ and changing primed into unprimed quantities and vice versa.

Thus,

$$\begin{aligned} x^0 &= \frac{x'^0+\beta x'^1}{\sqrt{1-\beta^2}} \\ x^1 &= \frac{\beta x'^0+x'^1}{\sqrt{1-\beta^2}} \\ x^2 &= x'^2 \\ x^3 &= x'^3 \end{aligned} \tag{1.59}$$

In terms of x, y, z, t the inverse transformations become

$$\begin{aligned} t &= \frac{t' + vx'/c^2}{\sqrt{1-\beta^2}} \\ x &= \frac{x' + vt'}{\sqrt{1-\beta^2}} \\ y &= y' \\ z &= z' \end{aligned} \tag{1.60}$$

The two sets of (x, y, z, t) and (x', y', z', t') coordinates that are related to each other by Eqs. (1.58) and (1.60) are referred to as Lorentz frames of reference.

Example 1.1

Assuming the invariance of the equations of wave-fronts in two inertial systems, that are moving with uniform velocity relative to each other, *i.e.*,

$$c^2t^2 - x^2 - y^2 - z^2 = c^2t'^2 - x'^2 - y'^2 - z'^2$$

derive Lorentz transformation.

Solution

Since the motion is in the x-direction, we assume that

$$y' = y, \quad z' = z$$

hold and so we have

$$c^2t^2 - x^2 = c^2t'^2 - x'^2 \tag{i}$$

The identical transformations $x' = x$ and $t' = t$ are trivial, implying that x, t have to transform separately. Since, at any instant, $x' = 0$ corresponds to $x = vt$, we write $x' = \gamma(x - vt)$ and the inverse transformation

$$x = \gamma'(x' + vt')$$

$$\Rightarrow \frac{x}{\gamma'} = \gamma x - v\gamma t + vt'$$

which gives

$$t' = \gamma\left[t - \frac{x}{v}\left(1 - \frac{1}{\gamma\gamma'}\right)\right] \tag{ii}$$

Substituting from Eq. (ii) into (i), one gets

$$\begin{aligned} c^2t'^2 - x'^2 &= \gamma^2t^2(c^2 - v^2) + 2\gamma^2 xt\left[v - \frac{c^2}{v}\left(1 - \frac{1}{\gamma\gamma'}\right)\right] \\ &\quad - \gamma^2x^2\left[1 - \frac{c^2}{v^2}\left(1 - \frac{1}{\gamma\gamma'}\right)^2\right] \\ &= c^2t^2 - x^2 \end{aligned}$$

which will be satisfied provided $\gamma = \pm\frac{1}{\sqrt{1-v^2/c^2}}$ and $\gamma' = \pm\frac{1}{\sqrt{1-v^2/c^2}}$.

Since $x' = x$ for $v = 0$, we take the positive sign of the square root. Thus, the transformations are

$$\begin{aligned} x' &= \gamma(x - vt) \\ y' &= y \\ z' &= z \\ t' &= \gamma(1 - vx/c^2) \end{aligned}$$

where $\gamma = (1 - v^2/c^2)^{-1/2}$.

Example 1.2

Show that D'Alembertian operator

$$\Box^2 \equiv \frac{\partial^2}{\partial x^2} + \frac{\partial^2}{\partial y^2} + \frac{\partial^2}{\partial z^2} - \frac{1}{c^2}\frac{\partial^2}{\partial t^2}$$

is invariant under Lorentz transformation.

Solution

The Lorentz transformations are

$$\begin{aligned} x' &= \gamma(x - vt) \\ y' &= y \\ z' &= z \\ t' &= \gamma(t - vx/c^2) \end{aligned} \tag{i}$$

From the set of Eq. (i), we get

$$\frac{\partial}{\partial x'} = \gamma\left(\frac{\partial}{\partial x} + \frac{v}{c^2}\frac{\partial}{\partial t}\right)$$
$$\frac{\partial}{\partial y'} = \frac{\partial}{\partial y}$$
$$\frac{\partial}{\partial z'} = \frac{\partial}{\partial z}$$
$$\frac{\partial}{\partial t'} = \gamma\left(v\frac{\partial}{\partial x} + \frac{\partial}{\partial t}\right)$$

Therefore,

$$\frac{\partial^2}{\partial x'^2} = \gamma^2\left(\frac{\partial^2}{\partial x^2} + \frac{2v}{c^2}\frac{\partial^2}{\partial x \partial t} + \frac{v^2}{c^4}\frac{\partial^2}{\partial t^2}\right)$$
$$\frac{\partial^2}{\partial y'^2} = \frac{\partial^2}{\partial y^2}, \quad \frac{\partial^2}{\partial z'^2} = \frac{\partial^2}{\partial z^2}$$
$$\frac{\partial^2}{\partial t'^2} = \gamma^2\left(v^2\frac{\partial^2}{\partial x^2} + 2v\frac{\partial^2}{\partial x \partial t} + \frac{\partial^2}{\partial t^2}\right)$$

so

$$\frac{\partial^2}{\partial x'^2} + \frac{\partial^2}{\partial y'^2} + \frac{\partial^2}{\partial z'^2} - \frac{1}{c^2}\frac{\partial^2}{\partial t'^2} = \frac{\partial^2}{\partial x^2} + \frac{\partial^2}{\partial y^2} + \frac{\partial^2}{\partial z^2} - \frac{1}{c^2}\frac{\partial^2}{\partial t^2}$$

1.7 KINEMATIC CONSEQUENCES OF LORENTZ TRANSFORMATION

1.7.1 Lorentz-Fitzgerald Length Contraction

That is a rod at rest in the system *S*, placed parallel to the *x*-axis and let its length as measured by an observer in *S* be $l = x_2 - x_1$, where x_1 and x_2 are the coordinates of the two ends of the rod in the system. Another observer in the system S' measures it at any instant, say t', what length of the rod will he measure. From Eq. (1.58), we get

$$x_1 = \frac{x_1' + vt'}{\sqrt{1 - v^2/c^2}} \tag{1.61}$$

$$x_2 = \frac{x_2' + vt'}{\sqrt{1 - v^2/c^2}} \tag{1.62}$$

Calling it $l' = x_2' - x_1'$, the length in the system S', one gets

$$l = l'(1 - v^2/c^2)^{-1/2} \tag{1.63}$$

$$\Rightarrow l' = l(1 - v^2/c^2)^{1/2} \tag{1.64}$$

The proper length of a rod is its length in a reference frame in which it is at rest. Thus, the proper length of a rod is the maximum and to any other observer, which is moving with velocity v, the rod appears to be contracted in the ratio $\sqrt{1 - v^2/c^2} : 1$. Furthermore, this effect is reciprocal.

1.7.2 Time Dilation or Apparent Slowing of Moving Clocks

Suppose an observer from system S' sends a light signal from the point $(x', 0, 0)$ at t_1' and a subsequent signal at time t_2'. How the interval $(t_2' - t_1')$ will appear to the observer in system S. From Eq. (1.58), the corresponding times are

$$\begin{aligned} t_1 &= \gamma(t_1' + vx'/c^2) \\ t_2 &= \gamma(t_2' + vx'/c^2) \\ \Rightarrow (t_2 - t_1) &= \gamma(t_2' - t_1') \end{aligned} \tag{1.65}$$

which indicates that the interval is increased in the ratio $1 : \sqrt{1 - v^2/c^2}$.

Calling, $(t_2 - t_1) = \tau,\ (t_2' - t_1') = \tau'$, one gets from Eq. (1.65)

$$\tau = \frac{\tau'}{\sqrt{1 - v^2/c^2}} \tag{1.66}$$

The time read by a clock moving with a given object is called the proper time for the object. It is obvious from Eq. (1.66) that the proper time of a moving object is always less than the corresponding interval in the rest frame. Thus, a clock will be found to run more and more slowly, the greater the relative motion between the clock and the observer. This kinematical effect of relativity is called *Time Dilation*.

Example 1.3

Show that the four-dimensional volume elementis $dx\,dy\,dz\,dt$ invariant under Lorentz transformation.

Solution

$$\begin{aligned} dx &= dx'(1 - v^2/c^2)^{-1/2} \\ dt &= dt'(1 - v^2/c^2)^{1/2} \end{aligned}$$

Now $$dx\,dy\,dz\,dt = J dx' dy' dz' dt'$$

where J is the Jacobian of the transformation and is given by

$$J = \frac{\partial(x, y, z, t)}{\partial(x', y', z', t')} = \begin{pmatrix} \frac{\partial x}{\partial x'} & \frac{\partial y}{\partial y'} & \frac{\partial z}{\partial z'} & \frac{\partial t}{\partial t'} \\ \cdots & \cdots & \cdots & \cdots \\ \cdots & \cdots & \cdots & \cdots \\ \frac{\partial x}{\partial t'} & \cdots & \cdots & \frac{\partial t}{\partial t'} \end{pmatrix}$$

It can be shown that $J = 1$, hence, the result.

1.8 LIGHT CONE

We study the concept of light cone and interval between two events in an imaginary four-dimensional space, called *Minkowski space*. The infinitesimal interval between two infinitely close events is defined by ds, whose square is

$$ds^2 = c^2dt^2 - dx^2 - dy^2 - dz^2 \tag{1.67}$$

As shown earlier in Eq. (1.18), the interval between two events is invariant and this invariance is an outcome of the principle of equivalence of all inertial frames. Since the invariant metric of Minkowski space is indefinite, *i.e.*, the quadratic form $c^2dt^2 - \sum_i (dx_i)^2$, may take either sign. Due to this indefiniteness, the properties of Minkowski space are different from those of Euclidean space. As human mind cannot visualize four-dimensions and these cannot be drawn on paper, we take up the simplest of all cases and take only two-dimensions, ct and x.

Let us consider an event whose world point is O, the origin of ct, x coordinates, Fig.1.3. We will investigate the relations other events bear to the given event O. Since the velocity of light is an upper limit on all velocities, the space–time is divisible into three regions. These are as follows:

(i) Light-like or $ds^2 = 0$.
(ii) Time-like or $ds^2 > 0$.
(iii) Space-like or $ds^2 < 0$.

Let us consider these regions one by one.

(i) **Light-like or $ds^2 = 0$:** The dotted lines $x = ct$ and $x = -ct$, represent respectively the propagation of light signals in opposite directions through the origin O, and these lines make an angle 45° of with the x-axis, *i.e.*, lines that satisfy

$$x^2 = c^2t^2 \tag{1.68}$$

Had we considered all the space coordinates x, y, z instead of x, we would have got the cone $x^2 + y^2 + z^2 = c^2t^2$,instead of light (or null) line. Light signals emitted at $t = 0$ from the origin, travel out along the dotted lines and all these events are on the light cone.

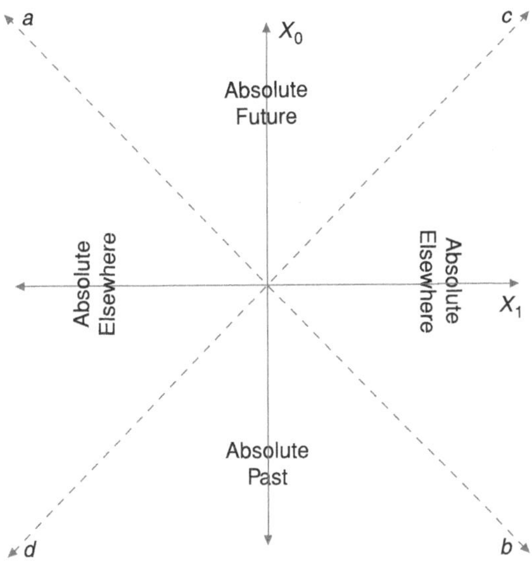

Figure 1.3 *Minkowski diagram of cone dividing space–time into past, future and elsewhere relative to an event O.*

(ii) **Time-like or $ds^2 > 0$:** Let us consider events whose world points lie within the region aOc. It is evident that for all points in this region $(c^2t^2 - x^2)$ is positive and the interval between any event in aOc and event O is time-like. As it is positive, all events in this region occur after the event O. Since the time-like character of an event is independent of the reference system, it is impossible to find a reference system in which any of the events in aOc, occurred before the event O. Therefore, the region aOc is the absolute future relative to event O.

Similarly, the region bOd is absolute past relative to O, *i.e.*, it is impossible to find a reference system in which any event of the region bOd, occurred after the event O.

(iii) **Space-like or $ds^2 < 0$:** Let us consider the regions dOc and bOc. The interval between any event in these regions and event O, is space-like, so any event in these regions and event O occur at different points in space in any reference system. Thus, the regions dOc and bOc are absolutely remote to the event O and are outside the light cone. For any event, one can find a reference system in these regions, where event O could occur either earlier, later and one system in which it occurred simultaneously with O.

1.9 FOUR-VECTORS

A four-vector is defined as a set of four variant scalars (x^0, x^1, x^2, x^3), which transform like (dx^0, dx^1, dx^2, dx^3) or, which is the same thing as the components of a point under rotation. Thus if the transformation is given by the equation

$$x'^{\mu} = \sum_{\nu=0}^{3} \alpha_{\nu}^{\mu} x^{\nu} \quad (\mu = 0,1,2,3) \tag{1.69}$$

then the four-vectors A and A'^{μ} are related as

$$A'^{\mu} = \sum_{\nu=0}^{3} \alpha^{\mu}_{\nu} A^{\nu} \quad (\mu = 0,1,2,3) \tag{1.70}$$

and the inverse transformation is

$$A^{\mu} = \sum_{\nu=0}^{3} \overline{\alpha^{\mu}_{\nu}} A'^{\nu} \quad (\mu = 0,1,2,3) \tag{1.71}$$

where $\overline{\alpha^{\mu}_{\nu}}$ are the coefficients of the inverse transformation.

The sum of two four-vectors is defined as

$$A + B = (A^0 + B^0, A^1 + B^1, A^2 + B^2, A^3 + B^3) \tag{1.72}$$

in analogy with the addition of three-dimensional vectors.

The scalar product of two four-vectors is

$$A \cdot B = A^0 B^0 - A^1 B^1 - A^2 B^2 - A^3 B^3 \tag{1.73}$$

As is evident, the scalar product of two 4-vectors is not a generalization of the scalar product of two 3-vectors, since the space-like components have a negative sign. It implies that the length of a 4-vector can be zero when neither of the components is zero. Such a 4-vector is called a *null vector*. Thus, from Eq. (1.73), when $A = B$, we have

$$A = \sqrt{A \cdot A} = 0$$

or

$$(A^0)^2 = (A^1)^2 + (A^2)^2 + (A^3)^2 \tag{1.74}$$

The displacement between two events on the path of a photon is a null vector.

If $(A^0)^2 > (A^1)^2 + (A^2)^2 + (A^3)^2$, the 4-vector A has a real length and is called *time-like*, whereas if $(A^0)^2 < (A^1)^2 + (A^2)^2 + (A^3)^2$, A is imaginary and the 4-vector A is called *space-like.*

The scalar product of two 4-vectors is invariant and there is a condition on the coefficients of the linear transformation, that is derived as follows:

Now

$$\sum_{\mu,\nu=0}^{3} g_{\mu\nu} A^{\mu} B^{\nu} = \sum_{\mu,\nu=0}^{3} g_{\mu\nu} A'^{\mu} B'^{\nu} \tag{1.75}$$

Substituting for A'^{μ} and B'^{ν} in Eq. (1.75) from Eq. (1.72), one gets

$$\sum_{\mu,\nu=0}^{3} g_{\mu\nu} A^{\mu} B^{\nu} = \sum_{\mu,\nu=0}^{3} g_{\mu\nu} \sum_{\rho=0}^{3} \alpha^{\mu}_{\rho} A^{\rho} \sum_{\sigma=0}^{3} \alpha^{\nu}_{\sigma} A^{\sigma} \tag{1.76}$$

Exchanging on the r.h.s. the dummy indices μ and ρ, ν and σ, one gets

$$\sum_{\mu,\nu=0}^{3} g_{\mu\nu} A^\mu B^\nu = \sum_{\mu,\nu=0}^{3} A^\mu B^\nu \sum_{\rho=\sigma=0}^{3} g_{\rho\sigma} \alpha_\mu^\rho \alpha_\nu^\sigma \tag{1.77}$$

$$\Rightarrow \sum_{\rho,\sigma=0}^{3} g_{\rho\sigma} \alpha_\mu^\rho \alpha_\nu^\sigma = g_{\mu\nu} \tag{1.78}$$

The square of the length of a 4-vector A is $A^\mu A_\mu$

$$A \cdot A = \sum_\mu A^\mu A_\mu = (A^0)^2 - \mathbf{A}^2 \tag{1.79}$$

A few examples of four-vectors are as follows:

(i) The four-vector $x^\mu\,(x^0, x^1, x^2, x^3)$. It gives rise to an invariant interval ds as

$$ds^2 = c^2(dt)^2 - [(dx^1)^2 + (dx^2)^2 + (dx^3)^2] \tag{1.80}$$

(ii) The four-momentum $p^\mu = (\mathcal{E}/c, \mathbf{p})$. The energy-momentum invariant is

$$p^\mu p_\mu = \frac{\mathcal{E}^2}{c^2} - p^2 \tag{1.81}$$

where $p = |\mathbf{p}|$.

(iii) The four-current $J^\mu = (c\rho, \mathbf{J})$
ρ is the electric charge density and $\mathbf{J}$ is the electric current density.

(iv) The four-potential $A^\mu = (\phi, \mathbf{A})$,
where ϕ and $\mathbf{A}$ are the scalar and vector potentials of electrodynamics.

(v) The four-gradient $\nabla^* = (\frac{1}{c}\frac{\partial}{\partial t}, -\nabla)$ and its square is

$$\begin{aligned}
\nabla^{*2} &= \sum_\mu \frac{\partial^\mu}{\partial x^\mu \partial x_\mu} \\
&= \frac{1}{c^2}\frac{\partial^2}{\partial t^2} - \frac{\partial^2}{\partial x^j \partial x_j} \\
&= \frac{1}{c^2}\frac{\partial^2}{\partial t^2} - \nabla^2
\end{aligned} \tag{1.82}$$

(vi) The four-wave vector $k^\mu = (\omega/c, \mathbf{k})$, where ω is the angular frequency and $\mathbf{k}$, the wave vector of a wave. It is a null vector since $k_\mu k^\mu = \omega^2/c^2 - k_x^2 - k_y^2 - k_z^2 = 0$.

Example 1.4

Prove that $\frac{\partial}{\partial x^\mu}$ and j^μ are four-vectors.

Solution

Let us take the case of $\frac{\partial}{\partial x^{\mu}}$. Expressing $\frac{\partial}{\partial x'^{\mu}}$ through the chain rule of differentiation, we get

$$\frac{\partial}{\partial x'^{\mu}} = \sum_{\nu} \frac{\partial x^{\nu}}{\partial x'^{\mu}} \frac{\partial}{\partial x^{\nu}} \tag{i}$$

The inverse transformation for x^{ν} is

$$x^{\nu} = \sum_{\mu} \overline{\alpha^{\nu}_{\mu}} x'^{\mu} \tag{ii}$$

$$\frac{\partial x^{\nu}}{\partial x'^{\mu}} = \overline{\alpha^{\nu}_{\mu}} \tag{iii}$$

Writing Eq. (i) in view of Eq. (iii), we get

$$\frac{\partial}{\partial x'^{\mu}} = \sum_{\nu} \overline{\alpha^{\nu}_{\mu}} \frac{\partial}{\partial x^{\nu}} \tag{iv}$$

It is evident that $\frac{\partial}{\partial x^{\mu}}$ transforms like a four-vector.

Next we show that j^{μ} is a four-vector. Assuming that a charge dq is at rest in a system S and contained in a volume dV, then

$$dq = \rho_0 dV \tag{v}$$

ρ_0 is the volume density of charge. Each fundamental particle like an electron, proton, etc., has a unit charge 'e', that has a discrete value. The total charge is found by the total number of particles. As the total number of particles cannot depend on the state of motion of the observer, the total charge is Lorentz invariant. Thus, the charge dq in system S when viewed from S' is

$$dq = \rho_0 dV = \rho dV' = dq' \tag{vi}$$

where ρ is the charge density in S'.

As
$$dV' = \sqrt{1-\beta^2}\, dV$$

so
$$\rho_0 dV = \rho dV'$$

$$= \rho dV \sqrt{1-\beta^2} \tag{vii}$$

$$\Rightarrow \rho = \frac{\rho_0}{\sqrt{1-\beta^2}} \tag{viii}$$

Thus, ρ transforms like, dx^0 *i.e.*, the time component of a four-vector.

The components of the current density $\mathbf{J} = \rho \mathbf{u}$ are

$$j^k = \rho u^k = \rho \frac{dx^k}{dt} \tag{ix}$$

It is obvious that j^k behaves like, dx^k $(k = 1,2,3)$, like the space component of a 4-vector. Multiplying ρ by c, so that it has the same dimensions as j_k, we infer that $j^\mu = (c\rho, \mathbf{J})$ is a 4-vector.

1.9.1 Applications of Energy-Momentum Invariant

We will treat the following applications:

- Kinematics of the decay of an unstable particle
- Centre of momentum system and threshold energy

Let us take these applications one by one.

Kinematics of the Decay of an Unstable Particle: The conservation of energy and momentum in the two body decay is expressed through a 4-vector equation

$$p^\mu = p^{1\mu} + p^{2\mu} \tag{1.83}$$

where $p^\mu, p^{1\mu}$, and $p^{2\mu}$ stand for the unstable particle, the decay particle no. 1 and decay particle no. 2, respectively.

Rewriting Eq. (1.83), we have

$$p^{2\mu} = p^\mu - p^{1\mu} \tag{1.84}$$

and forming the Lorentz invariant of the four-vectors on both the sides, one obtains

$$p^{2\mu} p_{2\mu} = p^\mu p_\mu + p^{1\mu} p_{1\mu} - 2p^\mu p_{1\mu} \tag{1.85}$$

Putting the values of these terms that are invariant as

$$\begin{aligned} p^{1\mu} p_{1\mu} &= m_1^2 c^2 \\ p^\mu p_\mu &= m_1^2 c^2 \\ p^\mu p_\mu &= M^2 c^2 \\ p^\mu p_{1\mu} &= M\mathcal{E}_1 \end{aligned} \tag{1.86}$$

The term $p^\mu p_{1\mu}$ is also Lorentz invariant and in the rest frame of *M*, its space part vanishes. $\mathcal{E}_1$ is the total energy of the particle of mass m_1.

From Eq. (1.85), we get

$$m_2^2 c^2 = M^2 c^2 + m_1^2 c^2 - 2M\mathcal{E}_1$$

$$\Rightarrow \mathcal{E}_1 = \frac{M^2 c^2 + m_1^2 c^2 - m_2^2 c^2}{2M} \tag{1.87}$$

Similarly,

$$\Rightarrow \mathcal{E}_2 = \frac{M^2 c^2 + m_2^2 c^2 - m_1^2 c^2}{2M} \tag{1.88}$$

Further, let us get the expressions for the kinetic energies of the decay products.

$$\begin{aligned}
T_1 &= E_1 - m_1c^2 \\
&= \frac{M^2c^2 + m_1^2c^2 - m_2^2c^2}{2M} - m_1c^2 \\
&= \frac{M^2 + m_1^2 - m_2^2 - 2Mm_1}{2M}c^2 \\
&= (M - m_1 - m_2)\left[\frac{M - m_1 + m_2}{2M}\right]c^2 \\
&= \Delta Mc^2\left[\frac{2M - 2m_1 - M + m_1 + m_2}{2M}\right] \\
&= \Delta Mc^2\left[1 - \frac{m_1}{M} - \frac{\Delta M}{2M}\right]
\end{aligned} \tag{1.89}$$

Analogously,

$$T_2 = \Delta Mc^2\left[1 - \frac{m_2}{M} - \frac{\Delta M}{2M}\right] \tag{1.90}$$

ΔM is the mass excess. $\Delta M/2M$ is the relativistic correction. For the sake of illustration, let us apply it to the decay

$$\begin{aligned}
&\pi^+ \to \mu^+ + v \\
&M = 139.5\,\text{MeV} \\
&m_\mu = 105.7\,\text{MeV} \\
&m_v \approx 0
\end{aligned}$$

The kinetic energy of the $\mu-$meson is

$$\begin{aligned}
T_1 &= 33.9\left[1 - \frac{105.7}{139.6} - \frac{33.9}{2(139.6)}\right] \\
&= 4.1\ \text{MeV}
\end{aligned}$$

It was the unique value of the $\mu-$meson kinetic energy from $\pi-\mu$ decay that led Powell and co-workers in 1947 to the discovery of $\pi-$meson through the nuclear emulsion technique.

Centre of Momentum System and Threshold Energy: To calculate the threshold energy for the production of a particle, we will make use of the invariance of the scalar product of two four-vectors in the lab as well as C. M. system. Thus,

$$(p^{1\mu} + p^{2\mu})(p_{1\mu} + p_{2\mu}) = (p'^{1\mu} + p'^{2\mu})(p'_{1\mu} + p'_{2\mu}) \tag{1.91}$$

The unprimed quantities refer to the lab system where momentum $p_2 = 0$ and the primed quantities on the r.h.s. pertain to the C. M. system where the spatial momenta $\mathbf{p}'_1 + \mathbf{p}'_2 = 0$.

Putting the values of different terms, we get

$$\frac{1}{c^2}(\mathcal{E}_1 + m_2c^2)^2 - p^2 = \frac{1}{c^2}(\mathcal{E}_1' + \mathcal{E}_2')^2$$

or

$$(\mathcal{E}_1 + m_2c^2)^2 - c^2p^2 = (\mathcal{E}_1' + \mathcal{E}_2')^2 \tag{1.92}$$

Taking $\mathcal{E}_1 = (m_1^2c^4 + c^2p^2)^{1/2}$, the total energy in the C. M. system in Eq. (1.92), becomes

$$\begin{aligned}\mathcal{E}' &= \mathcal{E}_1' + \mathcal{E}_2' \\ &= \left(m_1^2c^2 + m_2^2c^2 + 2m_2\mathcal{E}_1\right)^{1/2} c\end{aligned} \tag{1.93}$$

The separate energies $\mathcal{E}_1'$ and $\mathcal{E}_2'$ are determined from the scalar products like

$$p^{1\mu}(p_{1\mu} + p_{2\mu}) = p'^{1\mu}(p_{1\mu}' + p_{2\mu}') \tag{1.94}$$

Putting the values of different Lorentz invariants leads to the results

$$\mathcal{E}_1' = \frac{\mathcal{E}'^2 + m_1^2c^4 - m_2^2c^4}{2\mathcal{E}'} \tag{1.95}$$

$$\mathcal{E}_2' = \frac{\mathcal{E}'^2 + m_2^2c^4 - m_1^2c^4}{2\mathcal{E}'} \tag{1.96}$$

Let us apply the concept of the invariance of the scalar product of two 4-vectors to the calculation of the threshold energy for the production of particles.

In a reaction, the initial particles of mass m_1 and m_2 are transformed into two or more particles with masses m_i, $i = 3, 4,$ etc. Defining as ΔM, the difference between the sum of masses of the products and the reactants, one gets

$$\Delta M = (m_3 + m_4 + \cdots) - (m_1 + m_2) \tag{1.97}$$

If ΔM is positive, the reaction will not occur unless the projectile has certain minimum energy, called the threshold of the reaction. At threshold, the particles are produced with zero kinetic energy and this implies that

$$\mathcal{E}_{\text{th}}' = m_1c^2 + m_2c^2 + \Delta Mc^2 \tag{1.98}$$

Substituting the value of $\mathcal{E}_{\text{th}}'$ in Eq. (1.93), we get

$$(m_1 + m_2 + \Delta M)c^2 = \left(m_1^2c^2 + m_2^2c^2 + 2m_2\mathcal{E}_1\right)^{1/2} c \tag{1.99}$$

The incident kinetic energy of the projectile at threshold is

$$\mathcal{E}_{\text{th}} = \mathcal{E}_1 - m_1c^2 \tag{1.100}$$

Evaluating the value of $\mathcal{E}_1$ from Eq. (1.99), one gets

$$\mathcal{E}_1 = \Delta M\left[1+\frac{m_1}{m_2}+\frac{\Delta M}{2m_2}\right]c^2 + m_1c^2 \tag{1.101}$$

Therefore,

$$T_{\text{th}} = \Delta M\left[1+\frac{m_1}{m_2}+\frac{\Delta M}{2m_2}\right]c^2 \tag{1.102}$$

Illustrating it for the production of a proton–antiproton pair in proton-proton collisions

$$p + p \rightarrow p + p + p + \overline{p} \tag{1.103}$$

where p and $\overline{p}$ stand for the proton and antiproton.

The mass difference

$$\Delta Mc^2 = 2m_pc^2 = 1.877 \text{ BeV}$$

Hence, from Eq. (1.102)

$$T_{\text{th}} = 1.877[1+1+1] = 5.631 \text{ BeV}$$

1.10 RELATIVISTIC MASS

The rest mass of a particle is the mass as measured in the instantaneous rest frame of that particle. If a particle is moving with velocity $\boldsymbol{v}$ w.r.t. an observer, then the latter will measure its mass as

$$m = \frac{m_0}{\sqrt{1-v^2/c^2}} = \gamma m_0 \tag{1.104}$$

where m_0 is the rest mass.

Now

$$\begin{aligned} m &= \frac{m_0}{\sqrt{1-v^2/c^2}} \\ &\approx m_0\left(1+\frac{1}{2}\frac{v^2}{c^2}\right) \\ &= m_0 + \frac{1}{2}m_0\frac{v^2}{c^2} \end{aligned}$$

Multiplying through c^2, we obtain an expression that relates the relativistic energy to the rest mass energy plus the Newtonian kinetic energy of the particle

$$mc^2 = m_0c^2 + \frac{1}{2}m_0{}^2v^2 \tag{1.105}$$

1.10.1 Energy-Mass Equivalence

We set ourselves the task of determining the mass energy relationship. From Newton's law of force, we have

$$\mathbf{F} = \frac{d}{dt}(m\mathbf{u}) \tag{1.106}$$

Consider a particle of mass m, subjected to an accelerating force **F**, applied in the x-direction. The work done by the force in displacing the particle through distance x, is

$$\begin{aligned} T &= \int_0^x \mathrm{F}\,dx \\ &= \int_0^t F\frac{dx}{dt}\,dt \\ &= \int_0^t Fu\,dt \\ &= \int_0^t u\frac{dp}{dt}\,dt = \int_0^p u\,dp \end{aligned}$$

Putting $p = \frac{mu}{\sqrt{1-u^2/c^2}}$ and $udu = \frac{1}{2}d(u^2)$, we get

$$\begin{aligned} T &= \left.\frac{mu^2}{\sqrt{1-u^2/c^2}}\right|_0^u - \frac{m}{2}\int_0^u \frac{d(u^2)}{\sqrt{1-u^2/c^2}} \\ &= mc^2\left[\frac{u^2/c^2}{\sqrt{1-u^2/c^2}} + \sqrt{1-u^2/c^2}\right]_0^u \\ &= mc^2\left[\frac{1}{\sqrt{1-u^2/c^2}}\right]_0^u \\ &= \frac{mc^2}{\sqrt{1-u^2/c^2}} - mc^2 \\ &= m_r c^2 - mc^2 = (m_r - m)c^2 \end{aligned} \tag{1.107}$$

Equation (1.107) is the law of equivalence of mass and energy or the law of inertia of energy. This law forms a basis of the process of conversion of rest energy into other forms of energy, such as in a nuclear reactor or plasma. In this law of conservation of total relativistic energy, are fused together the two principles of classical physics, namely the law of conservation of energy and the law of conservation of mass. This law is an article of faith with the physicists.

Example 1.5

One of the earliest quantitative proofs of the validity of Einstein's mass-energy relation was provided by Cockroft and Walton, the first case of an artificial nuclear disintegration. They bombarded $_3Li^7$

with ${}_1H^1$ accelerated to energies of 0.1 to 0.7 MeV. The cloud chamber showed that two α-particles leave the point of disintegration and proceed with equal energies in opposite directions.

The reaction is represented as

$$ {}_3Li^7 + {}_1H^1 \rightarrow [{}_4Be^8] \rightarrow {}_2He^4 + {}_2He^4 + Q $$

where Q is the energy balance.

The masses are

$$ \begin{aligned} M({}_3Li^7) &= 7.01818 \text{ amu} \\ M({}_1H^1) &= 1.008142 \text{ amu} \\ M({}_2He^4) &= 4.00386 \text{ amu} \end{aligned} $$

Find the Q-value of the reaction.

Solution

The mass defect is

$$ \Delta M = .01860 \text{ amu} $$

and

$$ Q = \Delta Mc^2 = 17.32 \text{ MeV} $$

The experimental values of Q obtained from the energies of the incident protons and outgoing α-particles is 17.33 MeV. This release of energy was at the expense of mass.

1.11 ELECTROMAGNETIC FIELD TENSOR

The fields **E** and **B** are defined as

$$ \mathbf{E} = -\nabla\phi - \frac{1}{c}\frac{\partial \mathrm{A}}{\partial t} \tag{1.108} $$

$$ \mathbf{B} = \nabla \times \mathbf{A} \tag{1.109} $$

where ϕ and **A** are the scalar and vector potentials for the electromagnetic field.

Writing the expressions for the components of **E** and **B**, in terms of the four-potential $A_\mu = (\phi, -\mathbf{A})$, we get

$$ \begin{aligned} E_x &= -\frac{\partial \phi}{\partial x} - \frac{1}{c}\frac{\partial A_x}{\partial t} = \frac{\partial A_1}{\partial x^0} - \frac{\partial A_0}{\partial x^1} = \overset{*}{\nabla}_0 A_1 - \overset{*}{\nabla}_1 A_0 \\ E_y &= -\frac{\partial \phi}{\partial y} - \frac{1}{c}\frac{\partial A_y}{\partial t} = \frac{\partial A_2}{\partial x^0} - \frac{\partial A_0}{\partial x^2} = \overset{*}{\nabla}_0 A_2 - \overset{*}{\nabla}_2 A_0 \\ E_y &= -\frac{\partial \phi}{\partial z} - \frac{1}{c}\frac{\partial A_z}{\partial t} = \frac{\partial A_3}{\partial x^0} - \frac{\partial A_0}{\partial x^3} = \overset{*}{\nabla}_0 A_3 - \overset{*}{\nabla}_3 A_0 \end{aligned} \tag{1.110} $$

where use was made of the covariant components of the four-gradient operator

$$\overset{*}{\nabla}_0 = \frac{\partial}{\partial x^0}, \quad \overset{*}{\nabla}_1 = \frac{\partial}{\partial x^1}, \quad \overset{*}{\nabla}_2 = \frac{\partial}{\partial x^2}, \quad \overset{*}{\nabla}_3 = \frac{\partial}{\partial x^3} \tag{1.111}$$

and the components of covariant four-potential $A_\mu = (\phi, -\mathbf{A})$.

Furthermore, the components of **B** are

$$\begin{aligned} B_x &= \frac{\partial A_z}{\partial y} - \frac{\partial A_y}{\partial z} = \frac{\partial A_2}{\partial x^3} - \frac{\partial A_3}{\partial x^2} = \overset{*}{\nabla}_3 A_2 - \overset{*}{\nabla}_2 A_3 \\ B_y &= \frac{\partial A_x}{\partial z} - \frac{\partial A_z}{\partial x} = \frac{\partial A_3}{\partial x^1} - \frac{\partial A_1}{\partial x^3} = \overset{*}{\nabla}_1 A_3 - \overset{*}{\nabla}_3 A_1 \\ B_z &= \frac{\partial A_y}{\partial x} - \frac{\partial A_x}{\partial y} = \frac{\partial A_1}{\partial x^2} - \frac{\partial A_2}{\partial x^1} = \overset{*}{\nabla}_2 A_1 - \overset{*}{\nabla}_1 A_2 \end{aligned} \tag{1.112}$$

It is obvious from Eqs. (1.110) and (1.112) that the components of **E** and **B** are expressible as the covariant components of the anti-symmetric four-tensor

$$\begin{aligned} F_{\mu\nu} &= \overset{*}{\nabla}_\mu A_\nu - \overset{*}{\nabla}_\nu A_\mu \\ &= \frac{\partial A_\nu}{\partial x^\mu} - \frac{\partial A_\mu}{\partial x^\nu} \end{aligned} \tag{1.113}$$

$F_{\mu\nu}$ is anti-symmetric tensor, as

$$\begin{aligned} F_{\nu\mu} &= \frac{\partial A_\mu}{\partial x^\nu} - \frac{\partial A_\nu}{\partial x^\mu} \\ &= -F_{\mu\nu} \end{aligned} \tag{1.114}$$

and

$$F_{\mu\mu} = 0 \tag{1.115}$$

$F_{\mu\nu}$ is called the electromagnetic field tensor. Raising the indices μ and ν on both sides of the Eq. (1.113) and lowering them on the right hand side, we get the contravariant components of the electromagnetic field tensor

$$F^{\mu\nu} = \frac{\partial A^\nu}{\partial x_\mu} - \frac{\partial A^\mu}{\partial x_\nu} \tag{1.116}$$

Next we evaluate the various components of $F_{\mu\nu}$, making use of Eq. (1.113)

$$
\begin{aligned}
F_{01} &= -F_{10} = \frac{\partial A_1}{\partial x^0} - \frac{\partial A_0}{\partial x^1} = -\frac{\partial A_x}{c\,\partial t} - \frac{\partial \phi}{\partial x} = E_x \\
F_{02} &= -F_{20} = \frac{\partial A_2}{\partial x^0} - \frac{\partial A_0}{\partial x^2} = -\frac{\partial A_y}{c\,\partial t} - \frac{\partial \phi}{\partial y} = E_y \\
F_{03} &= -F_{30} = \frac{\partial A_3}{\partial x^0} - \frac{\partial A_0}{\partial x^3} = -\frac{\partial A_z}{c\,\partial t} - \frac{\partial \phi}{\partial z} = E_z \\
F_{12} &= -F_{21} = \frac{\partial A_2}{\partial x^1} - \frac{\partial A_1}{\partial x^2} = -\frac{\partial A_y}{\partial x} + \frac{\partial A_x}{\partial y} = -B_x \\
F_{13} &= -F_{31} = \frac{\partial A_3}{\partial x^1} - \frac{\partial A_1}{\partial x^3} = -\frac{\partial A_z}{\partial x} + \frac{\partial A_x}{\partial z} = B_y \\
F_{23} &= -F_{32} = \frac{\partial A_3}{\partial x^2} - \frac{\partial A_2}{\partial x^3} = -\frac{\partial A_z}{\partial y} + \frac{\partial A_y}{\partial x} = -B_z
\end{aligned}
\tag{1.117}
$$

The electromagnetic tensor $F_{\mu\nu}$ in tabular form, is

$$
F_{\mu\nu} = \begin{pmatrix} 0 & E_x & E_y & E_z \\ -E_x & 0 & -B_z & B_y \\ -E_y & B_z & 0 & -B_x \\ -E_z & -B_y & B_x & 0 \end{pmatrix} \tag{1.118}
$$

Noting that raising or lowering of the time index does not change the sign of a component, while raising or lowering of the space index reverses the sign of the component, the contravariant tensor $F^{\mu\nu}$, is

$$
F^{\mu\nu} = \begin{pmatrix} 0 & -E_x & -E_y & -E_z \\ E_x & 0 & -B_z & B_y \\ E_y & B_z & 0 & -B_x \\ E_z & -B_y & B & 0 \end{pmatrix} \tag{1.119}
$$

The space components of $F_{\mu\nu}$ are the magnetic field intensities. Since the components of **B** form a three-dimensional antisymmetric tensor of the second rank and it is only an axial vector that can form a three-dimensional antisymmetrical tensor of the second rank, **B** is an axial vector. However, the electric field intensities constitute. either a row or column vector in $F^{\mu\nu}$, therefore, **E** is a polar vector.

Example 1.6

The angle that **E** field makes with **B** field in a particular electromagnetic field is θ_0 and the angle is invariant to all observers. Find θ_0.

Solution

In any frame of reference, the angle between two vectors is

$$\cos\theta_0 = \frac{\mathbf{E}\cdot\mathbf{B}}{|\mathbf{E}||\mathbf{B}|}$$

$\mathbf{E}\cdot\mathbf{B}$ is invariant of electromagnetic field but $|\mathbf{E}||\mathbf{B}|$ is not. Therefore, θ_0 can be invariant angle if

$$\mathbf{E}\cdot\mathbf{B} = 0$$

or

$$\theta_0 = \pi/2$$

Example 1.7

Show that $\mathcal{E}^2 - |\mathbf{S}|^2$ is an invariant of the electromagnetic field where $\mathcal{E}$ is the energy density and $\mathbf{S}$ is the Poynting flux.

Solution

$$\mathcal{E} = \frac{1}{8\pi}(E^2 + B^2)$$

$$\mathbf{S} = \frac{1}{4\pi}\mathbf{E}\times\mathbf{B}$$

Now

$$\begin{aligned}\mathcal{E}^2 - |\mathbf{S}|^2 &= \frac{1}{64\pi^2}(E^2 + B^2)^2 - \frac{1}{16\pi^2}(\mathbf{E}\times\mathbf{B})^2 \\ &= \frac{1}{64\pi^2}\left[(E^2+B^2)^2 - 4(\mathbf{E}\times\mathbf{B})^2\right] \\ &= \frac{1}{64\pi^2}\left[E^4 + B^4 + 2E^2B^2 - 4(E^2B^2 - (\mathbf{E}\cdot\mathbf{B})^2\right] \\ &= \frac{1}{64\pi^2}\left[E^4 + B^4 - 2E^2B^2 + 4(\mathbf{E}\cdot\mathbf{B})^2\right] \\ &= \frac{1}{64\pi^2}\left[(E^2 - B^2)^2 + 4(\mathbf{E}\cdot\mathbf{B})^2\right]\end{aligned}$$

As $E^2 - B^2$ and $\mathbf{E}\cdot\mathbf{B}$ are invariants of the electromagnetic field, thus, the result follows.

1.12 COVARIANT FORM OF MAXWELL'S EQUATIONS

The origin of the special theory of relativity is the outcome of the conflict of the predictions of Galilean transformation and the observed constancy of the velocity of light. It required the modification of Newtonian mechanics in order to make it compatible with the special theory and resulted in the reformulation of the dynamical concepts. Maxwell's equations were invariant under Lorentz transformations and the theory of relativity led to a deeper understanding of the theory of electromagnetic fields. As discussed earlier, when expressed in four-dimensional form, the electric and magnetic field intensities turn out to be the components of the electromagnetic field tensor.

We proceed to express Maxwell's equations in the four-dimensional form. The fundamental microscopic form of Maxwell's equations, is

$$\begin{gathered}\nabla\times\mathbf{E}+\frac{1}{c}\frac{\partial\mathbf{B}}{\partial t}=0\\ \nabla\cdot\mathbf{B}=0\\ \nabla\times\mathbf{B}-\frac{1}{c}\frac{\partial\mathbf{E}}{\partial t}=4\pi\mathbf{J}\\ \nabla\cdot\mathbf{E}=4\pi\rho\end{gathered}\tag{1.120}$$

Let us consider the inhomogeneous set of equations

$$\begin{gathered}\nabla\times\mathbf{B}-\frac{1}{c}\frac{\partial\mathbf{E}}{\partial t}=4\pi\mathbf{J}\\ \nabla\cdot\mathbf{E}=4\pi\rho\end{gathered}\tag{1.121}$$

Expressed in the form of components, these equations become

$$\begin{gathered}0+\frac{\partial E_x}{\partial x^1}+\frac{\partial E_y}{\partial x^2}+\frac{\partial E_z}{\partial x^3}=\frac{4\pi}{c}j^0\\ -\frac{\partial E_x}{\partial x^0}+0+\frac{\partial B_z}{\partial x^2}-\frac{\partial B_y}{\partial x^3}=\frac{4\pi}{c}j^1\\ -\frac{\partial E_y}{\partial x^0}-\frac{\partial B_z}{\partial x^1}+0+\frac{\partial B_x}{\partial x^3}=\frac{4\pi}{c}j^2\\ -\frac{\partial E_z}{\partial x^0}+\frac{\partial B_y}{\partial x^1}-\frac{\partial B_x}{\partial x^2}+0=\frac{4\pi}{c}j^3\end{gathered}\tag{1.122}$$

Introducing in the left-hand members, a set of dependent variables, defined as

$$\begin{array}{llll}F^{00}=0 & F^{01}=-E_x & F^{02}=-E_y & F^{03}=-E_z\\ F^{10}=E_x & F^{11}=0 & F^{12}=-B_z & F^{13}=B_y\\ F^{20}=E_y & F^{21}=B_z & F^{22}=0 & F^{23}=-B_x\\ F^{30}=E_z & F^{31}=-B_y & F^{32}=B_x & F^{33}=0\end{array}\tag{1.123}$$

Putting $j^\mu=(j^0, j^1, j^2, j^3)$, Eqs. (1.122) become

$$\sum\frac{\partial F^{\mu\nu}}{\partial x^\nu}=-\frac{4\pi}{c}j^\mu\quad(\mu=0,1,2,3)\tag{1.124}$$

where

$$(F^{\mu\nu}) = \begin{pmatrix} 0 & -E_x & -E_y & -E_z \\ E_x & 0 & -B_z & B_y \\ E_y & B_z & 0 & -B_x \\ E_z & -B_y & B_x & 0 \end{pmatrix} \tag{1.125}$$

The pair of homogeneous Maxwell's equations can be written as a single equation in terms of tensor $F_{\mu\nu}$, as

$$\frac{\partial F_{\mu\nu}}{\partial x^\rho} + \frac{\partial F_{\nu\rho}}{\partial x^\mu} + \frac{\partial F_{\rho\mu}}{\partial x^\nu} = 0 \tag{1.126}$$

The Eqs. (1.124) and (1.126) are the Maxwell's equations in the covariant form.

Example 1.8

Show by explicit examination of components that the equations

$$F_{\alpha\beta,\gamma} + F_{\beta\gamma,\alpha} + F_{\gamma\alpha,\beta} = 0$$

reduce to Maxwell's homogeneous equations

$$\nabla \times \mathbf{E} + \frac{1}{c}\frac{\partial \mathbf{B}}{\partial t} = 0$$

$$\nabla \cdot \mathbf{B} = 0$$

Solution

It may be noted that the indices in each of the terms in the equation

$$F_{\alpha\beta,\gamma} + F_{\beta\gamma,\alpha} + F_{\gamma\alpha,\beta} = 0 \tag{i}$$

form a cyclic transposition of the sequence of α, β, γ. This equation is a set of four equations, the first of which is obtained at α, β, γ equal to 0,1,2; the second at α, β, γ equal to 1,2,3; the third at α, β, γ equal to 2,3,0 and the fourth at α, β, γ equal to 3,0,1. The equation obtained at any other combination of the three non-coinciding indices reduces to any one of those four equations.

Let us write Eq. (i) for $\alpha = 0, \beta = 1, \gamma = 2$, making use of the expression of $F_{\mu\nu}$,

$$F_{\mu\nu} = \begin{pmatrix} 0 & E_x & E_y & E_z \\ -E_x & 0 & -B_z & B_y \\ -E_y & B_z & 0 & -B_x \\ -E_z & -B_y & B_x & 0 \end{pmatrix} \tag{ii}$$

$$\frac{\partial F_{01}}{\partial x^2} + \frac{\partial F_{12}}{\partial x^0} + \frac{\partial F_{20}}{\partial x^1} = 0 \tag{iii}$$

Putting the values of the tensor components from Eq. (ii) and coordinate x^μ, we get

$$\frac{\partial E_x}{\partial y} - \frac{1}{c}\frac{\partial B_z}{\partial t} - \frac{\partial E_y}{\partial x} = 0$$

or

$$\frac{\partial E_y}{\partial x} - \frac{\partial E_x}{\partial y} = -\frac{1}{c}\frac{\partial B_z}{\partial t} \tag{iv}$$

This is the z-component of Equation $\nabla \times \mathbf{E} + \frac{1}{c}\frac{\partial \mathbf{B}}{\partial t} = 0$.

Similarly, for $\alpha = 2,\ \beta = 3,\ \gamma = 0$, and $\alpha = 3,\ \beta = 0,\ \gamma = 1$ will give the x- and y-components of the same equation.

Analogouly, for $\alpha = 1,\ \beta = 2,\ \gamma = 3$, we get

$$\frac{\partial F_{12}}{\partial x^3} + \frac{\partial F_{23}}{\partial x^2} + \frac{\partial F_{31}}{\partial x^2} = 0$$

or

$$-\frac{\partial B_z}{\partial z} - \frac{\partial B_x}{\partial x} - \frac{\partial B_y}{\partial y} = 0$$

which is the equation $\nabla \cdot \mathbf{B} = 0$.

1.13 PHOTONS AND NEUTRINOS

Light has a dual character, behaving as a wave as inferred from the phenomenon of interference and diffraction and as corpuscles from the photoelctric as well as Compton effects. However, this dual nature is a complex momentary behaviour, that is, one cannot consider it to behave as waves and quanta simultaneously. When it is considered as a quanta, corpuscle or photon, it is attributed energy $\mathcal{E}$ and momentum **p**, whereas when considered as a wave, it is associated with frequency ν and wavelength λ. The special theory of relativity, helps in establishing the relationship between the wave and corpuscular attributes.

The momentum **p** and energy of a particle are

$$\mathbf{p} = m_r\mathbf{u} = \frac{m\mathbf{u}}{\sqrt{1 - u^2/c^2}} \tag{1.127}$$

$$\mathcal{E} = m_r c^2 = \frac{mc^2}{\sqrt{1 - u^2/c^2}} \tag{1.128}$$

The particle velocity **u** can never exceed c, since for limiting value c, the mass and energy of the particle become infinite. However, in the case of velocity $u \to c$, if the rest mass of the particle tends to zero, the mass and energy of the particle, will remain finite. Such a particle of zero rest mass is Einstein's light quantum or photon that moves with velocity c. Besides the photon, there is another particle, called neutrino (which has an anti-particle, called anti-neutrino) that has zero rest mass. It was hypothesized in 1932 by W. Pauli, on the basis of laws of conservation of energy and momentum, in the case of the continuous distribution of electron energy in $\beta - \text{decay}$. However, it was discovered experimentally by Reines and Cowan in 1956. It has a spin of $\hbar/2$, whereas photon has spin $\hbar$.

Putting $u = c$, for a photon, we get

$$p = mu = \frac{\mathcal{E}}{c^2}u = \frac{\mathcal{E}}{c^2}c = \frac{\mathcal{E}}{c} \tag{1.129}$$

In the corresponding wave picture, the wave number k, is

$$k = \frac{1}{\lambda\!\!\!^{-}} = \frac{2\pi}{\lambda} = \frac{2\pi v}{v\lambda} = \frac{\omega}{c} \tag{1.130}$$

From the comparison of Eqs. (1.129) and (1.130), it emerges that momentum **p** corresponds to the wave vector k and the energy $\mathcal{E}$ to the angular frequency ω. As **p** is a 3-vector, for getting the desired relationship between the corpuscular and wave pictures, we assume that **k** is a 3-vector with components. k_x, k_y, k_z that determine the direction of propagation of light waves.

If there exists an invariant relationship between the 4-vectors $\mathcal{E}/c$, p_x, p_y, p_z, and ω/c, k_x, k_y, k_z, it has to be of the form

$$\frac{\mathcal{E}}{\omega} = \frac{p_x}{k_x} = \frac{p_y}{k_y} = \frac{p_z}{k} = \hbar, \text{ say} \tag{1.131}$$

As the above relation, Eq. (1.131) has to be independent of coordinate axes, it has to be an absolute constant. It is called Planck's constant as it was introduced by him in black body radiation.

Thus, from Eq. (1.131), we have

$$\mathcal{E} = \hbar\omega \tag{1.132a}$$

$$p_x = \hbar k_x \tag{1.132b}$$

$$p_y = \hbar k_y \tag{1.132c}$$

$$p_z = \hbar k_z \tag{1.132d}$$

or

$$|\mathbf{p}| = \hbar k = \frac{h}{\lambda} = \frac{hv}{c} \tag{1.133}$$

Equation (1.133) is consistent with Eq. (1.129). A photon has energy hv and momentum hv/c and the same ray of light when considered as a wave has frequency v and wavelength λ, the energy and momentum hv/c.

Problems

1.1 Two electron beams are travelling in opposite directions with velocities $v = 0.9c$ relative to the laboratory system. Find the relative velocity of the electrons according to Newtonian dynamics. Further, find the velocity measured by the observer moving with one of the electron beams. (1.8*c*, 0.994*c*)

1.2 The length of a moving rod measured by an inertial observer is $3\sqrt{2}$ cm. The rod is moving with velocity 0.8*c* and is making an angle of $45°$ with the direction of motion. Find the proper length of rod. ($\sqrt{34}$ cm)

1.3 Show that the electromagnetic wave equation

$$\frac{\partial^2\phi}{\partial x^2}+\frac{\partial^2\phi}{\partial y^2}+\frac{\partial^2\phi}{\partial z^2}-\frac{1}{c^2}\frac{\partial^2\phi}{\partial t^2}=0$$

is invariant under Lorentz transformation. ϕ stands for either a Cartesian component of vector potential **A** or scalar potential.

1.4 The average lifetime of a $\pi-$meson in its own frame of reference is 26.0×10^{-9}s. If the $\pi-$meson moves with speed $0.95c$ with respect to earth, what is its lifetime as measured by an observer at rest on earth. What is the average distance it travels before decaying as observed by an observer at rest on earth? (83.3 ns, 240 m)

1.5 Momentum is conserved in a collision of two particles according to an observer in uniform motion, say on train. Show that the law of conservation holds for a ground observer.

1.6 In the Compton effect, a gamma-ray of wavelength λ strikes a free but stationary electron. After the collision, the photon is scattered at an angle θ and its scattered wavelength is λ'. The electron recoils at an angle ϕ. Show that the change $(\lambda'-\lambda)$ in photon wavelength for $\theta=\pi/2$ is h/mc, where h is Planck constant and m is the mass of the electron.

1.7 A free neutron undergoes a $\beta-$decay into a proton, an electron and an anti-neutrino, according to the reaction

$$n\rightarrow p+e^-+\bar{v}$$

The decay products have a total kinetic energy of 0.781 ± 0.005 MeV. Show that the observation is compatible with the mass-energy equivalence.

1.8 The frames S and S' are moving with velocity v with respect to each other and a particle has velocities u and u' in these frames. Show that

$$\sqrt{1-u'^2/c^2}=\frac{\sqrt{(1-v^2/c^2)(1-u^2/c^2)}}{1-\frac{u_x v}{c^2}}$$

1.9 Show that the energy-momentum relationship

$$\mathcal{E}^2=m^2c^4+p^2c^2$$

follows from

$$\mathcal{E}=\gamma mc^2$$
$$p=\gamma mv$$

1.10 The solar energy reaching the earth's atmosphere, is at the rate of 1.35×10^3 watt/m^2. Calculate the decrease in the mass of sun per second. (4.4×10^9 kg)

1.11 'The existence of rigid bodies is in contradiction to the principle of relativity.' Comment.

1.12 (a) Show that $\mathbf{E}\cdot\mathbf{B}$ and $\mathbf{E}^2-\mathbf{B}^2$ are invariants of an electromagnetic field.

(b) Show that the statement that **E** and **B** are perpendicular have absolute significance, *i.e.*, if they are at right angles for one observer, they will be at right angles for all observers.

1.13 Show that the law of conservation of charge, *i.e.*, continuity equation is contained in the inhomogeneous pair of Maxwell's equation

$$\sum_{\nu=0}^{3}\frac{\partial F^{\mu\nu}}{\partial x^\nu}=-\frac{4\pi}{c}j^\mu \quad (\mu=0,1,2,3)$$

1.14 Show that a photon in free space cannot lead to a pair-production, but it can in the vicinity of another atom.

2

Tensor Analysis and Riemannian Geometry

Part I Line Element

2.1 RIEMANNIAN SPACE

In the Euclidean space of three dimensions, each point is specified by three coordinates (x^1, x^2, x^3). The distance ds between two neighbouring points (x^1, x^2, x^3) and $(x^1 + dx^1, x^2 + dx^2, x^3 + dx^3)$ is given by

$$ds^2 = (dx^1)^2 + (dx^2)^2 + (dx^3)^2 \tag{2.1}$$

We may extend the concept of Cartesian space in three dimensions to n-dimensional space. Each point will be designated by n coordinates $(x^1, x^2, \ldots, x^n)$, which are shown collectively by (x). Further, we assume that the distance between any two neighbouring points is given by

$$ds^2 = \sum_{\mu,\nu=1}^{n} g_{\mu\nu}(x) dx^u dx^\nu \quad (\mu, \nu = 1, 2, 3, \ldots, n) \tag{2.2}$$

where $g_{\mu\nu}(x)$ are functions of the coordinates (x) and may vary from point to point. These are the components of a metric tensor. The $g_{\mu\nu}$'s are symmetric in indices, μ and ν, *i.e.*,

$$g_{\mu\nu} = g_{\nu\mu} \tag{2.3}$$

The quadratic differential, Eq. (2.2) is called a Riemannian metric and any space characterized by such a metric is called a Riemannian space. ds is called an *interval* or the *line element*. The three dimensional Euclidean space is a special case of the Riemannian space where $g_{ij} = \delta_{ij}$. The functions $g_{\mu\nu}$, n^2 in number, are real and need not be positive; but it is always assumed that the determinant

$$(g_{\mu\nu}) = \begin{vmatrix} g_{11} & g_{12} & \cdots & g_{1n} \\ g_{21} & g_{22} & \cdots & g_{2n} \\ \cdots & \cdots & \cdots & \cdots \\ g_{n1} & g_{n2} & \cdots & g_{nn} \end{vmatrix} \neq 0 \tag{2.4}$$

Minkowski's space–time continuum is a special case of Riemannian space of four-dimensions.

2.2 TRANSFORMATION OF COORDINATES

Let us consider a transformation from one coordinate system $(x^1, x^2, \ldots, x^n)$ to another $(x'^1, x'^2, \ldots, x'^n)$, where

$$x'^1 = f^1(x^1, x^2, \ldots, x^n) \tag{2.5a}$$

$$x'^2 = f^2(x^1, x^2, \ldots, x^n) \tag{2.5b}$$

....

$$x'^\mu = f^\mu(x^1, x'^2, \ldots, x'^n) \quad (\mu = 1, 2, \ldots, n) \tag{2.5c}$$

The functions $f^\mu(x)$ in terms of which the new set of coordinates, are defined are single-valued, continuous, differentiable functions of coordinates $(x^1, x^2, x^3, \ldots, x^n)$. It is essential that these functions be independent and a necessary and sufficient condition for that is that the Jacobian determinant obtained from the partial derivatives $\partial x'^\mu / \partial x^\mu$ is non-zero, *i.e.*,

$$\begin{vmatrix} \dfrac{\partial x'^1}{\partial x^1} & \dfrac{\partial x'^1}{\partial x^2} & \cdots & \dfrac{\partial x'^1}{\partial x^\mu} \\ \dfrac{\partial x'^2}{\partial x^1} & \dfrac{\partial x'^2}{\partial x^2} & \cdots & \dfrac{\partial x'^2}{\partial x^\mu} \\ \cdots & \cdots & \cdots & \cdots \\ \cdots & \cdots & \cdots & \cdots \\ \dfrac{\partial x'^n}{\partial x^1} & \dfrac{\partial x'^n}{\partial x^2} & \cdots & \dfrac{\partial x'^n}{\partial x^\mu} \end{vmatrix} \neq 0 \tag{2.6}$$

The differentials $(dx^1, dx^2, \ldots, dx^n)$ are transformed to $(dx'^1, dx'^2, \ldots, dx'^n)$ as

$$dx'^1 = \frac{\partial x'^1}{\partial x^1} dx^1 + \frac{\partial x'^1}{\partial x^2} dx^2 + \cdots + \frac{\partial x'^1}{\partial x^n} dx^n \quad \text{etc.}$$

or

$$dx'^\mu = \sum_{\nu=1}^{n} \frac{\partial x'^\mu}{\partial x^\nu} dx^\nu \quad (\mu = 1, 2 \ldots n) \tag{2.7}$$

Example 2.1

The line element in Euclidean space is given by $ds^2 = dx^2 + dy^2 + dz^2$. Transform it in polar and cylindrical coordinates.

Solution

The line element in Cartesian coordinates is given by

$$ds^2 = dx^2 + dy^2 + dz^2$$

Comparing it with $$ds^2 = \sum_{i,j=1}^{3} g_{ij} dx^i dx^j$$

one gets
$$g_{11} = g_{22} = g_{33} = 1$$
$$g_{ij} = 0 \quad \text{for } i \neq j$$

Now
$$g'_{ij} = \sum_{a,b=1}^{3} \frac{\partial x^a}{\partial x'^i} \frac{\partial x^b}{\partial x'^j} g_{ab}$$
$$= \sum_{a=1}^{3} g_{aa} \frac{\partial x^a}{\partial x'^i} \frac{\partial x^a}{\partial x'^j}$$
$$= \sum_{a=1}^{3} \frac{\partial x^a}{\partial x'^i} \frac{\partial x^a}{\partial x'^j} \quad \text{since } g_{aa} = 1 \quad (a = 1, 2, 3) \qquad \text{(i)}$$

(i) Let us determine the polar form of the given line element.

Let $$x^1 = x, \quad x^2 = y, \quad x^3 = z$$

and $$x'^1 = r, \quad x'^2 = \theta, \quad x'^3 = \phi$$

then
$$x = r \sin\theta \cos\phi$$
$$y = r \sin\theta \sin\phi$$
$$z = r \cos\theta$$

Then from Eq. (i), we get

$$g'_{11} = \sum_{a=1}^{3} \frac{\partial x^a}{\partial x'^1} \frac{\partial x^a}{\partial x'^1}$$
$$= \frac{\partial x^1}{\partial x'^1} \frac{\partial x^1}{\partial x'^1} + \frac{\partial x^2}{\partial x'^1} \frac{\partial x^2}{\partial x'^1} + \frac{\partial x^3}{\partial x'^1} \frac{\partial x^3}{\partial x'^1}$$
$$= \left(\frac{\partial x}{\partial r}\right)^2 + \left(\frac{\partial y}{\partial r}\right)^2 + \left(\frac{\partial z}{\partial r}\right)^2$$
$$= (\sin\theta \cos\phi)^2 + (\sin\theta \sin\phi)^2 + (\cos\phi)^2$$
$$= 1$$

$$\begin{aligned}
g'_{22} &= \sum_{a=1}^{3} \frac{\partial x^a}{\partial x'^2} \frac{\partial x^a}{\partial x'^2} \\
&= \sum_{a=1}^{3} \left(\frac{\partial x^a}{\partial x'^2} \right)^2 \\
&= \left(\frac{\partial x}{\partial \theta} \right)^2 + \left(\frac{\partial y}{\partial \theta} \right)^2 + \left(\frac{\partial z}{\partial \theta} \right)^2 \\
&= (r\cos\theta\cos\phi)^2 + (r\cos\theta\sin\phi)^2 + (-r\sin\theta)^2 \\
&= r^2
\end{aligned}$$

$$\begin{aligned}
g'_{33} &= \sum_{a=1}^{3} \frac{\partial x^a}{\partial x'^3} \frac{\partial x^a}{\partial x'^3} \\
&= \sum_{a=1}^{3} \left(\frac{\partial x^a}{\partial x'^3} \right)^2 \\
&= \left(\frac{\partial x}{\partial \phi} \right)^2 + \left(\frac{\partial y}{\partial \phi} \right)^2 + \left(\frac{\partial z}{\partial \phi} \right)^2 \\
&= (-\sin\theta\sin\phi)^2 + (r\sin\theta\cos\phi)^2 + 0 \\
&= r^2 \sin^2\theta
\end{aligned}$$

Further,

$$\begin{aligned}
g'_{12} &= \sum_{a=1}^{3} \frac{\partial x^a}{\partial x'^1} \frac{\partial x^a}{\partial x'^2} \\
&= \frac{\partial x}{\partial r}\frac{\partial x}{\partial \theta} + \frac{\partial y}{\partial r}\frac{\partial y}{\partial \theta} + \frac{\partial z}{\partial r}\frac{\partial z}{\partial \theta} \\
&= r\sin\theta\cos\theta\cos^2\phi + r\sin\theta\cos\theta\sin^2\phi - \sin\theta\cos\theta \\
&= 0
\end{aligned}$$

Similarly,

$$g'_{23} = 0$$
$$g'_{31} = 0$$

Therefore,

$$\begin{aligned}
ds^2 &= \sum_{a,b=1}^{3} g'_{ab} dx'^a dx'^b \\
&= \sum_{a=1}^{3} g'_{aa} dx'^a dx'^a \\
&= g'_{11}(dx'^1)^2 + g'_{22}(dx'^2)^2 + g'_{33}(dx'^3)^2 \\
&= dr^2 + r^2 d\theta^2 + r^2 \sin^2\theta \, d\phi^2
\end{aligned}$$

(ii) Let us find the line element in cylindrical coordinates.

Now
$$x^1 = x, \quad x^2 = y, \quad x^3 = z$$
$$x'^1 = r, \quad x'^2 = \theta, \quad x'^3 = z$$

which gives
$$x = r\cos\theta$$
$$x = r\sin\theta$$
$$z = z$$

From Eq. (i), we get
$$g'_{11} = \sum_{a=1}^{3} \frac{\partial x^a}{\partial x'^1}\frac{\partial x^a}{\partial x'^1}$$
$$= \sum_{a=1}^{3} \left(\frac{\partial x^a}{\partial x'^1}\right)^2$$
$$= \left(\frac{\partial x}{\partial r}\right)^2 + \left(\frac{\partial y}{\partial r}\right)^2 + \left(\frac{\partial z}{\partial r}\right)^2$$
$$= (\cos\theta)^2 + (\sin\theta)^2 + 0$$
$$= 1$$
$$g'_{22} = \sum_{a=1}^{3} \frac{\partial x^a}{\partial x'^2}\frac{\partial x^a}{\partial x'^2}$$
$$= \sum_{a=1}^{3} \left(\frac{\partial x^a}{\partial x'^2}\right)^2$$
$$= \left(\frac{\partial x}{\partial \theta}\right)^2 + \left(\frac{\partial y}{\partial \theta}\right)^2 + \left(\frac{\partial z}{\partial \theta}\right)^2$$
$$= (-r\sin\theta)^2 + (r\cos\theta)^2 + 0$$
$$= r^2$$
$$g'_{33} = \sum_{a=1}^{3} \left(\frac{\partial x^a}{\partial x'^3}\right)^2$$
$$= \left(\frac{\partial x}{\partial z}\right)^2 + \left(\frac{\partial y}{\partial z}\right)^2 + \left(\frac{\partial z}{\partial z}\right)^2$$
$$= 0 + 0 + 1$$
$$= 1$$

Similarly,
$$g'_{12} = \sum_{a=1}^{3} \frac{\partial x^a}{\partial x'^1}\frac{\partial x^a}{\partial x'^2}$$
$$= \frac{\partial x}{\partial r}\frac{\partial x}{\partial \theta} + \frac{\partial y}{\partial r}\frac{\partial y}{\partial \theta} + \frac{\partial z}{\partial r}\frac{\partial z}{\partial \theta}$$
$$= \cos\theta(-r\sin\theta) + \sin\theta(r\cos\theta) + 0$$
$$= 0$$
$$g'_{23} = 0$$
$$g'_{31} = 0$$

Therefore,
$$\begin{aligned} ds^2 &= g'_{ab}dx'^a dx'^b \\ &= \sum_{a=1}^{3} g'_{aa}dx'^a dx'^b \\ &= 1drdr + r^2 d\theta d\theta + 1dzdz \\ &= dr^2 + r^2 d\theta^2 + dz^2 \end{aligned}$$

Example 2.2

Find the fundamental tensor g_{ij} for the three-dimensional Euclidean space in terms of

(a) Cartesian Coordinates,

(b) Spherical Polar Coordinates, and

(c) Cylindrical Coordinates.

Solution

(a) **Cartesian Coordinates:** In a three-dimensional space with Cartesian coordinates (x, y, z), the distance ds between two points (x, y, z) and $(x + dx, y + dy, z + dz)$ is

$$ds^2 = dx^2 + dy^2 + dz^2 \tag{i}$$

Comparing it with
$$ds^2 = \sum_{i,j=1}^{3} g_{ij}dx^i dx^j$$

we get
$$(g_{ij}) = \begin{pmatrix} 1 & 0 & 0 \\ 0 & 1 & 0 \\ 0 & 0 & 1 \end{pmatrix} \tag{ii}$$

and
$$g \equiv \det(g_{ij}) = \begin{vmatrix} 1 & 0 & 0 \\ 0 & 1 & 0 \\ 0 & 0 & 1 \end{vmatrix} = 1$$

(b) **Spherical Polar Coordinates:** In terms of spherical polar coordinates (r, θ, ϕ) in a three-dimensional space, the distance between two points (r, θ, ϕ) and $(r + dr, \theta + d\theta, \phi + d\phi)$ is

$$ds^2 = dr^2 + r^2 d\theta^2 + r^2 \sin^2\theta d\phi^2 \tag{iii}$$

Comparing Eq. (iii) with $ds^2 = \Sigma_{i,j=1}^{3} g_{ij}dx^i dx^j$

we get
$$g_{11} = 1, \quad g_{22} = r^2, \quad g_{33} = r^2 \sin^2\theta$$
$$g_{ij} = 0 \quad \text{for } i \neq j$$

with
$$(g_{ij}) = \begin{pmatrix} 1 & 0 & 0 \\ 0 & r^2 & 0 \\ 0 & 0 & r^2\sin^2\theta \end{pmatrix} \tag{iv}$$

and
$$g \equiv \det(g_{ij}) = r^4 \sin^2\theta$$

(c) **Cylindrical Coordinates:** In cylindrical coordinates (r, θ, z) in three-dimensional space, the distance between two points (r, θ, z) and $(r+dr, \theta+d\theta, z+dz)$ is

$$ds^2 = dr^2 + r^2 d\theta^2 + dz^2 \tag{v}$$

Comparing it with $ds^2 = \sum_{i,j=1}^{3} g_{ij} dx^i dx^j$

we get

$$g_{11} = 1, \quad g_{22} = r^2, \quad g_{33} = 1$$
$$g_{ij} = 0 \quad \text{for } i \neq j$$

with

$$(g_{ij}) = \begin{pmatrix} 1 & 0 & 0 \\ 0 & r^2 & 0 \\ 0 & 0 & 1 \end{pmatrix} \tag{vi}$$

and

$$g \equiv \det g_{ij} = r^2$$

2.3 CONTRAVARIANT AND COVARIANT VECTORS

2.3.1 Contravariant Vector

Let us consider a set of n quantities $(A^1, A^2, \ldots, A^n)$ where each $A^\mu (\mu = 1, 2, \ldots, n)$ is some known function of the coordinates $x^\nu (\nu = 1, 2, \ldots, n)$. If the quantities A^μ are transformed into a new set $A'^\mu(x')$ according to the law,

$$A'^\mu = \sum_{\nu=1}^{n} \frac{\partial x'^\mu}{\partial x^\nu} A^\nu(x) \quad (\mu = 1, 2, \ldots, n) \tag{2.8}$$

then the n quantities are the components of a contravariant vector, Eq. (2.8). The law of transformation is the same as that for the differentials of the coordinates. A superscript is used to indicate a contravariant vector. Evidently the differentials $(dx^1, dx^2, \ldots, dx^n)$ of the coordinates constitute the components of a contravariant vector.

2.3.2 Covariant Vector

There is another kind of vector called covariant vector and we derive the law of transformation for its components. Assuming that ϕ is an invariant function of position, *i.e.*, it has a fixed value at each point, independent of the coordinate system, *i.e.*, $\phi(x) = \phi'(x')$, then the n quantities

$$\left(\frac{\partial \phi}{\partial x^1}, \frac{\partial \phi}{\partial x^2}, \ldots, \frac{\partial \phi}{\partial x^n} \right) \tag{2.9}$$

are transformed into the set

$$\left(\frac{\partial \phi}{\partial x'^1}, \frac{\partial \phi}{\partial x'^2}, \ldots, \frac{\partial \phi}{\partial x'^n}\right) \tag{2.10}$$

according to the equation

$$\frac{\partial \phi}{\partial x'^1} = \frac{\partial \phi}{\partial x^1}\frac{\partial x^1}{\partial x'^1} + \frac{\partial \phi}{\partial x^2}\frac{\partial x^2}{\partial x'^1} + \cdots + \frac{\partial \phi}{\partial x^n}\frac{\partial x^n}{\partial x'^1} \text{ etc.} \tag{2.11}$$

or

$$\frac{\partial \phi}{\partial x'^\mu} = \sum_{\nu=1}^{n} \frac{\partial x^\nu}{\partial x'^\mu}\frac{\partial \phi}{\partial x^\nu} \tag{2.12}$$

Defining $A_\nu = \frac{\partial \phi}{\partial x^\nu}$, $A'_\nu = \frac{\partial \phi}{\partial x'^\nu}$, we rewrite Eq. (2.12) as

$$A'_\mu = \sum_{\nu=1}^{n} \frac{\partial x^\nu}{\partial x'^\mu} A_\nu \tag{2.13}$$

A set of n quantities $(A_1, A_2, \ldots, A_n)$ transformed according to Eq. (2.13) under the coordinate transformations, are the components of a covariant vector. A subscript is used to denote a covariant vector.

2.3.3 Relation Between Contravariant and Covariant Vectors

As we have seen, the laws of transformation for the contravariant and covariant vectors are different. However, there exists an important relationship between them and we will show that the sum of products, $\sum_{\mu=1}^{n} A'^\mu B'_\mu$ is a scalar invariant.

Thus,

$$\begin{aligned}
\sum_{\mu=1}^{n} A'^\mu B'_\mu &= \sum_{\mu=1}^{n}\left(\sum_{\nu=1}^{n}\frac{\partial x'^\mu}{\partial x^\nu}A^\nu\right)\left(\sum_{\rho=1}^{n}\frac{\partial x^\rho}{\partial x'^\mu}B_\rho\right) \\
&= \sum_{\nu=1}^{n}\sum_{\rho=1}^{n}\frac{\partial x^\rho}{\partial x^\nu}A^\nu B_\rho \\
&= \sum_{\nu=1}^{n}\delta^\rho_\nu A^\nu B_\rho \\
&= \sum_{\nu=1}^{n} A^\nu B_\nu
\end{aligned} \tag{2.14}$$

The sum is called the inner product of a contravariant with covariant vector.

2.4 SUMMATION CONVENTION

A summation convention due to Einstein is used in writing mathematical expressions. According to this convention, if a suffix (a superscript or a subscript) is repeated twice in a term, that term is to be summed over the range of values of the suffix. This implies that the summation sign is dispensed with. Thus, according to this convention, Eq. (2.2) is rewritten as

$$ds^2 = g_{\mu\nu} dx^\mu dx^\nu \tag{2.15}$$

As the suffices μ,ν occur twice, summation over these from 1 to n is implied in the equation. The repeated index is called a dummy index. Obviously a dummy index can be replaced by any other index as it will not appear in the final result. No dummy index is to occur more than twice and any two dummy indices can be interchanged.

If in any term an index appears only once, it has a definite value between 1 and n. It is called a free index. Another important symbol is Kronecker delta or Kronecker symbol δ^μ_ν, such that

$$\delta^\mu_\nu = \begin{cases} 1 & \text{for } \mu = \nu \\ 0 & \text{for } \mu \neq \nu \end{cases} \quad (\mu,\nu = 1,2,\ldots,n) \tag{2.16}$$

Since x^μ and x^ν are independent of each other

$$\frac{\partial x^\mu}{\partial x^\nu} = \delta^\mu_\nu \tag{2.17}$$

Furthermore,

$$\sum_\nu \delta^\mu_\nu A^\nu = \delta^\mu_1 A^1 + \delta^\mu_2 A^2 + \cdots + \delta^\mu_n A^n$$

$$= A^\mu \tag{2.18}$$

Also

$$\delta^\mu_\nu \delta^\nu_\sigma = \delta^\mu_\sigma \tag{2.19}$$

and

$$\delta^\mu_\mu = \delta^1_1 + \delta^2_2 + \cdots + \delta^n_n$$
$$= n \tag{2.20}$$

Example 2.3

Prove that Kronecker delta δ^μ_ν is a tensor.

Solution

There are two ways to show that δ^μ_ν is a tensor. One approach is by combining δ^μ_ν with the components of two vectors to form a scalar. Thus,

$$A^\nu B_\mu \delta^\mu_\nu = A^\mu B_\mu = \mathbf{A} \cdot \mathbf{B}$$

An alternative is by checking the transformation properties. Thus,

$$\frac{\partial x^{\mu'}}{\partial x^\alpha} \frac{\partial x^\beta}{\partial x^{\nu'}} \delta^\alpha_\beta = \frac{\partial x^{\mu'}}{\partial x^\alpha} \frac{\partial x^\alpha}{\partial x^{\nu'}} = \delta^{\mu'}_{\nu'}$$

where the last equality follows from the fact that $\frac{\partial x^{\mu'}}{\partial x^{\alpha}}$ and $\frac{\partial x^{\alpha}}{\partial x^{\nu'}}$ are matrix inverse of each other.

2.5 THE METRIC

In the flat space–time of special relativity, consider the events given by the coordinates. The infinitesimal interval between them is given by

$$ds^2 = dt^2 - dx^2 - dy^2 - dz^2 \tag{2.21}$$

The interval ds^2, often called the metric, contains the information about how the given space–time deviates from the flat space–time. As we will see, the interval describes the geometry as modified by the gravitational field.

In calculus, the interval ds^2 is often called a line element. Let us consider the line elements for different coordinate systems. For the Cartesian coordinates, it is

$$ds^2 = dx^2 + dy^2 + dz^2 \tag{2.22}$$

For spherical coordinates, it is

$$ds^2 = dr^2 + r^2 d\theta^2 + r^2 \sin^2\theta d\phi^2 \tag{2.23}$$

whereas for cylindrical coordinates, it is given by

$$ds^2 = dr^2 + r^2 d\phi^2 + dz^2 \tag{2.24}$$

These and other line elements can be written compactly by writing the coordinates with indices and assuming the summation convention. Thus, the line element is written as,

$$ds^2 = g_{\mu\nu}(x) dx^\mu dx^\nu \tag{2.25}$$

where $g_{\mu\nu}(x)$ are the components of a second rank tensor, called the metric. The matrix elements representing the tensor, are the coefficient functions that multiply the differentials in the line element. For example, a metric representing ordinary three dimensional space, has its components arranged in a matrix as

$$g_{ij} = \begin{pmatrix} g_{11} & g_{12} & g_{13} \\ g_{21} & g_{22} & g_{23} \\ g_{31} & g_{32} & g_{33} \end{pmatrix} \tag{2.26}$$

The metric for the Cartesian system, Eq. (2.22), is

$$g_{ij} = \begin{pmatrix} 1 & 0 & 0 \\ 0 & 1 & 0 \\ 0 & 0 & 1 \end{pmatrix} \tag{2.27}$$

When dealing with space–time, the time coordinate is labeled as x^0 and the matrix representation of the metric is given by the matrix

$$g_{\mu\nu} = \begin{pmatrix} g_{00} & g_{01} & g_{02} & g_{03} \\ g_{10} & g_{11} & g_{12} & g_{13} \\ g_{20} & g_{21} & g_{22} & g_{23} \\ g_{30} & g_{31} & g_{32} & g_{33} \end{pmatrix} \tag{2.28}$$

In the spherical coordinate system, the coordinates (r,θ,ϕ) are identified as (x^1,x^2,x^3) and the matrix Eq. (2.23), is written as

$$g_{ij} = \begin{pmatrix} 1 & 0 & 0 \\ 0 & r^2 & 0 \\ 0 & 0 & r^2\sin^2\theta \end{pmatrix} \tag{2.29}$$

Analogously for the cylindrical coordinates, the matrix for the metric becomes

$$g_{ij} = \begin{pmatrix} 1 & 0 & 0 \\ 0 & r^2 & 0 \\ 0 & 0 & 1 \end{pmatrix} \tag{2.30}$$

For the metrics we considered in Eqs.. (2.27), (2.29) and (2.30), the matrices are diagonal, and so may be the situation in many cases. However, this is not always the case. For example, a metric that arises in the study of gravitational radiation, called the Bondi metric, has off-diagonal elements. In this metric, the coordinates are (u,r,θ,ϕ), with the following line element

$$\begin{aligned} ds^2 &= \left(\frac{f}{r}e^{2\beta} - g^2r^2e^{2\alpha}\right)du^2 + 2e^{2\beta}dudr + 2gr^2e^{2\alpha}dud\theta \\ &\quad - r^2(e^{2\alpha}d\theta^2 + e^{-2\alpha}\sin^2\theta d\phi^2) \end{aligned} \tag{2.31}$$

where f,g,α,β are functions of the coordinates (u,r,θ,ϕ). The marix representation of the metric is

$$g_{\mu\nu} = \begin{pmatrix} g_{uu} & g_{ur} & g_{u\theta} & g_{u\phi} \\ g_{ru} & g_{rr} & g_{r\theta} & g_{r\phi} \\ g_{\theta u} & g_{\theta r} & g_{\theta\theta} & g_{\theta\phi} \\ g_{\phi u} & g_{\phi r} & g_{\phi\theta} & g_{\phi\phi} \end{pmatrix} \tag{2.32}$$

Furthermore, the fact that the metric is symmetric, *i.e.*, $g_{\mu\nu} = g_{\nu\mu}$, helps us to write down the components of the metric associated with the mixed terms in the line element. Thus, in this case, we have,

$$\begin{aligned} 2e^{2\beta}dudr &= e^{2\beta}dudr + e^{2\beta}drdu = g_{ur}dudr + g_{ru}drdu \\ 2gr^2e^{2\alpha}dud\theta &= gr^2e^{2\alpha}dud\theta + gr^2e^{2\alpha}d\theta du = g_{u\theta}dud\theta + g_{\theta u}d\theta du \end{aligned} \tag{2.33}$$

Thus, the matrix is

$$g_{\mu\nu} = \begin{pmatrix} \frac{f}{r}e^{2\beta} - g^2r^2e^{2\alpha} & e^{2\beta} & gr^2e^{2\alpha} & 0 \\ e^{2\beta} & 0 & 0 & 0 \\ gr^2e^{2\alpha} & 0 & -r^2e^{2\alpha} & 0 \\ 0 & 0 & 0 & -r^2e^{-2\alpha}\sin^2\theta \end{pmatrix} \tag{2.34}$$

The metric is a coordinate dependent function as is evident from all the cases discussed so far. Furthermore, it may be noted that different sign conventions are used for time and space coordinates. Consider the case of flat Minkowski space, written in spherical coordinates. It may be written either as

$$ds^2 = dt^2 - dr^2 - r^2 d\theta^2 - r^2 \sin^2\theta d\phi^2 \tag{2.35}$$

or

$$ds^2 = -dt^2 + dr^2 + r^2 d\theta^2 + r^2 \sin^2\theta d\phi^2 \tag{2.36}$$

In Eq. (2.35), the signature of space is chosen to be $+---$ and in Eq. (2.36), these are $-+++$. Although any choice is valid but whichever is selected at the beginning, one has to stick with it throughout the solution. These are cases of pseudo-Euclidean space.

By convention, the metric of flat Minkowski space is denoted by $\eta_{\mu\nu}$ as

$$\eta_{\mu\nu} = \begin{pmatrix} 1 & 0 & 0 & 0 \\ 0 & -1 & 0 & 0 \\ 0 & 0 & -1 & 0 \\ 0 & 0 & 0 & -1 \end{pmatrix} \tag{2.37}$$

provided that

$$ds^2 = dt^2 - dx^2 - dy^2 - dz^2$$

2.6 THE METRIC AS A TENSOR

So far we considered the metric as a matrix of the coefficients from the line element. But metric is a symmetric second rank tensor. It is formally written as the sum over tensor products of basis one forms

$$g = g_{\mu\nu} dx^\mu \otimes dx^\nu \tag{2.38}$$

Further, it may be stated that the metric has an inverse which is written as $g^{\mu\nu}$ through the relationship

$$g_{\mu\nu} g^{\nu\rho} = \delta_\mu^\rho \tag{2.39}$$

where δ_μ^ρ is the Kronecker delta function.

In the case of a diagonal metric, it is relatively easier to find the inverse. Let us illustrate it by taking the metric in spherical coordinates, Eq. (2.29), $g_{ij} = 0$ for $i \neq j$. The non-zero components are

$$g^{11} g_{11} = 1 \Rightarrow g^{11} = 1$$

$$g^{22} g_{22} = g^{22} r^2 = 1 \Rightarrow g^{22} = \frac{1}{r^2}$$

$$g^{33} g_{33} = g^{33} r^2 \sin^2\theta = 1 \Rightarrow g^{33} = \frac{1}{r^2 \sin^2\theta}$$

The components in the matrix form become

$$g^{ij} = \begin{pmatrix} 1 & 0 & 0 \\ 0 & \dfrac{1}{r^2} & 0 \\ 0 & 0 & \dfrac{1}{r^2 \sin^2 \theta} \end{pmatrix} \tag{2.40}$$

2.6.1 Levi–Civita Symbol

Named after the Italian mathematician and physicist Tullio Levi–Civita, the Levi–Civita symbol, also called the permutation symbol, is used in tensor calculus.

In three dimensions, it is defined as follows:

$$\varepsilon_{ijk} = \begin{cases} +1 & \text{if } i,\ j,\ k \text{ is an even permutation of } 1,2,3 \\ -1 & \text{if } i,\ j,\ k \text{ is an odd permutation of } 1,2,3 \\ 0 & \text{if any two indices are equal} \end{cases} \tag{2.41}$$

For illustration, in linear algebra, the determinant of a 3×3 matrix A is written as

$$\det A = \sum_{i=1}^{3} \sum_{j=1}^{3} \sum_{k=1}^{3} \varepsilon_{ijk} a_{1i} a_{2j} a_{3k} \tag{2.42}$$

and the cross product of two vectors **a** and **b** can be written as a determinant

$$\mathbf{a} \times \mathbf{b} = \begin{vmatrix} \mathbf{e}_1 & \mathbf{e}_2 & \mathbf{e}_3 \\ a_1 & a_2 & a_3 \\ b_1 & b_2 & b_3 \end{vmatrix} = \sum_{i=1}^{3} \sum_{j=1}^{3} \sum_{k=1}^{3} \varepsilon_{ijk} \mathbf{e}_i a_j b_k \tag{2.43}$$

where $\mathbf{e}_1$, $\mathbf{e}_2$, $\mathbf{e}_3$ are the unit vectors along the three axes 1, 2, 3. These are the base vectors and satisfy the following relations

$$\begin{aligned} \mathbf{e}_i \cdot \mathbf{e}_j &= \delta_{ij} \\ \mathbf{e}_i \times \mathbf{e}_j &= \sum_k \varepsilon_{ijk} \mathbf{e}_k \end{aligned} \tag{2.44}$$

where δ_{ij} is the Kronecker delta.

ε_{ijk} is a pseudotensor because under an orthogonal transformation of Jacobian determinant -1 (a rotation followed by a reflection), it acquires a negative sign.

Example 2.4

Levi–Civita symbol $\varepsilon^{\mu\nu\rho\sigma}$ in 4-dimensions is defined as follows:

$$\varepsilon^{\mu\nu\rho\sigma} = \begin{cases} 0 & \text{if any two indices are the same} \\ +1 & \text{if } \mu\nu\rho\sigma \text{ is an even permutation of } 0,1,2,3 \\ -1 & \text{if } \mu\nu\rho\sigma \text{ is an odd permutation of } 0,1,2,3 \end{cases}$$

Evaluate $\varepsilon_{\mu\nu\rho\sigma}\varepsilon^{\mu\nu\rho\sigma}$.

Solution

In an orthonormal frame

$$\varepsilon^{\mu\nu\rho\sigma} = \eta^{\mu\alpha}\eta^{\nu\beta}\eta^{\rho\gamma}\eta^{\sigma\delta}\varepsilon_{\alpha\beta\gamma\delta}$$

For non-zero components of $\varepsilon^{\mu\nu\rho\sigma}$, precisely one of the indices μ, ν, ρ, σ must equal zero. Since $\eta^{00} = 1$ and $\eta^{ij} = -1$, this introduces one minus sign in what is an identity transformation. Therefore,

$$\varepsilon_{\mu\nu\rho\sigma} = -\varepsilon^{\mu\nu\rho\sigma}$$

$$\varepsilon_{\mu\nu\rho\sigma}\varepsilon^{\mu\nu\rho\sigma} = -\sum \left|\varepsilon_{\mu\nu\rho\sigma}\right|^2$$

$$= -\sum \left|\varepsilon_{0,1,2,3}\right|^2$$

where the sum is over the permutations of 0, 1, 2, 3. As there are $4! = 24$ permutations, thus,

$$\varepsilon_{\mu\nu\rho\sigma}\varepsilon^{\mu\nu\rho\sigma} = -24$$

2.7 CONTRAVARIANT, COVARIANT, AND MIXED TENSORS

2.7.1 Contravariant Tensor

A set of n^2 functions $A^{\mu\nu}$ are components of a contravariant tensor of rank two, if they transform according to the law

$$A'^{\mu\nu} = \frac{\partial x'^{\mu}}{\partial x^{\alpha}}\frac{\partial x'^{\nu}}{\partial x^{\beta}}A^{\alpha\beta} \tag{2.45}$$

under coordinate transformation. $A'^{\mu\nu}$ are the components of the tensor in the primed coordinate system. The contravariant vector A^{μ} is called contravariant tensor of rank one. Analogously, one can define contravariant tensors of higher rank as

$$A'^{\mu\nu\rho\ldots} = \frac{\partial x'^{\mu}}{\partial x^{\alpha}}\frac{\partial x'^{\nu}}{\partial x^{\beta}}\frac{\partial x'^{\rho}}{\partial x^{\gamma}}\cdots A^{\alpha\beta\gamma\ldots} \tag{2.46}$$

The contravariant indices are always written as superscripts.

2.7.2 Covariant Tensor

A set of n^2 functions $A_{\mu\nu}$ are components of a covariant tensor of rank two if they transform as

$$A'_{\mu\nu} = \frac{\partial x^{\alpha}}{\partial x'^{\mu}} \frac{\partial x^{\beta}}{\partial x'^{\nu}} A_{\alpha\beta} \tag{2.47}$$

under coordinate transformation. Covariant tensors of higher rank have the corresponding transformation law

$$A'_{\mu\nu\rho\ldots} = \frac{\partial x^{\alpha}}{\partial x'^{\mu}} \frac{\partial x^{\beta}}{\partial x'^{\nu}} \frac{\partial x^{\gamma}}{\partial x'^{\rho}} \cdots A_{\alpha\beta\gamma\ldots} \tag{2.48}$$

The covariant vector A_{μ} is a covariant tensor of rank one. The covariant character is always specified by a lower suffix.

2.7.3 Mixed Tensor

A set of n^2 functions A^{μ}_{ν} are the components of a contravariant tensor of rank one and covariant tensor of rank one (or a mixed tensor of rank two) if they transform according to the law

$$A'^{\mu}_{\nu} = \frac{\partial x'^{\mu}}{\partial x^{\alpha}} \frac{\partial x^{\beta}}{\partial x'^{\nu}} A^{\alpha}_{\beta} \tag{2.49}$$

under coordinate transformations. It is obvious that the quantities A^{μ}_{ν} have the transformation character of a contravariant vector for the index μ and of covariant vector for the index ν; *i.e.*, both types of transformation laws are operative. For that reason A^{μ}_{ν} is called a mixed tensor.

A tensor of rank one is an n-component vector and a tensor of rank two can be written as a square matrix of order n. Thus, a contravariant tensor of rank two is written as

$$A^{\mu\nu} = \begin{pmatrix} A^{11} & A^{12} & \ldots & A^{1n} \\ A^{21} & A^{22} & \ldots & A^{2n} \\ \ldots & \ldots & \ldots & \ldots \\ A^{n1} & A^{n2} & \ldots & A^{nn} \end{pmatrix} \tag{2.50}$$

It may be remarked that a scalar and a vector are particular cases of a tensor. A scalar is a tensor of rank zero and thus has $n^0 = 1$ component. A vector is a tensor of rank one and has n-components in an n-dimensional space.

It is pertinent to bring out the difference between the contravariant and covariant tensors geometrically. A contravariant tensor is represented by components *in the directions of coordinate increase* , whereas a covariant tensor is represented by components *in the directions orthogonal to constant coordinate surfaces*. By way of physical examples, one can mention that velocity and acceleration vectors, being represented by components in the directions of coordinate increase, are contravariant vectors. As a gradient vector is represented in terms of components in the directions orthogonal to the constant coordinate surface, it is a covariant vector. In the case of a Cartesian coordinate system, the coordinate direction x^i coincides with the direction orthogonal to the constant surface, so there is no distinction between the contravariant and covariant vectors in this case. It is called a *Cartesian* tensor.

Example 2.5

Show that in a Cartesian system, there is no distinction between the contravariant and covariant components of a vector.

Solution

Let A^i and A_i be the contravariant and covariant components of a vector with respect to a Cartesian coordinate system. These are related to each other by the relation

$$A_i = A^j g_{ji} \tag{i}$$

For a Cartesian system

$$(g_{ji}) = (g_{ji}) = \begin{pmatrix} 1 & 0 & 0 \\ 0 & 1 & 0 \\ 0 & 0 & 1 \end{pmatrix} \tag{ii}$$

Therefore, Eq. (i) may be written as

$$\begin{aligned} A_i &= A^i g_{ii} \quad \text{(no summation over } i\text{)} \\ &= A^i \end{aligned} \tag{iii}$$

Thus, in a Cartesian system, the contravariant and covariant components of a vector are identical.

Example 2.6

Show that $\sqrt{g}\ dx^1 dx^2 \ldots dx^n$ is an invariant.

Solution

$g_{\mu\nu}$ is a second rank covariant tensor, so

$$g'_{\mu\nu} = \frac{\partial x^\alpha}{\partial x'^\mu} \frac{\partial x^\beta}{\partial x'^\nu} g_{\alpha\beta} \quad (\mu, \nu = 1, 2, \ldots, n) \tag{i}$$

Taking determinant of both sides,

$$|g'_{\mu\nu}| = |g_{\alpha\beta}| \left| \frac{\partial x^\alpha}{\partial x'^\mu} \right| \left| \frac{\partial x^\beta}{\partial x'^\nu} \right|$$

or

$$g' = g \left| \frac{\partial x}{\partial x'} \right|^2 \tag{ii}$$

But

$$\left| \frac{\partial x}{\partial x'} \right| dx'^1 dx'^2 \ldots dx'^n = dx^1 dx^2 \ldots dx^n \tag{iii}$$

Eliminating $|\frac{\partial x}{\partial x'}|$ from Eqs. (ii) and (iii), we get

$$\frac{\sqrt{g'}}{\sqrt{g}} dx'^1 dx'^2 \dots dx'^n = dx^1 dx^2 \dots dx^n$$

or

$$\sqrt{g'} dx'^1 dx'^2 \dots dx'^n = \sqrt{g} dx^1 dx^2 \dots dx^n \tag{iv}$$

which shows that $\sqrt{g} dx^1 dx^2 \dots dx^n$ is an invariant.

2.7.4 Rank of a Tensor

The rank of a tensor is equal to the number of indices for a component. An arbitrary tensor of rank r in an n-dimensional space has n^r components. Consider a tensor $A^{\rho}_{\mu\nu\lambda}$. As it has one upper and three lower indices, it is a tensor of 4th rank. Its transformation is given by

$$A'^{\rho}_{\mu\nu\lambda} = \frac{\partial x'^{\rho}}{\partial x^{\alpha}} \frac{\partial x^{\beta}}{\partial x'^{\mu}} \frac{\partial x^{\gamma}}{\partial x'^{\nu}} \frac{\partial x^{\delta}}{\partial x'^{\lambda}} A^{\alpha}_{\beta\gamma\delta} \tag{2.51}$$

2.7.5 Equality and Null Tensor

Two tensors are said to be equal if they have the same contravariant rank and covariant rank and every component of one is equal to the corresponding component of the other.

A tensor of rank r in an n-dimensional space has n^r components and if all these components are zero, it is said to be a null tensor. It is evident from Eq. (2.51), that if the components of a tensor vanish in one coordinate system, they will vanish in any other coordinate system.

Two tensors are said to be of the *same type* if they have the same contravariant rank and same covariant rank.

2.7.6 Addition and Subtraction of Tensors

The sum and difference of two tensors of same type, is a tensor of the same rank and type. Thus, if $A^{\mu\nu}_{\lambda}$ and $B^{\mu\nu}_{\lambda}$ are two such tensors, then

$$C^{\mu\nu}_{\lambda} = A^{\mu\nu}_{\lambda} + B^{\mu\nu}_{\lambda} \tag{2.52}$$

$$D^{\mu\nu}_{\lambda} = A^{\mu\nu}_{\lambda} - B^{\mu\nu}_{\lambda} \tag{2.53}$$

are also tensors of the same rank and type.

2.7.7 Symmetric and Antisymmetric Tensors

A tensor is said to be symmetric if two contravariant or covariant indices can be interchanged without changing the tensor. Thus,

$$A^{\mu\nu} = A^{\nu\mu} \tag{2.54}$$

In the primed system, it becomes

$$A'^{\mu\nu} = \frac{\partial x'^{\mu}}{\partial x^{\alpha}}\frac{\partial x'^{\nu}}{\partial x^{\beta}}A^{\alpha\beta}$$
$$= \frac{\partial x'^{\nu}}{\partial x^{\beta}}\frac{\partial x'^{\mu}}{\partial x^{\alpha}}A^{\beta\alpha}$$

since $A^{\alpha\beta} = A^{\beta\alpha}$

therefore, $A'^{\mu\nu} = A'^{\nu\mu}$ (2.55)

implying thereby that symmetry remains unaltered in the other system.

Similarly, if $A_{\mu\nu} = A_{\nu\mu}$

Then
$$A'_{\mu\nu} = \frac{\partial x^{\alpha}}{\partial x'^{\mu}}\frac{\partial x^{\beta}}{\partial x'^{\nu}}A_{\alpha\beta}$$
$$= \frac{\partial x^{\beta}}{\partial x'^{\nu}}\frac{\partial x^{\alpha}}{\partial x'^{\mu}}A^{\beta\alpha}$$
$$= A'_{\nu\mu} \tag{2.56}$$

One cannot define symmetry when one index is contravariant and the other is covariant. However, Kronecker delta which is a mixed tensor, is symmetrical with respect to its indices.

A tensor is said to be antisymmetric (or skew-symmetric) when on interchanging two contravariant or covariant indices, it changes in sign but not in magnitude. Thus,

$$A^{\mu\nu} = -A^{\nu\mu} \tag{2.57}$$

On transforming to a primed system, one gets

$$A'^{\mu\nu} = \frac{\partial x'^{\mu}}{\partial x^{\alpha}}\frac{\partial x'^{\nu}}{\partial x^{\beta}}A^{\alpha\beta}$$
$$= \frac{\partial x'^{\nu}}{\partial x^{\beta}}\frac{\partial x'^{\mu}}{\partial x^{\alpha}}A^{\beta\alpha}$$

Since $A^{\alpha\beta} = -A^{\beta\alpha}$

therefore, $A'^{\mu\nu} = -A'^{\nu\mu}$ (2.58a)

Similarly, if $A_{\mu\nu} = -A_{\nu\mu}$

then $A'_{\mu\nu} = -A'_{\nu\mu}$ (2.58b)

Any second rank tensor $A_{\mu\nu}$ or $B^{\mu\nu}$ can be written as a sum of a symmetric and antisymmetric part

$$A_{\mu\nu} = \frac{1}{2}(A_{\mu\nu} + A_{\nu\mu}) + \frac{1}{2}(A_{\mu\nu} - A_{\nu\mu}) \tag{2.59}$$

2.8 MULTIPLICATION OF TENSORS—INNER AND OUTER PRODUCTS AND CONTRACTION

2.8.1 Inner Product

The product $A^\mu B_\mu$ is the inner (or scalar) product of two vectors. As shown earlier, Eq. (2.14), it is invariant under coordinate transformation. It is the sum of diagonal terms, as

$$A^\mu B_\mu = \sum_{\mu=1}^{4} A^\mu B_\mu = A^1 B_1 + A^2 B_2 + A^3 B_3 + A^4 B_4 \tag{2.60}$$

2.8.2 Outer Product

The product of two tensors is a tensor whose rank is the sum of the ranks of the two tensors. For example, if a tensor $A^{\mu_1\mu_2\cdots\mu_n}_{\nu_1\nu_2\cdots\nu_m}$ (which is covariant of order m and contravariant of order n) and a tensor $B^{\alpha_1\alpha_2\cdots\alpha_p}_{\beta_1\beta_2\cdots\beta_q}$ (which is covariant of order q and contravariant of order p) are multiplied, the product is a tensor, which is covariant tensor of order $m + q$ and contravariant of order $n + p$. This product is called *open product* or *outer product* of the two tensors.

Without any loss of generality, we confine ourselves to space–time four-dimensional manifold and consider two four-vectors A^μ and B_ν. The outer product of these four-vectors is $A^\mu B_\nu$, which is a tensor of rank two. This is easily seen as follows:

$$A'^\mu = \frac{\partial x'^\mu}{\partial x^\alpha} A^\alpha \tag{2.61}$$

$$B'_\nu = \frac{\partial x^\beta}{\partial x'^\nu} B_\beta \tag{2.62}$$

Now
$$A'^\mu B'_\nu = \frac{\partial x'^\mu}{\partial x^\alpha} \frac{\partial x^\beta}{\partial x'^\nu} A^\alpha B_\beta \tag{2.63}$$

It is obvious that $A^\mu B_\nu$ transforms like the components of a second rank tensor, contravariant in μ and covariant in ν.

The outer product is defined for any types of two tensors, the rank of the resulting tensor is the sum of the individual ranks of the tensors.

2.8.3 Contraction

Contraction is a method of producing from a mixed tensor, a tensor of *rank two lower*, by equating covariant and contravariant indices. For the sake of illustration, consider a mixed tensor $A^\lambda_{\mu\nu\rho}$. One can form a contracted tensor by equating the indices λ and ρ, *i.e.*, putting $\lambda = \rho$ and getting $A^\rho_{\mu\nu\rho}$, which is a tensor of rank two lower, since ρ is a dummy index. Further, we show that $A^\rho_{\mu\nu\rho}$ is a tensor of rank two. Thus, taking the original tensor $A^\lambda_{\mu\nu\rho}$, one obtains,

$$A'^\lambda_{\mu\nu\rho} = \frac{\partial x'^\lambda}{\partial x^\alpha} \frac{\partial x^\beta}{\partial x'^\mu} \frac{\partial x^\gamma}{\partial x'^\nu} \frac{\partial x^\delta}{\partial x'^\rho} A^\alpha_{\beta\gamma\delta} \tag{2.64}$$

Putting $\lambda = \rho$, we get

$$A'^{\rho}_{\mu\nu\rho} = \frac{\partial x'^{\rho}}{\partial x^{\alpha}} \frac{dx^{\beta}}{dx'^{\mu}} \frac{\partial x^{\gamma}}{\partial x'^{\nu}} \frac{\partial x^{\delta}}{\partial x'^{\rho}} A^{\alpha}_{\beta\gamma\delta} \tag{2.65}$$

However,
$$\frac{\partial x'^{\rho}}{\partial x^{\alpha}} \frac{\partial x^{\delta}}{\partial x'^{\rho}} = \frac{\partial x^{\delta}}{\partial x^{\alpha}} = \delta^{\delta}_{\alpha}$$

Therefore, rewriting Eq. (2.65), we get

$$\begin{aligned} A'^{\rho}_{\mu\nu\rho} &= \frac{\partial x^{\beta}}{\partial x'^{\mu}} \frac{dx^{\gamma}}{\partial x'^{\nu}} \delta^{\delta}_{\alpha} A^{\alpha}_{\beta\gamma\delta} \\ &= \frac{\partial x^{\beta}}{\partial x'^{\mu}} \frac{dx^{\gamma}}{\partial x'^{\nu}} A^{\alpha}_{\beta\gamma\alpha} \end{aligned} \tag{2.66}$$

which shows that $A^{\rho}_{\mu\nu\rho}$ is a second rank covariant tensor. The indices ρ and α are dummy and do not appear in the final result.

Similarly, one can show that A^{μ}_{μ} is an invariant. Thus,

$$\begin{aligned} A'^{\mu}_{\mu} &= \frac{\partial x'^{\mu}}{\partial x^{\alpha}} \frac{\partial x^{\beta}}{\partial x'^{\mu}} A^{\alpha}_{\beta} \\ &= \delta^{\beta}_{\alpha} A^{\alpha}_{\beta} = A^{\alpha}_{\alpha} \end{aligned} \tag{2.67}$$

Analogously, one can prove that $A^{\mu}B_{\mu}$, $A^{\mu\nu}_{\mu\nu}$, $A^{\mu}_{\nu}B^{\nu}_{\mu}$ etc., are invariants.

Summarily, when a contravariant index and a covariant index are same, the corresponding contravariant and covariant character cancel out. If all the upper suffixes cancel out with the lower ones, the resulting expression is an invariant. The indices that are contracted must be of opposite character. From a general mixed tensor of rank *r*, one contraction gives a tensor of rank $(r-2)$.

2.9 QUOTIENT LAW OF TENSORS

We do come across certain quantities whose transformation properties are not known, *i.e.*, it is not known whether these quantities are tensors or not. One has to establish a criterion by which on can find whether these quantities are tensors. The tensorial nature of a given set of quantities can be established with the help of quotient law, according to which *a quantity whose inner product with an arbitrary covariant (or contravariant) vector or tensor, is a tensor, is itself a tensor.* In order to illustrate the law, we consider the following examples.

Let $X(\mu\nu)$ be a set of quantities whose tensorial nature is not known. Further, let A^{ν} be a contravariant vector, chosen arbitrarily, whose inner product $X(\mu\nu)A^{\nu}$ is a contravariant vector. Under coordinate transformation, we get

$$X'(\mu\nu)A'^{\nu} = \frac{\partial x'^{\mu}}{\partial x^{\alpha}} \left\{ X(\alpha\beta)A^{\beta} \right\} \tag{2.68}$$

because the transformation law for contravariant vector is

$$A'^{\nu} = \frac{\partial x'^{\nu}}{\partial x^{\alpha}} A^{\alpha}$$

Since A^{β} is a contravariant vector, we get from reverse transformation

$$A^{\beta} = \frac{\partial x^{\beta}}{\partial x'^{\nu}} A'^{\nu} \tag{2.69}$$

Substituting for A^{β} from Eq. (2.69) into Eq. (2.68) and transposing, one gets

$$A'^{\nu} = \left\{ X'(\mu\nu) - \frac{\partial x'^{\mu}}{\partial x^{\alpha}} \frac{\partial x^{\beta}}{\partial x'^{\nu}} X(\alpha\beta) \right\} = 0$$

Since A'^{ν} is arbitrary, we can choose it in such a way that only one of its components is non-zero, while all others are zero; and this process can be repeated for all the components of A'^{ν}. The above equation then can only be satisfied if the quantity in the bracket is zero. Thus,

$$X'(\mu\nu) = \frac{\partial x'^{\mu}}{\partial x^{\alpha}} \frac{\partial x^{\beta}}{\partial x'^{\nu}} X(\alpha\beta) \tag{2.70}$$

This is obviously the law of transformation for a mixed tensor of rank two, contravariant in μ and covariant in ν. Thus, $X(\mu\nu) = X^{\mu}_{\nu}$.

Similarly, if A^{ν} is a contravariant vector and $X(\mu\nu)$ any quantity such that its inner product with A^{ν} *i.e.*, $X(\mu\nu)A^{\nu}$ is a covariant vector. Since $X(\mu\nu)A^{\nu}$ is a covariant vector, we have

$$X'(\mu\nu)A'^{\nu} = \frac{\partial x^{\alpha}}{\partial x'^{\mu}} X(\alpha\beta)A^{\beta} \tag{2.71}$$

because a covariant vector transforms according to the law

$$A'_{\mu} = \frac{\partial x^{\alpha}}{\partial x'^{\mu}} A_{\alpha}$$

As A^{ν} is a contravariant vector, we get by reverse transformation

$$A^{\beta} = \frac{\partial x^{\beta}}{\partial x'^{\nu}} A'^{\nu} \tag{2.72}$$

Substituting Eq. (2.72) into Eq. (2.71), one gets

$$A'^{\nu} \left\{ X'(\mu\nu) - \frac{\partial x^{\alpha}}{\partial x'^{\mu}} \frac{\partial x^{\beta}}{\partial x'^{\nu}} X(\alpha\beta) \right\} = 0$$

Since A'^{ν} is arbitrary, it leads to

$$X'(\mu\nu) - \frac{\partial x^{\alpha}}{\partial x'^{\mu}} \frac{\partial x^{\beta}}{\partial x'^{\nu}} X(\alpha\beta) = 0$$

or

$$X'(\mu\nu) = \frac{\partial x^{\alpha}}{\partial x'^{\mu}} \frac{\partial x^{\beta}}{\partial x'^{\nu}} X(\alpha\beta) \tag{2.73}$$

This shows that $X(\mu\nu)$ is a covariant tensor.

Example 2.7

$A^{\rho}_{\mu\nu}$ is an arbitrary tensor and its inner product with $X^{\mu\nu\lambda}$

$$X^{\mu\nu\lambda} A^{\rho}_{\mu\nu} = B^{\lambda\rho}$$

gives a contravariant tensor of rank two. Show that $X^{\mu\nu\lambda}$ is a third rank contravariant tensor.

Solution

Given that
$$X^{\mu\nu\lambda} A^{\rho}_{\mu\nu} = B^{\lambda\rho} \tag{i}$$

which in primed coordinate system becomes

$$\begin{aligned} X'^{\mu\nu\lambda} A'^{\rho}_{\mu\nu} &= B'^{\lambda\rho} \\ &= \frac{\partial x'^{\lambda}}{\partial x^{\alpha}} \frac{\partial x'^{\rho}}{\partial x^{\beta}} B^{\alpha\beta} \\ &= \frac{\partial x'^{\lambda}}{\partial x^{\alpha}} \frac{\partial x'^{\rho}}{\partial x^{\beta}} \left(X^{\gamma\delta\alpha} A^{\beta}_{\gamma\delta} \right) \end{aligned} \tag{ii}$$

According to reverse transformation $A^{\beta}_{\gamma\delta}$ is

$$A^{\beta}_{\gamma\delta} = \frac{\partial x^{\beta}}{\partial x'^{\sigma}} \frac{\partial x'^{\mu}}{\partial x^{\gamma}} \frac{\partial x'^{\nu}}{\partial x^{\delta}} A'^{\sigma}_{\mu\nu} \tag{iii}$$

Substituting Eq. (iii) in the r.h.s. of Eq. (ii), we get

$$\begin{aligned} X'^{\mu\nu\lambda} A'^{\rho}_{\mu\nu} &= \frac{\partial x'^{\lambda}}{\partial x^{\alpha}} \frac{\partial x'^{\rho}}{\partial x^{\beta}} \frac{\partial x^{\beta}}{\partial x'^{\sigma}} \frac{\partial x'^{\mu}}{\partial x^{\gamma}} \frac{\partial x'^{\nu}}{\partial x'^{\delta}} X^{\gamma\delta\alpha} A'^{\sigma}_{\mu\nu} \\ &= \frac{\partial x'^{\mu}}{\partial x^{\gamma}} \frac{\partial x'^{\nu}}{\partial x^{\delta}} \frac{\partial x'^{\lambda}}{\partial x^{\alpha}} \frac{\partial x'^{\rho}}{\partial x^{\beta}} \frac{\partial x^{\beta}}{\partial x^{\sigma}} X^{\gamma\delta\alpha} A'^{\sigma}_{\mu\nu} \end{aligned}$$

However
$$\frac{\partial x'^{\rho}}{\partial x^{\beta}} \frac{\partial x^{\beta}}{\partial x'^{\sigma}} = \frac{\partial x'^{\rho}}{\partial x'^{\sigma}} = \delta^{\rho}_{\sigma}$$

Therefore,
$$\begin{aligned} X'^{\mu\nu\lambda} A'^{\rho}_{\mu\nu} &= \frac{\partial x'^{\mu}}{\partial x^{\gamma}} \frac{\partial x'^{\nu}}{\partial x^{\delta}} \frac{\partial x'^{\lambda}}{\partial x^{\alpha}} X^{\gamma\delta\alpha} \left(\delta^{\rho}_{\sigma} A'^{\sigma}_{\mu\nu} \right) \\ &= \frac{\partial x'^{\mu}}{\partial x^{\gamma}} \frac{\partial x'^{\nu}}{\partial x^{\delta}} \frac{\partial x'^{\lambda}}{\partial x^{\alpha}} X^{\gamma\delta\alpha} A'^{\rho}_{\mu\nu} \end{aligned}$$

or
$$\left\{ X'^{\mu\nu\lambda} - \frac{\partial x'^{\mu}}{\partial x^{\gamma}} \frac{\partial x'^{\nu}}{\partial x^{\delta}} \frac{\partial x'^{\lambda}}{\partial x^{\alpha}} X^{\gamma\delta\alpha} \right\} A'^{\rho}_{\mu\nu} = 0$$

Since $A'^{\rho}_{\mu\nu}$ is arbitrary, this will be satisfied if

$$X'^{\mu\nu\lambda} = \frac{\partial x'^{\mu}}{\partial x^{\gamma}} \frac{\partial x'^{\nu}}{\partial x^{\delta}} \frac{\partial x'^{\lambda}}{\partial x^{\alpha}} X^{\gamma\delta\alpha}$$

which shows that $X^{\mu\nu\lambda}$ is a contravariant tensor of rank three.

2.10 FUNDAMENTAL TENSORS: $g_{\mu\nu}$, $g^{\mu\nu}$, AND g^{μ}_{ν}

The interval between two nearby points is an intrinsic property of the points and is given by the expression

$$ds^2 = g_{\mu\nu} dx^\mu dx^\nu \tag{2.74}$$

Now, since dx^μ and dx^ν are contravariant vectors, ds^2 is an invariant; on the basis of quotient law, one can infer that $g_{\mu\nu}$ is a covariant tensor of rank two. $g_{\mu\nu}$ is called the *fundamental tensor or the metric tensor*. That $g_{\mu\nu}$ is a tensor can be construed from its transformation law, we get

$$ds^2 = g'_{\alpha\beta} dx'^\alpha dx'^\beta \tag{2.75}$$

Applying the inverse transformation to dx^μ and dx^ν, one gets

$$dx^\mu = \frac{\partial x^\mu}{\partial x'^\alpha} dx'^\alpha \tag{2.76}$$

$$dx^\nu = \frac{\partial x^\nu}{\partial x'^\beta} dx'^\beta \tag{2.77}$$

On substituting Eqs. (2.76) and (2.77) into Eq. (2.75), we get

$$g'_{\alpha\beta} dx'^\alpha dx'^\beta = g_{\mu\nu} \frac{\partial x^\mu}{\partial x'^\alpha} \frac{\partial x^\nu}{\partial x'^\beta} dx'^\alpha dx'^\beta \tag{2.78}$$

As dx'^α and dx'^β are arbitrary contravariant vectors, we obtain

$$g'_{\alpha\beta} = \frac{\partial x^\mu}{\partial x'^\alpha} \frac{\partial x^\nu}{\partial x'^\beta} g_{\mu\nu} \tag{2.79}$$

Thus, $g_{\alpha\beta}$ transforms like a covariant tensor of rank two, as we had expected.

Let us consider the remaining fundamental tensors $g^{\mu\nu}$ and g^{ν}_{μ}. In the determinant

$$|g| = \begin{vmatrix} g_{11} & g_{12} & \cdots & g_{1n} \\ g_{21} & g_{22} & \cdots & g_{2n} \\ \cdots & \cdots & \cdots & \cdots \\ g_{n1} & g_{n2} & \cdots & g_{nn} \end{vmatrix} \tag{2.80}$$

$g^{\mu\nu}$ is defined as

$$g^{\mu\nu} = \frac{\text{cofactor of } g_{\mu\nu} \text{ in } |g|}{|g|} \tag{2.81}$$

Thus,

$$\begin{aligned} g_{\mu\sigma} g^{\nu\sigma} &= g_{\mu 1} g^{\nu 1} + g_{\mu 2} g^{\nu 2} + \cdots + g_{\mu n} g^{\nu n} \\ &= \frac{1}{|g|} (g_{\mu 1} C_{\nu 1} + g_{\mu 2} C_{\nu 2} + \cdots + g_{\mu n} C_{\nu n}) \end{aligned}$$

where C stands for the cofactor. Therefore,

$$g_{\mu\sigma} g^{\nu\sigma} = \frac{1}{|g|} \times 0 \quad \text{if } \mu \neq \nu$$
$$g_{\mu\sigma} g^{\nu\sigma} = \frac{1}{|g|} \times |g| \quad \text{if } \mu = \nu \tag{2.82}$$

or
$$g_{\mu\sigma} g^{\nu\sigma} = \delta^{\nu}_{\mu} \tag{2.83}$$

Evidently, g^{ν}_{μ} has the property of substitution operator δ^{ν}_{μ}.

Now
$$g^{\nu}_{\mu} A^{\mu} = A^{\nu} \tag{2.84}$$

shows that g^{ν}_{μ} multiplied by a contravariant vector gives a contravariant vector, so from quotient law, we infer that δ^{ν}_{μ} is a mixed tensor of rank two; contravariant in ν and covariant in μ.

Further, let us derive. g'^{ν}_{μ}. Thus,

$$\begin{aligned} g'^{\nu}_{\mu} &= \frac{\partial x^{\alpha}}{\partial x'^{\mu}} \frac{\partial x'^{\nu}}{\partial x^{\beta}} g^{\beta}_{\alpha} \\ &= g_{\alpha}{}^{\beta} \frac{\partial x'^{\nu}}{\partial x^{\beta}} \frac{\partial x^{\alpha}}{\partial x'^{\mu}} \\ &= \frac{\partial x'^{\nu}}{\partial x^{\alpha}} \frac{\partial x^{\alpha}}{\partial x'^{\mu}} \\ &= 0 \quad \text{if } \mu \neq \nu \\ &= 1 \quad \text{if } \mu = \nu \end{aligned}$$

So
$$g'^{\nu}_{\mu} = g^{\nu}_{\mu} \quad (\mu, \nu = 1, 2, \ldots, n) \tag{2.85}$$

Thus, the components of the mixed tensor δ^{ν}_{μ} have the same constant value in all coordinate systems.

Lastly, let us consider the fundamental tensor $g^{\mu\nu}$ and prove that $g^{\mu\nu}$ are the components of a contravariant tensor of rank two. Let A^{μ} be an arbitrary contravariant vector so that its inner product with a covariant vector $g_{\mu\nu}$ gives say, a covariant vector B_{σ}. Thus,

$$g_{\mu\sigma} A^{\mu} = B_{\sigma} \tag{2.86}$$

Then
$$g^{\nu\sigma} B_{\sigma} = g^{\nu\sigma} g_{\mu\sigma} A^{\mu} = g^{\nu}_{\mu} A^{\mu} = A^{\nu} \tag{2.87}$$

Since the inner product of $g^{\nu\sigma}$ with a covariant vector B_{σ}, gives a contravariant vector A^{ν}, so $g^{\nu\sigma}$ must be a contravariant tensor of rank two.

All the three fundamental tensors $g_{\mu\nu}, g^{\mu\nu}$ and g^{μ}_{ν} have importance in the general theory of relativity.

2.11 RAISING AND LOWERING OF INDICES

The operations of raising and lowering the indices of a tensor are performed with the fundamental tensors $g_{\mu\nu}$ and $g^{\mu\nu}$. The operation of raising a suffix changes a covariant suffix into a contravariant one and vice versa. It consists of forming inner product of the given tensor with either of the tensors $g_{\mu\nu}, g^{\mu\nu}$ and the resulting tensor is not considered a new tensor but rather the same tensor in a new form and is denoted by the same letter.

Let us consider the first rank tensors A^{μ} and A_{μ}. The lowering of a suffix is defined as

$$A_{\mu} = g_{\mu\nu} A^{\nu} \tag{2.88}$$

and raising as

$$A^{\mu} = g^{\mu\nu} A_{\nu} \tag{2.89}$$

Similarly, for a tensor of rank two $A^{\rho\nu}$, we get

$$A_{\mu}^{\ \rho} = g_{\mu\nu} A^{\rho\nu} \tag{2.90}$$

In a straight forward manner, one can verify that the entities obtained through raising or lowering operations, with the fundamental tensors are themselves tensors. Thus, from Eq. (2.88), we obtain

$$\begin{aligned}
A'_{\mu} &= g'_{\mu\nu} A'^{\nu} \\
&= \left(\frac{\partial x^{\alpha}}{\partial x'^{\mu}} \frac{\partial x^{\beta}}{\partial x'^{\nu}} g^{\mu\nu} \right) \left(\frac{\partial x'^{\nu}}{\partial x^{\rho}} A^{\rho} \right) \\
&= \frac{\partial x^{\alpha}}{\partial x'^{\mu}} \frac{\partial x^{\beta}}{\partial x'^{\nu}} \frac{\partial x'^{\nu}}{\partial x^{\rho}} g_{\alpha\beta} A^{\rho} \\
&= \frac{\partial x^{\alpha}}{\partial x'^{\mu}} \frac{\partial x^{\beta}}{\partial x^{\rho}} g_{\alpha\beta} A^{\rho} \\
&= \frac{\partial x^{\alpha}}{\partial x'^{\mu}} \delta^{\beta}_{\rho} g_{\alpha\beta} A^{\rho} \\
&= \frac{\partial x^{\alpha}}{\partial x'^{\mu}} g_{\alpha\beta} A^{\beta} \\
&= \frac{\partial x^{\alpha}}{\partial x'^{\mu}} A_{\alpha}
\end{aligned} \tag{2.91}$$

Thus, A_{μ} transforms like a covariant vector.

2.11.1 Associate Tensors—Conjugate Tensors

Tensors obtained through the operations of raising or lowering the indices, are called *associate tensors*.

For each index of a tensor, one introduces the *conjugate tensor* where the index is transferred to its counterpart (covariant/countervariant) using the relations

$$A^{\mu} = g^{\mu\nu} A_{\nu} \tag{2.92a}$$

$$A_{\mu} = g_{\mu\nu} A^{\nu} \tag{2.92b}$$

Conjugate tensors are also called associate tensors.

In Euclidean geometry, the metric is given by

$$ds^2 = g_{\mu\nu} dx^{\mu} dx^{\nu}$$

and in Cartesian coordinates, it becomes

$$ds^2 = (dx^1)^2 + (dx^2)^2 + (dx^3)^2 \tag{2.93}$$

Obviously, the components of the metric tensor $g_{\mu\nu}$, are

$$g_{11} = g_{22} = g_{33} = 1 \qquad g_{ij} = 0 \quad \text{if } i \neq j$$

and hence the determinant

$$|g| = \begin{vmatrix} g_{11} & g_{12} & g_{13} \\ g_{21} & g_{22} & g_{23} \\ g_{31} & g_{32} & g_{33} \end{vmatrix} = \begin{vmatrix} 1 & 0 & 0 \\ 0 & 1 & 0 \\ 0 & 0 & 1 \end{vmatrix} = 1$$

The corresponding components of the contravariant tensor $g^{\mu\nu}$ are now given by

$$g^{11} = g^{22} = g^{33} = 1, \qquad g^{\mu\nu} = 0 \quad \text{if } \mu \neq \nu \tag{2.94}$$

Thus,
$$g_{\mu\nu} = g^{\mu\nu} = \delta_\mu^\nu \tag{2.95}$$

Thus, the operation of raising or lowering the index, leaves the value of the component unaltered.

2.11.2 Magnitude of a Vector

The magnitude l of a vector A^μ, is defined by the scalar product $A^\mu A_\mu$, as

$$l^2 = A^\mu A_\mu = A_\mu A^\mu = g_{\mu\nu} A^\mu A^\nu = g^{\mu\nu} A_\mu A_\nu \tag{2.96}$$

and it is invariant under coordinate transformation.

2.11.3 Angle Between Two Vectors

In vector algebra, the angle between two vectors is defined by their scalar product, which is given by the product of the lengths of individual vectors and the cosine of the angle between them, *i.e.*,

$$\mathbf{A.B} = |\mathbf{A}||\mathbf{B}|\cos\theta$$

where **A** and **B** are three dimensional vectors and $|\mathbf{A}|$ and $|\mathbf{B}|$ are their lengths and θ is the angle between them. Analogously if A_μ and B_μ are two vectors, then the angle between them is given by

$$\begin{aligned}\cos\theta &= \frac{\text{scalar product of } A^\mu \text{ and } B_\mu}{\text{length of } A_\mu \times \text{length of } B_\mu} \\ &= \frac{A^\mu B_\mu}{\sqrt{(A^\mu A_\mu)(B^\mu B_\mu)}} \\ &= \frac{g^{\mu\nu} A_\mu B_\nu}{\sqrt{(g^{\alpha\beta} A_\alpha B_\beta)(g^{\rho\sigma} A_\rho B_\sigma)}}\end{aligned} \tag{2.97}$$

The angle θ defined by Eq. (2.97) is an invariant quantity. It is real when the metric is positive definite, and when real, it has the usual geometrical meaning even in non-Euclidean space. In three-dimensional case, it is invariant only for transformations in three dimensions and not under Lorentz transformations.

Problems

2.1 The Cartesian components of the velocity vector are

$$u_x = \frac{dx}{dt}, \quad u_y = \frac{dy}{dt}$$

Show that its components in the spherical polar coordinates are

$$\dot{r} = \frac{x\dot{x} + y\dot{y}}{r}, \quad \dot{\theta} = \frac{x\dot{y} + \dot{x}y}{r^2}$$

2.2 The Cartesian components of the acceleration vector are

$$a_x = \frac{d^2x}{dt^2}, \quad a_y = \frac{d^2y}{dt^2} \; a_z = \frac{d^2z}{dt^2}$$

Show that its components in the spherical polar coordinates are

$$a_r = \ddot{r} - r\dot{\theta}^2 - r\sin^2\theta\,\dot{\varphi}^2$$
$$a_\theta = \ddot{\theta} - \sin\theta\cos\varphi\dot{\varphi}$$
$$a_\varphi = \ddot{\varphi} + \frac{2}{r}\dot{r}\,\dot{\varphi} + 2\cot\theta\dot{\theta}\dot{\varphi}$$

2.3 Show that in spherical polar coordinates r, θ, φ

(a) $divA^i = \frac{\partial A^r}{\partial r} + \frac{\partial A^\theta}{\partial \theta} + \frac{\partial A^\varphi}{\partial \varphi} + \frac{2A^r}{r} + \cot\theta\,\varphi$

(b) $divA_i = \frac{\partial A_r}{\partial r} + \frac{1}{r^2}\frac{\partial A_\theta}{\partial \theta} + \frac{1}{r^2\sin^2\theta}\frac{\partial A_\varphi}{\partial \varphi} + \frac{2A_r}{r} + \frac{\cot\theta}{r^2}A_\theta$

(c) $\nabla^2\phi = \frac{\partial^2\phi}{\partial r^2} + \frac{1}{r^2}\frac{\partial^2\phi}{\partial \theta^2} + \frac{1}{r^2\sin^2\theta}\frac{\partial^2\phi}{\partial \varphi^2} + \frac{2}{r}\frac{\partial\phi}{\partial r} + \frac{\cot\theta}{r^2}\frac{\partial\phi}{\partial\theta}$

where ϕ is a scalar field.

2.4 Show that

(a) $A_{\mu\nu}B^{\mu\nu} = A^{\mu\nu}B_{\mu\nu}$

(b) $A_{\mu\alpha}B^{\nu\alpha} = A_\mu{}^\alpha B^\nu_\alpha$

2.5 If $A_{\mu\nu}$ is a symmetric covariant tensor, define the conjugate tensor $A^{\mu\nu}$ and prove that it is a tensor.

Part II Geodesic Curves—Covariant Differentiation

2.12 MANIFOLDS

Manifold is a mathematical concept for describing curved space–time. It is a continuous space of points that may be curved or more complicated globally but locally it appears like flat space. As examples of a manifold, one can consider the surface of a sphere or the surface of earth. Globally, the earth is a curved surface and if one draws a triangle with its base on the equator and sides going from the equator to the north pole, the sum of the three angles will turn out to be greater than 180°. The formulae of Euclidean geometry will not apply, although locally the surface of earth is flat and Euclidean geometry will apply.

A differentiated manifold is a space that is continuous and differentiable. In physics, we describe space–time by a differentiable manifold.

One cannot cover a manifold by a single coordinate system, but is covered with a set of open sets U_μ called coordinate patches. Each coordinate patch can be mapped to flat Euclidean space, Fig. 2.1.

2.13 COVARIANT DERIVATIVE

Let us consider the problem of taking the derivative of a vector. In Cartesian coordinates, it is relatively straight forward as it is obtained by differentiating the components of the vector, since the base vectors are constant. In curved space, the base vectors may vary from point to point, so while taking the derivative of a vector, the base vectors are to be differentiated as well.

To compute the derivative of vector **A**, we expand the vector in a basis and then differentiate, such as

$$\frac{\partial \mathbf{A}}{\partial x} = \frac{\partial}{\partial x^\mu}(A^\nu e_\nu) = \frac{\partial A^\nu}{\partial x^\mu} e_\nu + A^\nu \frac{\partial e_\nu}{\partial x^\mu} \tag{2.98}$$

In Cartesian coordinates, the second term is zero, since the base vector e_ν is constant.

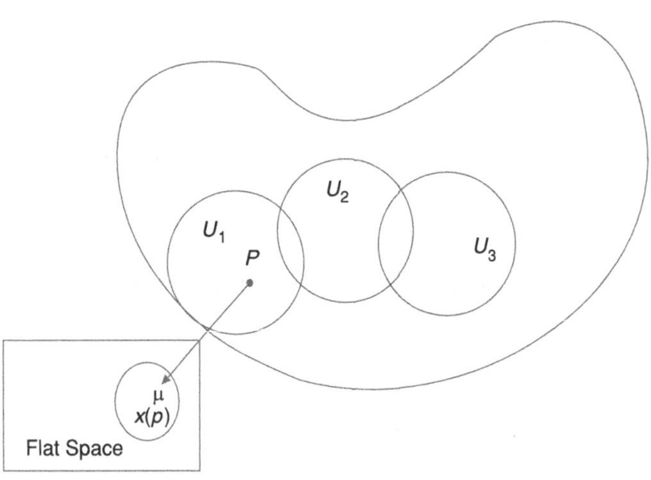

Figure 2.1 *A point P in coordinate patch U_1 in the manifold, is mapped to coordinate $x^\mu(p)$ in the flat Euclidean space.*

Let us consider the vector **A** in spherical polar coordinates. Cartesian coordinates expressed in spherical coordinates, are

$$\begin{aligned} x &= r\sin\theta\cos\phi \\ y &= r\sin\theta\sin\phi \\ z &= r\cos\theta \end{aligned} \tag{2.99}$$

Firstly, we are to work out the base vectors in spherical coordinates in terms of Cartesian base vectors. This is accomplished through the use of a transformation matrix. Taking (x, y, z) as the unprimed system and (r, θ, ϕ) as the primed system, the transformation matrix is

$$\Lambda^{i}_{\ j'} = \begin{pmatrix} \dfrac{\partial x}{\partial r} & \dfrac{\partial y}{\partial r} & \dfrac{\partial z}{\partial r} \\ \dfrac{\partial x}{\partial \theta} & \dfrac{\partial y}{\partial \theta} & \dfrac{\partial z}{\partial \theta} \\ \dfrac{\partial x}{\partial \phi} & \dfrac{\partial y}{\partial \phi} & \dfrac{\partial z}{\partial \phi} \end{pmatrix} \tag{2.100}$$

Using Eq. (2.99), we get the transformation matrix

$$\Lambda^{i}_{\ j'} = \begin{pmatrix} \sin\theta\cos\phi & \sin\theta\sin\phi & \cos\theta \\ r\cos\theta\cos\phi & r\cos\theta\sin\phi & -r\sin\theta \\ -r\sin\theta\sin\phi & r\sin\theta\cos\phi & 0 \end{pmatrix} \tag{2.101}$$

The base vectors for spherical coordinates as expansions in terms of the Cartesian bases are

$$\begin{pmatrix} e_r \\ e_\theta \\ e_\phi \end{pmatrix} = \begin{pmatrix} \sin\theta\cos\phi & \sin\theta\sin\phi & \cos\theta \\ r\cos\theta\cos\phi & r\cos\theta\sin\phi & -r\sin\theta \\ -r\sin\theta\sin\phi & r\sin\theta\cos\phi & 0 \end{pmatrix} \begin{pmatrix} e_x \\ e_y \\ e_z \end{pmatrix} \tag{2.102}$$

or

$$\begin{aligned} e_r &= \sin\theta\cos\phi\, e_x + \sin\theta\sin\phi\, e_y + \cos\theta\, e_z \\ e_\theta &= r\cos\theta\cos\phi\, e_x + r\cos\theta\sin\phi\, e_y - r\sin\theta\, e_z \\ e_\phi &= -r\sin\theta\sin\phi\, e_x + r\sin\theta\cos\phi\, e_y \end{aligned} \tag{2.103}$$

Now let us differentiate these base vectors. Taking the case of e_r, we obtain

$$\begin{aligned} \frac{\partial e_r}{\partial r} &= \frac{\partial}{\partial r}(\sin\theta\cos\phi\, e_x + \sin\theta\sin\phi\, e_y + \cos\theta\, e_z) \\ &= 0 \end{aligned}$$

$$\begin{aligned} \frac{\partial e_r}{\partial \theta} &= \frac{\partial}{\partial \theta}(\sin\theta\cos\phi\, e_x + \sin\theta\sin\phi\, e_y + \cos\theta\, e_z) \\ &= \cos\theta\cos\phi\, e_x + \cos\theta\sin\phi\, e_y - \sin\theta\, e_z \\ &= \frac{1}{r} e_\theta \end{aligned} \tag{2.104}$$

$$\frac{\partial e_r}{\partial \phi} = \frac{\partial}{\partial \phi}(\sin\theta\cos\phi\, e_x + \sin\theta\sin\phi\, e_y + \cos\theta\, e_z)$$
$$= -\sin\theta\sin\phi e_x + \sin\theta\cos\phi e_y$$
$$= \frac{1}{r} e_\phi$$

Thus, we see when differentiating the base vectors, the relationship that results gives the derivative of a base vector in terms of a weighted sum. The sum is an expression in terms of base vectors with integrating coefficients which we denote by $\Gamma^{\mu}_{\nu\rho}$

$$\frac{\partial e_\mu}{\partial x^\nu} = \Gamma^{\rho}_{\mu\nu} e_\rho \tag{2.105}$$

The coefficients $\Gamma^{\mu}_{\nu\rho}$ are functions of the coordinates, as is the case in the example above. Considering the Eqs. (2.104), we identify the results as follows:

$$\frac{\partial e_r}{\partial \theta} = \frac{1}{r} e_\theta \Rightarrow \Gamma^{\theta}_{r\theta} = \frac{1}{r} \tag{2.106}$$

$$\frac{\partial e_r}{\partial \phi} = \frac{1}{r} e_\phi \Rightarrow \Gamma^{\phi}_{r\phi} = \frac{1}{r} \tag{2.107}$$

Let us consider another derivative

$$\frac{\partial e_\phi}{\partial \theta} = \frac{\partial}{\partial \theta}(-r\sin\theta\sin\phi e_x + r\sin\theta\cos\phi e_y)$$
$$= -r\cos\theta\sin\phi e_x + r\cos\theta\cos\phi e_y$$
$$= \cos\theta(-r\sin\phi e_x + r\cos\phi e_y)$$
$$= \frac{\cos\theta}{\sin\theta}(-r\sin\theta\sin\phi e_x + r\sin\theta\cos\phi e_y)$$
$$= \cot\theta e_\phi \tag{2.108}$$

Comparing Eq. (2.108) with Eq. (2.105), we obtain

$$\Gamma^{\phi}_{\phi\theta} = \cot\theta \tag{2.109}$$

The coefficient functions $\Gamma^{\rho}_{\mu\nu}$ are known as *Christoffel symbols* or as *affine connection*.

In essence, Christoffel symbols represent correction terms. A derivative operator applying to a tensor should yield another tensor. As we see in the treatment above, in curved space, the derivative of a vector involves the derivatives of the base vectors as well. Another reason for the Christoffel symbols is that the partial derivative of a tensor is not a tensor. We will see in the next section that Christoffel symbols act as correction terms and the covariant derivative of a tensor transforms as a tensor.

2.14 CHRISTOFFEL'S 3-INDEX SYMBOLS AND THEIR TRANSFORMATION LAW

Let us introduce two expressions of great importance. These are Christoffel's symbols of the first and second kind and are defined as follows

$$\Gamma_{\sigma,\mu\nu} \equiv [\mu\nu,\sigma] \equiv \frac{1}{2}\left(\frac{\partial g_{\mu\sigma}}{\partial x^\nu}+\frac{\partial g_{\nu\sigma}}{\partial x^\mu}-\frac{\partial g_{\mu\nu}}{\partial x^\sigma}\right) \tag{2.110}$$

$$\Gamma^\sigma_{\mu\nu} \equiv \begin{Bmatrix}\sigma\\ \mu\nu\end{Bmatrix} \equiv g^{\sigma\rho}[\mu\nu,\rho] = \frac{1}{2}g^{\sigma\rho}\left(\frac{\partial g_{\mu\rho}}{\partial x^\nu}+\frac{\partial g_{\nu\rho}}{\partial x^\mu}-\frac{\partial g_{\mu\nu}}{\partial x^\rho}\right) \tag{2.111}$$

The first one, *i.e.*, $\Gamma_{\sigma,\mu\nu}$ is called Christoffel's symbol of the first kind, whereas the second one, *i.e.*, $\Gamma^\sigma_{\mu\nu}$ is called Christoffel's symbol or bracket of the second kind.

Since the metric tensor is symmetric, it is obvious from the above equations that the Christoffel symbols have the following symmetry

$$[\mu\nu,\sigma] = [\nu\mu,\sigma]$$

$$\begin{Bmatrix}\sigma\\ \mu\nu\end{Bmatrix} = \begin{Bmatrix}\sigma\\ \nu\mu\end{Bmatrix}$$

implying thereby that the 3-index symbols are symmetrical with respect to μ and ν.

Further, one sees that

$$\Gamma^\sigma_{\mu\nu} = g^{\sigma\rho}[\mu\nu,\rho] \tag{2.112}$$

and

$$\Gamma_{\sigma,\mu\nu} = g_{\sigma\rho}\begin{Bmatrix}\rho\\ \mu\nu\end{Bmatrix} \tag{2.113}$$

The Eq. (2.112) is clear from Eq. (2.111). However, Eq. (2.113) is proved as follows. Multiply Eq. (2.112) by $g_{\sigma\alpha}$ and we get

$$g_{\sigma\alpha}\Gamma^\sigma_{\mu\nu} = g_{\sigma\alpha}g^{\sigma\rho}[\mu\nu,\rho]$$

$$= g^\rho_\alpha[\mu\nu,\rho] = \begin{Bmatrix}\alpha\\ \mu\nu\end{Bmatrix}$$

This is the same as Eq. (2.113).

Example 2.8

Show that

$$[\mu\nu,\rho]+[\rho\nu,\mu] = \frac{\partial g_{\mu\rho}}{\partial x^\nu}.$$

Solution

$$[\mu\nu,\rho]+[\rho\nu,\mu]=\frac{1}{2}\left[\frac{\partial g_{\mu\rho}}{\partial x^{\nu}}+\frac{\partial g_{\nu\rho}}{\partial x^{\mu}}+\frac{\partial g_{\mu\nu}}{\partial x^{\rho}}\right]+\frac{1}{2}\left[\frac{\partial g_{\rho\mu}}{\partial g^{\nu}}+\frac{\partial g_{\nu\mu}}{\partial x^{\rho}}-\frac{\partial g_{\rho\nu}}{\partial x^{\mu}}\right]$$
$$=\frac{\partial g_{\mu\rho}}{\partial x^{\nu}}$$

The Christoffel symbols are not tensors and this can be inferred from their transformation law. Let us consider the transformation of the fundamental tensor under coordinate transformation.

Since $g_{\mu\nu}$ is a covariant tensor of rank two, it will transform as

$$g'_{\mu\nu}=\frac{\partial x^{\alpha}}{\partial x'^{\mu}}\frac{\partial x^{\beta}}{\partial x'^{\nu}}g_{\alpha\beta} \tag{2.114}$$

Differentiating Eq. (2.114) with respect to x'^{ρ}, we get

$$\frac{\partial g'_{\mu\nu}}{\partial x'^{\rho}}=\left[\frac{\partial^2 x^{\alpha}}{\partial x'^{\rho}\partial x'^{\mu}}\frac{\partial x^{\beta}}{\partial x'^{\nu}}+\frac{\partial x^{\alpha}}{\partial x'^{\mu}}\frac{\partial^2 x^{\beta}}{\partial x'^{\rho}\partial x'^{\nu}}\right]g_{\alpha\beta}+\frac{\partial x^{\alpha}}{\partial x'^{\mu}}\frac{\partial x^{\beta}}{\partial x'^{\nu}}\frac{\partial g_{\alpha\beta}}{\partial x^{\gamma}}\frac{\partial x^{\gamma}}{\partial x'^{\rho}} \tag{2.115}$$

where use was made of the result

$$\frac{\partial g_{\alpha\beta}}{\partial x'^{\rho}}\frac{\partial g_{\alpha\beta}}{\partial x^{\gamma}}\frac{\partial x^{\gamma}}{\partial x'^{\rho}}$$

Interchanging the dummy indices α and β in the second term in the bracket on the r.h.s. of Eq. (2.115), and using $g_{\alpha\beta}=g_{\beta\alpha}$, we obtain

$$\frac{\partial g'_{\mu\nu}}{\partial x'^{\rho}}=\left(\frac{\partial^2 x^{\alpha}}{\partial x'^{\rho}\partial x'^{\mu}}\frac{\partial x^{\beta}}{\partial x'^{\nu}}+\frac{\partial^2 x^{\alpha}}{\partial x'^{\rho}\partial x'^{\nu}}\frac{\partial x^{\beta}}{\partial x'^{\mu}}\right)g_{\alpha\beta}+\frac{\partial x^{\alpha}}{\partial x'^{\mu}}\frac{\partial x^{\beta}}{\partial x'^{\nu}}\frac{\partial g_{\alpha\beta}}{\partial x^{\gamma}}\frac{\partial x^{\gamma}}{\partial x'^{\rho}} \tag{2.116}$$

Further, permuting the indices $\mu\nu\rho$ cyclically, we get two more equations, and interchanging the dummy indices α,β,γ in the last term of Eq. (2.116), so as to have $\frac{\partial x^{\alpha}}{\partial x'^{\mu}}\frac{\partial x^{\beta}}{\partial x'^{\nu}}\frac{\partial x^{\gamma}}{\partial x'^{\rho}}$ as the coefficient of the last term, we obtain

$$\frac{\partial g'_{\nu\rho}}{\partial x'^{\mu}}=\left[\frac{\partial^2 x^{\alpha}}{\partial x'^{\mu}\partial x'^{\nu}}\frac{\partial x^{\beta}}{\partial x'^{\rho}}+\frac{\partial^2 x^{\alpha}}{\partial x'^{\mu}\partial x'^{\rho}}\frac{\partial x^{\beta}}{\partial x'^{\nu}}\right]g_{\alpha\beta}+\frac{\partial x^{\alpha}}{\partial x'^{\mu}}\frac{\partial x^{\beta}}{\partial x'^{\nu}}\frac{\partial x^{\gamma}}{\partial x'^{\rho}}\frac{\partial g_{\beta\gamma}}{\partial x^{\alpha}} \tag{2.117}$$

$$\frac{\partial g'_{\rho\mu}}{\partial x'^{\nu}}=\left[\frac{\partial^2 x^{\alpha}}{\partial x'^{\nu}\partial x'^{\rho}}\frac{\partial x^{\beta}}{\partial x'^{\mu}}+\frac{\partial^2 x^{\alpha}}{\partial x'^{\nu}\partial x'^{\mu}}\frac{\partial x^{\beta}}{\partial x'^{\rho}}\right]g_{\alpha\beta}+\frac{\partial x^{\alpha}}{\partial x'^{\mu}}\frac{\partial x^{\beta}}{\partial x'^{\nu}}\frac{\partial x^{\gamma}}{\partial x'^{\rho}}\frac{\partial g_{\gamma\alpha}}{\partial x^{\beta}} \tag{2.118}$$

Adding Eqs. (2.117) and (2.118) and subtracting Eq. (2.116) from the sum and dividing by 2, we get

$$\frac{1}{2}\left(\frac{\partial g'_{\nu\rho}}{\partial x'^{\mu}}+\frac{\partial g'_{\rho\mu}}{\partial x'^{\nu}}-\frac{\partial g'_{\mu\nu}}{\partial x'^{\rho}}\right)=\frac{\partial^2 x^{\alpha}}{\partial x'^{\mu}\partial x'^{\nu}}\frac{\partial x^{\beta}}{\partial x'^{\rho}}g_{\alpha\beta}+\frac{1}{2}\left(\frac{\partial g_{\gamma\alpha}}{\partial x^{\beta}}+\frac{\partial g_{\beta\gamma}}{\partial x^{\alpha}}-\frac{\partial g_{\alpha\beta}}{\partial x^{\gamma}}\right)\frac{\partial x^{\alpha}}{\partial x'^{\mu}}\frac{\partial x^{\beta}}{\partial x'^{\nu}}\frac{\partial x^{\gamma}}{\partial x'^{\rho}} \tag{2.119}$$

Rewriting it in terms of Christoffel's 3-index symbols, we get

$$\Gamma'_{\rho,\mu\nu} = \frac{\partial x^{\alpha}}{\partial x'^{\mu}}\frac{\partial x^{\beta}}{\partial x'^{\nu}}\frac{\partial x^{\gamma}}{\partial x'^{\rho}}\Gamma_{\gamma,\alpha\beta} + \frac{\partial^2 x^{\alpha}}{\partial x'^{\mu}\partial x'^{\nu}}\frac{\partial x^{\beta}}{\partial x'^{\rho}} g_{\alpha\beta} \tag{2.120}$$

This is the transformation law for the Christoffel's 3-index symbol of the first kind and it shows clearly that it is not a tensor. The first term in Eq. (2.120), does transform like a covariant tensor of third rank but the presence of the second term shows that Christoffel symbol of the first kind is not a tensor.

Further, we consider the transformation law for the 3-index symbol of the second kind. Employing the law of transformation for the covariant tensor of rank two

$$g'^{\lambda\rho} = \frac{\partial x'^{\lambda}}{\partial x^{\sigma}}\frac{\partial x'^{\rho}}{\partial x^{\varepsilon}} g^{\sigma\varepsilon} \tag{2.121}$$

and multiplying both sides of Eq. (2.120) with the corresponding sides of Eq. (2.121), we get

$$\begin{aligned} g'^{\lambda\rho}\Gamma'_{\rho,\mu\nu} &= \Gamma'^{\lambda}_{\mu\nu} \\ &= \frac{\partial x^{\alpha}}{\partial x'^{\mu}}\frac{\partial x^{\beta}}{\partial x'^{\nu}}\frac{\partial x^{\gamma}}{\partial x'^{\rho}}\frac{\partial x'^{\lambda}}{\partial x^{\sigma}}\frac{\partial x^{\rho}}{\partial x^{\varepsilon}} g^{\sigma\varepsilon}\Gamma_{\gamma,\alpha\beta} + \frac{\partial^2 x^{\alpha}}{\partial x'^{\mu}\partial x'^{\nu}}\frac{\partial x^{\beta}}{\partial x'^{\rho}}\frac{\partial x'^{\lambda}}{\partial x^{\sigma}}\frac{\partial x'^{\rho}}{\partial x^{\varepsilon}} g^{\sigma\varepsilon} g_{\alpha\beta} \\ &= \frac{\partial x^{\alpha}}{\partial x'^{\mu}}\frac{\partial x^{\beta}}{\partial x'^{\nu}}\frac{\partial x'^{\lambda}}{\partial x^{\sigma}}\delta^{\gamma}_{\varepsilon} g^{\sigma\varepsilon}\Gamma_{\gamma,\alpha\beta} + \frac{\partial^2 x^{\alpha}}{\partial x'^{\mu}\partial x'^{\nu}}\frac{\partial x'^{\lambda}}{\partial x^{\sigma}}\delta^{\beta}_{\varepsilon} g^{\sigma\varepsilon} g_{\alpha\beta} \\ &= \frac{\partial x^{\alpha}}{\partial x'^{\mu}}\frac{\partial x^{\beta}}{\partial x'^{\nu}}\frac{\partial x'^{\lambda}}{\partial x^{\sigma}}\Gamma^{\sigma}_{\alpha\beta} + \frac{\partial^2 x^{\alpha}}{\partial x'^{\mu}\partial x'^{\nu}}\frac{\partial x'^{\lambda}}{\partial x^{\alpha}} \end{aligned} \tag{2.122}$$

Analogously, the presence of second term in Eq. (2.122) shows clearly that the 3-index symbols of second kind also are not tensors. However, it may be remarked that for linear transformations of the type $x^{\mu} = a^{\mu}_{\nu}x'_{\nu} + b^{\mu}$, then $\frac{\partial^2 x^{\alpha}}{\partial x'^{\mu}\partial x'^{\nu}} = 0$, *i.e.*, the second term vanishes, implying thereby that Christoffel's 3-index symbols are tensors relative to linear transformations. Such quantities are called 'pseudo-tensors'.

Example 2.9

Find the fundamental metric tensor and Christoffel's symbols of the first and second kind for surface of a sphere in a two-dimensional space.

Solution

The surface of a sphere of fixed radius *a* is given by

$$ds^2 = a^2 d\theta^2 + a^2 \sin^2\theta d\phi^2$$

$$g_{11} = a^2$$

$$g_{22} = a^2 \sin^2\theta$$

$$g_{12} = g_{21} = 0$$

Thus, $$|g| = \begin{vmatrix} a^2 & 0 \\ 0 & a^2 \sin^2\theta \end{vmatrix} = a^4 \sin^2\theta$$

The contravariant components of the metric tensor are

$$g^{11} = \frac{a^2 \sin^2\theta}{|g|} = \frac{1}{a^2}$$

$$g^{22} = \frac{a^2}{|g|} = \frac{1}{a^2 \sin^2\theta}$$

$$g^{12} = g^{21} = 0$$

The Christoffel symbol of the first kind is from Eq. (2.106)

$$[\mu\nu, \rho] = \frac{1}{2}\left(\frac{\partial g_{\mu\rho}}{\partial x^\nu} + \frac{\partial g_{\nu\rho}}{\partial x^\mu} - \frac{\partial g_{\mu\nu}}{\partial x^\rho}\right) \tag{i}$$

As the indices μ, ν, ρ can take only two values 1 and 2, therefore, we have the following three categories of Christoffel symbols of the first kind.

(i) $\mu = \nu = \rho$. The Eq. (i) reduces to

$$[\mu\mu, \mu] = \frac{1}{2}\left(\frac{\partial g_{\mu\mu}}{\partial x^\mu} + \frac{\partial g_{\mu\mu}}{\partial x^\mu} - \frac{\partial g_{\mu\mu}}{\partial x^\mu}\right) = \frac{1}{2}\frac{\partial g_{\mu\mu}}{\partial x^\mu}$$

Therefore,

$$\frac{\partial g_{11}}{\partial \theta} = 0$$

$$\frac{\partial g_{22}}{\partial \phi} = 0$$

Thus, $$[\mu\mu, \mu] = 0 \quad (\mu = 1, 2) \tag{ii}$$

(ii) $\mu = \nu \neq \rho$. The Eq. (i) reduces to

$$[\mu\mu, \rho] = \frac{1}{2}\left(\frac{\partial g_{\mu\rho}}{\partial x^\mu} + \frac{\partial g_{\mu\rho}}{\partial x^\mu} - \frac{\partial g_{\mu\mu}}{\partial x^\rho}\right) = -\frac{\partial g_{\mu\mu}}{\partial x^\rho}$$

Thus, $$[11, 2 = 0]$$

$$[22,1] = -\frac{1}{2}\frac{\partial a^2 \sin^2\theta}{\partial \theta} = -a^2 \sin\theta \cos\theta \tag{iii}$$

(iii) $\mu = \rho \neq \nu$. The Eq. (i) reduces to

$$[\mu\nu,\rho] = \frac{1}{2}\left(\frac{\partial g_{\mu\mu}}{\partial x^{\nu}} + \frac{\partial g_{\nu\mu}}{\partial x^{\mu}} - \frac{\partial g_{\mu\nu}}{\partial x^{\mu}}\right) = \frac{1}{2}\frac{\partial g_{\mu\mu}}{\partial x^{\mu}}$$

Therefore,

$$[12,1] = [21,1] = \frac{1}{2}\frac{\partial g_{11}}{\partial \phi} = 0$$

$$[21,2] = [12,2] = -\frac{1}{2}\frac{\partial g_{11}}{\partial \theta} = -a^2 \sin\theta\cos\theta \qquad \text{(iv)}$$

Writing collectively, Christoffel index of the first kind from Eqs. (ii), (iii) and (iv), we get

$$\begin{aligned}
&[\mu\mu,\mu] = 0 \quad (\mu = 1,\ 2)\\
&[11,2] = 0\\
&[22,1] = -a^2\ \sin\theta\ \cos\theta\\
&[12,1] = [21,1] = 0\\
&[21,2] = [12,2] = a^2\ \sin\theta\ \cos\theta
\end{aligned}$$

The Christoffel indices of the second kind as defined in Eq. (2.111), are

$$\Gamma^{\sigma}_{\mu\nu} = \frac{1}{2}g^{\sigma\rho}\left(\frac{\partial g_{\mu\rho}}{\partial x^{\nu}} + \frac{\partial g_{\nu\rho}}{\partial x^{\mu}} - \frac{\partial g_{\mu\nu}}{\partial x^{\rho}}\right)$$

As the metric tensor is diagonal in this case, the 3-index of the second kind becomes

$$\Gamma^{\sigma}_{\mu\nu} = g^{\sigma\sigma}[\mu\nu,\sigma]$$

where σ is a free index. Thus, making use of the contravariant components of the metric tensor and the Christoffel indices of the first kind, we get

$$\begin{aligned}
&\Gamma^{\sigma}_{\sigma\sigma} = 0 \quad (\sigma = 1,\ 2)\\
&\Gamma^{1}_{12} = \Gamma^{1}_{21} = 0\\
&\Gamma^{2}_{12} = \Gamma^{2}_{21} = \frac{a^2 \sin\theta\cos\theta}{a^2 \sin^2\theta} = \cot\theta\\
&\Gamma^{2}_{11} = 0\\
&\Gamma^{2}_{22} = \frac{-a^2 \sin\theta\cos\theta}{a^2} = -\sin\theta\cos\theta
\end{aligned}$$

The Christoffel indices of the second kind are independent of the radius of the sphere.

2.15 GEODESICS

In Euclidean space, the shortest distance between any two points is a straight line. However, in Riemannian space, this is no longer true and the shortest distance between any two points, is a curved path. This point is best illustrated by considering the surface of a sphere, as the distance between any two points on it, is the smaller arc on its surface of the great circle passing through these points. Unlike Euclidean space, we cannot employ the path of a light ray, for measuring the shortest distance between say earth and a star, since a ray of light follows a curved path in the gravitational field; as a matter of fact this bending of light in gravitational field constituted a major proof of Einstein's general theory of relativity. In view of this argument, we adopt the concept of a geodesic.

Geodesic is defined as a curve such that for arbitrary displacement, the length of the curve remains unchanged provided the end points are kept fixed. That is to say that $\int_A^B ds$ is stationary or

$$\delta \int_A^B ds = 0 \tag{2.123}$$

where δ is the variation from the actual world line between the two points A and B to any other in its vicinity. Thus a geodesic is a path of extremum distance between any two points on it.

Now ds for Riemannian space is given by

$$ds^2 = g_{\mu\nu} dx^\mu dx^\nu \tag{2.124}$$

Therefore, taking its variation, we get

$$\begin{aligned} 2ds\delta(ds) &= dx^\mu dx^\nu \delta g_{\mu\nu} + g_{\mu\nu} dx^\mu \delta(dx^\nu) + g_{\mu\nu} dx^\nu \delta(dx^\mu) \\ &= dx^\mu dx^\nu \frac{\partial g_{\mu\nu}}{\partial x^\rho} \delta x^\rho + g_{\mu\nu} dx^\mu \delta(dx^\nu) + g_{\mu\nu} dx^\nu \delta(dx^\mu) \end{aligned} \tag{2.125}$$

or

$$\delta(ds) = \frac{1}{2}\left\{\frac{dx^\mu}{ds}\frac{dx^\nu}{ds}\frac{\partial g_{\mu\nu}}{\partial x^\rho}\delta x^\rho + g_{\mu\nu}\frac{dx^\mu}{ds}\frac{d}{ds}(\delta x^\nu) + g_{\mu\nu}\frac{dx^\nu}{ds}\frac{d}{ds}(\delta x^\mu)\right\} ds \tag{2.126}$$

Thus, rewriting Eq. (2.123) in view of Eq. (2.126), we get

$$\frac{1}{2}\int\left\{\frac{dx^\mu}{ds}\frac{dx^\nu}{ds}\frac{\partial g_{\mu\nu}}{\partial x^\rho}\delta x^\rho + g_{\mu\nu}\frac{dx^\mu}{ds}\frac{d}{ds}(\delta x^\nu) + g_{\mu\nu}\frac{dx^\nu}{ds}\frac{d}{ds}(\delta x^\mu)\right\} ds = 0 \tag{2.127}$$

Changing the dummy indices ν and μ to ρ, respectively in the last two terms, we obtain

$$\frac{1}{2}\int\left\{\frac{dx^\mu}{ds}\frac{dx^\nu}{ds}\frac{\partial g_{\mu\nu}}{\partial x^\rho}\delta x^\rho + \left(g_{\mu\rho}\frac{dx^\mu}{ds} + g_{\rho\nu}\frac{dx^\nu}{ds}\right)\frac{d}{ds}(\delta x^\rho)\right\} ds = 0 \tag{2.128}$$

Integrating the second term by parts and using the fact that variation is zero at the fixed ends A and B, one gets

$$\frac{1}{2}\int\left\{\frac{dx^{\mu}}{ds}\frac{dx^{\nu}}{ds}\frac{\partial g_{\mu\nu}}{\partial x^{\rho}}-\frac{d}{ds}\left(g_{\mu\rho}\frac{dx^{\mu}}{ds}+g_{\rho\nu}\frac{dx^{\nu}}{ds}\right)\right\}\delta x^{\rho}\,ds=0 \tag{2.129}$$

As the variation δx^{ρ} is arbitrary, Eq. (2.129) will hold only if its coefficient in the integral vanishes at all the points of the path, *i.e.*,

$$\frac{1}{2}\left\{\frac{dx^{\mu}}{ds}\frac{dx^{\nu}}{ds}\frac{\partial g_{\mu\nu}}{\partial x^{\rho}}-\frac{d}{ds}\left(g_{\mu\rho}\frac{dx^{\mu}}{ds}+g_{\rho\nu}\frac{dx^{\nu}}{ds}\right)\right\}=0 \tag{2.130}$$

or

$$\frac{1}{2}\frac{dx^{\mu}}{ds}\frac{dx^{\nu}}{ds}\frac{\partial g_{\mu\nu}}{\partial x^{\rho}}-\frac{1}{2}\frac{dg_{\mu\rho}}{ds}\frac{dx^{\mu}}{ds}-\frac{1}{2}g_{\mu\rho}\frac{d^{2}x^{\mu}}{ds^{2}}-\frac{1}{2}\frac{dg_{\rho\nu}}{ds}\frac{dx^{\nu}}{ds}-\frac{1}{2}g_{\rho\nu}\frac{d^{2}x^{\nu}}{ds^{2}}=0 \tag{2.131}$$

Since

$$\frac{dg_{\mu\rho}}{ds}=\frac{\partial g_{\mu\rho}}{\partial x^{\nu}}\frac{dx^{\nu}}{ds}$$

$$\frac{dg_{\rho\nu}}{ds}=\frac{\partial g_{\rho\nu}}{\partial x^{\mu}}\frac{dx^{\mu}}{ds}$$

we write Eq. (2.131) as

$$\frac{1}{2}\frac{dx^{\mu}}{ds}\frac{dx^{\nu}}{ds}\left[\frac{\partial g_{\mu\nu}}{\partial x^{\rho}}-\frac{\partial g_{\mu\rho}}{\partial x^{\nu}}-\frac{\partial g_{\rho\nu}}{\partial x^{\mu}}\right]-g_{\sigma\rho}\frac{d^{2}x^{\sigma}}{ds^{2}}=0 \tag{2.132}$$

Multiplying throughout by $g^{\rho\lambda}$, one gets

$$\frac{1}{2}\frac{dx^{\mu}}{ds}\frac{dx^{\nu}}{ds}g^{\rho\lambda}\left[\frac{\partial g_{\mu\nu}}{\partial x^{\rho}}-\frac{\partial g_{\mu\rho}}{\partial x^{\nu}}-\frac{\partial g_{\rho\nu}}{\partial x^{\mu}}\right]-\delta^{\lambda}_{\sigma}\frac{d^{2}x^{\sigma}}{ds^{2}}=0$$

or

$$\frac{d^{2}x^{\lambda}}{ds^{2}}+\frac{1}{2}g^{\rho\lambda}\left[\frac{\partial g_{\mu\rho}}{\partial x^{\nu}}+\frac{\partial g_{\rho\nu}}{\partial x^{\mu}}-\frac{\partial g_{\mu\nu}}{\partial x^{\rho}}\right]\frac{dx^{\mu}}{ds}\frac{dx^{\nu}}{ds}=0 \tag{2.133}$$

Introducing Christoffel's 3-index symbols of the first and second kind, Eq. (2.133) is rewritten as

$$\frac{d^{2}x^{\lambda}}{ds^{2}}+\begin{Bmatrix}\lambda\\ \mu\nu\end{Bmatrix}\frac{dx^{\mu}}{ds}\frac{dx^{\nu}}{ds}=0 \tag{2.134}$$

For $\lambda = 1, 2, 3, 4$ this gives four equations determining a geodesic.

Example 2.12

Show that geodesics in three-dimensional Euclidean space are straight lines.

Solution

The differential equations of a geodesic are

$$\frac{d^2x^\lambda}{ds^2}+\begin{Bmatrix}\lambda\\ \mu\nu\end{Bmatrix}\frac{dx^\mu}{ds}\frac{dx^\nu}{ds}=0$$

In case of Euclidean space, $g_{\mu\nu}$ = constant for all values of μ and ν.

Thus, $\{^{\lambda}_{\mu\nu}\}=0$ for all values of $\lambda,\ \mu,\ \nu$.

The differential equation becomes $$\frac{d^2x^\lambda}{ds^2}=0$$

Integrating it once, we get

$$\frac{dx^\lambda}{ds}=a^\lambda$$

Integrating it again, we get

$$x^\lambda=a^\lambda s+b^\lambda$$

where a^λ and b^λ are constants of integration. This equation is of the type $y = mx + c$ and represents a straight line. Thus geodesics are straight lines in case of Euclidean space.

Example 2.11

Obtain geodesic equations from Lagrangian equations of motion.

Solution

$$ds^2=g_{\mu\nu}dx^\mu dx^\nu$$

or

$$\left(\frac{ds}{dt}\right)^2=\dot{s}^2=g_{\mu\nu}\frac{dx^\mu}{dt}\frac{dx^\nu}{dt}$$

$$=g_{\mu\nu}\,\dot{x}^\mu\dot{x}^\nu$$

The dot denotes differentiation with respect to parameter t.

Therefore, $$ds=\sqrt{[g_{\mu\nu}\dot{x}^\mu\dot{x}^\nu]}\,dt$$

Calling $I = \sqrt{[g_{\mu\nu}\dot{x}^{\mu}\dot{x}^{\nu}]}$, we have

$$ds = Idt$$

or

$$\frac{ds}{dt} = I \equiv \sqrt{[g_{\mu\nu}\dot{x}^{\mu}\dot{x}^{\nu}]}$$

Now

$$\frac{\partial I}{\partial x^{\rho}} = \frac{1}{2\dot{s}}\frac{\partial g_{\mu\nu}}{\partial x^{\rho}}\dot{x}^{\mu}\,\dot{x}^{\nu}$$

and

$$\frac{\partial I}{\partial \dot{x}^{\rho}} = \frac{1}{2\dot{s}}[g_{\mu\rho}\dot{x}^{\mu} + g_{\nu\rho}\dot{x}^{\nu}]$$

$$= \frac{1}{\dot{s}} g_{\mu\rho}\,\dot{x}^{\mu}$$

also

$$\frac{d}{dt}\left(\frac{\partial I}{\partial \dot{x}^{\rho}}\right) = -\frac{1}{\dot{s}^2}\ddot{s}\,g_{\mu\rho}\,\dot{x}^{\mu} + \frac{1}{\dot{s}}\frac{\partial g_{\mu\rho}}{\partial x^{\nu}}\dot{x}^{\nu}\,\dot{x}^{\mu} + \frac{1}{\dot{s}}g_{\mu\rho}\,\ddot{x}^{\mu}$$

For a geodesic, $\int_A^B ds$ is stationary, *i.e.*, $\delta\int_A^B ds = 0$.

This is obtained from Eular–Lagrange equations

$$\frac{d}{dt}\left(\frac{\partial I}{\partial \dot{x}^{\rho}}\right) - \frac{\partial I}{\partial x^{\rho}} = 0$$

Substituting for the terms $\frac{\partial I}{\partial \dot{x}^{\rho}}$ and $\frac{\partial I}{\partial x^{\rho}}$, one obtains

$$-\frac{1}{\dot{s}^2}\ddot{s}g_{\mu\rho}\,\dot{x}^{\mu} + \frac{1}{\dot{s}}\frac{\partial g_{\mu\rho}}{\partial x^{\nu}}\dot{x}^{\nu}\,\dot{x}^{\mu} + \frac{1}{\dot{s}}g_{\mu\rho}\,\ddot{x}^{\mu} - \frac{1}{2\dot{s}}\frac{\partial g_{\mu\nu}}{\partial x^{\rho}}\dot{x}^{\mu}\,\dot{x}^{\nu} = 0$$

Taking $s = t$, we get

$$\dot{s} = \frac{ds}{dt} = 1; \quad \ddot{s} = 0$$

Therefore,

$$\frac{1}{2}\frac{\partial g_{\mu\rho}}{\partial x^{\nu}}\dot{x}^{\nu}\,\dot{x}^{\mu} + \frac{1}{2}\frac{\partial g_{\mu\rho}}{\partial x^{\nu}}\dot{x}^{\nu}\,\dot{x}^{\mu} + g_{\mu\rho}\,\ddot{x}^{\mu} - \frac{1}{2}\frac{\partial g_{\mu\nu}}{\partial x^{\rho}}\dot{x}^{\mu}\,\dot{x}^{\nu} = 0$$

or

$$\frac{1}{2}\frac{\partial g_{\mu\rho}}{\partial x^{\nu}}\dot{x}^{\nu}\,\dot{x}^{\mu} + \frac{1}{2}\frac{\partial g_{\mu\rho}}{\partial x^{\nu}}\dot{x}^{\nu}\,\dot{x}^{\mu} + g_{\sigma\rho}\,\ddot{x}^{\sigma} - \frac{1}{2}\frac{\partial g_{\mu\nu}}{\partial x^{\rho}}\dot{x}^{\mu}\dot{x}^{\nu} = 0$$

This is written as

$$g_{\sigma\rho}\frac{d^2x^\sigma}{ds^2}+\frac{1}{2}\left(\frac{\partial g_{\mu\rho}}{\partial x^\nu}+\frac{\partial g_{\nu\rho}}{\partial x^\mu}-\frac{\partial g_{\mu\nu}}{\partial x^\rho}\right)\frac{dx^\mu}{dt}\frac{dx^\nu}{dt}=0$$

Multiplying throughout by $g^{\rho\lambda}$, one gets

$$\frac{d^2x^\lambda}{ds^2}+\frac{1}{2}g^{\rho\lambda}\left(\frac{\partial g_{\mu\rho}}{\partial x^\nu}+\frac{\partial g_{\nu\rho}}{\partial x^\mu}-\frac{\partial g_{\mu\nu}}{\partial x^\rho}\right)\frac{dx^\mu}{dt}\frac{dx^\nu}{dt}=0$$

or

$$\frac{d^2x^\lambda}{ds^2}+\begin{Bmatrix}\lambda\\ \mu\nu\end{Bmatrix}\frac{dx^\mu}{dt}\frac{dx^\nu}{dt}=0$$

for each value of $\lambda = 1, 2, 3, 4$. These are the geodesic equations obtained by the variational technique.

2.16 COVARIANT DIFFERENTIATION OF VECTORS

2.16.1 Covariant Derivative of Covariant Vector

We have shown earlier that the derivative of an invariant is a covariant vector. But, in general, the derivative of a vector is not a tensor. This is seen easily as follows.

Let A^μ be a contravariant vector transforming according to the law

$$A'^\mu=\frac{\partial x'^\mu}{\partial x^\nu}A^\nu \tag{2.135}$$

We seek its derivative with respect to x'^ρ

$$\begin{aligned}\frac{\partial A'^\mu}{\partial x'^\rho}&=\frac{\partial}{\partial x'^\rho}\left(\frac{\partial x'^\mu}{\partial x^\nu}A^\nu\right)\\&=\frac{\partial}{\partial x^\lambda}\left(\frac{\partial x'^\mu}{\partial x^\nu}A^\nu\right)\frac{\partial x^\lambda}{\partial x'^\rho}\\&=\left[\frac{\partial^2 x'^\mu}{\partial x^\lambda\partial x^\nu}A^\nu+\frac{\partial x'^\mu}{\partial x^\nu}\frac{\partial A^\mu}{\partial x^\lambda}\right]\frac{\partial x^\lambda}{\partial x'^\rho}\\&=\frac{\partial x'^\mu}{\partial x^\nu}\frac{\partial x^\lambda}{\partial x'^\rho}\frac{\partial A^\nu}{\partial x^\lambda}+A^\nu\frac{\partial^2 x'^\mu}{\partial x^\lambda\partial x^\nu}\frac{\partial x^\lambda}{\partial x'^\rho}\end{aligned} \tag{2.136}$$

The presence of the term $A^\nu\frac{\partial^2 x'^\mu}{\partial x^\lambda\partial x^\nu}\frac{\partial x^\lambda}{\partial x'^\rho}$ shows that $\frac{\partial A'^\mu}{\partial x'^\rho}$ does not transform like the components of a tensor.

Next the question arises how we can find quantities that are obtained through partial differentiation and transform like tensors. Since dx^μ is contravariant and ds is invariant, then dx^μ/ds is a contravariant vector. The inner product of dx^μ/ds with a covariant vector A_μ *i.e.*, $A_\mu dx^\mu/ds$ is invariant. The rate of change of this expression per unit interval along any given curve must be independent of the coordinate system, *i.e.*,

$$\frac{d}{ds}\left(A_\mu \frac{dx^\mu}{ds}\right) = \text{invariant} \tag{2.137}$$

This amounts to the assumption that we keep the same absolute curve, irrespective of the choice of coordinate system. Thus this result will be of practical importance when applied to a curve, which is defined independently of the coordinate system. Geodesic being such a curve, we apply this to a geodesic. Therefore,

$$\frac{\partial A_\mu}{dx^\nu}\frac{dx^\nu}{ds}\frac{dx^\mu}{ds} + A_\mu \frac{d^2x^\mu}{ds^2} = \text{invariant} \tag{2.138}$$

The equation of a geodesic is given by Eq. (2.134)

$$\frac{d^2x^\lambda}{ds^2} + \begin{Bmatrix} \lambda \\ \mu\nu \end{Bmatrix} \frac{dx^\mu}{ds}\frac{dx^\nu}{ds} = 0 \tag{2.139}$$

Thus we get

$$A_\mu \frac{d^2x^\mu}{ds^2} = A_\lambda \frac{d^2x^\lambda}{ds^2} = -A_\lambda \begin{Bmatrix} \lambda \\ \mu\nu \end{Bmatrix} \frac{dx^\mu}{ds}\frac{dx^\nu}{ds} \tag{2.140}$$

Substituting Eq. (2.140) in Eq. (2.138), we get

$$\left(\frac{\partial A_\mu}{dx^\nu} - \begin{Bmatrix} \lambda \\ \mu\nu \end{Bmatrix} A_\lambda\right)\frac{dx^\mu}{ds}\frac{dx^\nu}{ds} = \text{invariant} \tag{2.141}$$

As dx^μ/ds and dx^ν/ds are contravariant vectors, we infer from the quotient law that the expression in the parenthesis is a covariant tensor of rank two. Denoting it by $A_{\mu;\nu}$, we obtain

$$A_{\mu;\nu} = \frac{\partial A_\mu}{dx^\nu} - \begin{Bmatrix} \lambda \\ \mu\nu \end{Bmatrix} A_\lambda \tag{2.142}$$

$A_{\mu;\nu}$ is the covariant derivative of the covariant vector A_μ with respect to the coordinate x^ν. If the Christoffel's 3-index symbols vanish as these do for a linear coordinate transformation; the covariant derivative of a four-vector is the same as the ordinary derivative of A_μ with respect to x^ν *i.e.*, $\partial A_\mu/\partial x^\nu$.

2.16.2 Covariant Derivative of a Contravariant Vector

Now $$A_\rho = g_{\rho\sigma} A^\sigma \tag{2.143}$$

Thus,
$$\begin{aligned}
A_{\rho;\nu} &= \frac{\partial A_\rho}{\partial x^\nu} - \begin{Bmatrix} \alpha \\ \rho\nu \end{Bmatrix} A_\alpha \\
&= \frac{\partial}{\partial x^\nu}(g_{\rho\sigma} A^\sigma) - \begin{Bmatrix} \alpha \\ \rho\nu \end{Bmatrix} g_{\alpha\sigma} A^\sigma \\
&= g_{\rho\sigma} \frac{\partial A^\sigma}{\partial x^\nu} + A^\sigma \frac{\partial g_{\rho\sigma}}{\partial x^\nu} - [\rho\nu, \sigma] A^\sigma \\
&= g_{\rho\sigma} \frac{\partial A^\sigma}{\partial x^\nu} + A^\sigma \frac{\partial g_{\rho\sigma}}{\partial x^\nu} - \frac{1}{2}\left(\frac{\partial g_{\rho\sigma}}{\partial x^\nu} - \frac{\partial g_{\rho\nu}}{\partial x^\sigma} + \frac{\partial g_{\nu\sigma}}{\partial x^\rho} \right) A^\sigma \\
&= g_{\rho\sigma} \frac{\partial A^\sigma}{\partial x^\nu} + A^\sigma \frac{1}{2}\left(\frac{\partial g_{\rho\sigma}}{\partial x^\nu} + \frac{\partial g_{\rho\nu}}{\partial x^\sigma} - \frac{\partial g_{\nu\alpha}}{\partial x^\rho} \right) \\
&= g_{\rho\sigma} \frac{\partial A^\sigma}{\partial x^\nu} + [\nu\sigma, \rho] A^\sigma
\end{aligned} \tag{2.144}$$

Multiplying throughout by $g^{\mu\rho}$, one obtains

$$\begin{aligned}
A^\mu{}_{;\nu} &= g^{\mu\rho} g_{\rho\sigma} \frac{\partial A^\sigma}{\partial x^\nu} + g^{\mu\rho}[\nu\sigma, \rho] A^\sigma \\
&= \frac{\partial A^\mu}{\partial x^\nu} + \begin{Bmatrix} \mu \\ \nu\sigma \end{Bmatrix} A^\sigma
\end{aligned} \tag{2.145}$$

This gives the covariant derivative of a contravariant vector.

2.17 COVARIANT DERIVATIVES OF TENSORS

In the earlier section, we dealt with the covariant derivatives of covariant and contravariant vectors and learnt that these are tensors of second rank and are respectively given by the relations

$$A_{\mu;\nu} = \frac{\partial A_\mu}{\partial x^\nu} - \begin{Bmatrix} \lambda \\ \mu\nu \end{Bmatrix} A_\lambda \tag{2.146}$$

and $$A^\mu_{;\nu} = \frac{\partial A^\mu}{\partial x^\nu} + \begin{Bmatrix} \lambda \\ \mu\nu \end{Bmatrix} A \tag{2.147}$$

Now we seek the covariant derivative of a tensor of higher rank. Let us obtain the covariant derivative of a tensor of second rank $A_{\mu\nu}$. If λ^μ denotes the unit tangent vector along the geodesic at the point $\lambda^\mu = \frac{dx^\mu}{ds}$, then $A_{\mu\nu}\lambda^\mu\lambda^\nu$ is a scalar invariant. We calculate its derivative $\frac{d}{ds}(A_{\mu\nu}\lambda^\mu\lambda^\nu)$.

Thus,
$$\frac{d}{ds}(A_{\mu\nu}\lambda^\mu\lambda^\nu)=\frac{dA_{\mu\nu}}{ds}\lambda^\mu\lambda^\nu+A_{\mu\nu}\frac{d\lambda^\mu}{ds}\lambda^\nu+A_{\mu\nu}\lambda^\mu\frac{d\lambda^\nu}{ds}$$
$$=\frac{dA_{\mu\nu}}{dx^\rho}\frac{dx^\rho}{ds}\lambda^\mu\lambda^\nu+A_{\mu\nu}\frac{d^2x^\mu}{ds^2}\lambda^\nu+A_{\mu\nu}\lambda^\mu\frac{d^2x^\nu}{ds^2}$$
$$=\frac{dA_{\mu\nu}}{dx^\rho}\lambda^\rho\lambda^\mu\lambda^\nu+A_{\sigma\nu}\frac{d^2x^\sigma}{ds^2}\lambda^\nu+A_{\mu\sigma}\frac{d^2x^\sigma}{ds^2}\lambda^\mu$$
$$=\frac{dA_{\mu\nu}}{dx^\rho}\lambda^\rho\lambda^\mu\lambda^\nu-\begin{Bmatrix}\sigma\\ \mu\rho\end{Bmatrix}A_{\sigma\nu}\lambda^\mu\lambda^\rho\lambda^\nu-\begin{Bmatrix}\sigma\\ \nu\rho\end{Bmatrix}A_{\mu\sigma}\lambda^\nu\lambda^\rho\lambda^\mu$$
$$=\left[\frac{\partial A_{\mu\nu}}{\partial x^\rho}-\begin{Bmatrix}\sigma\\ \mu\rho\end{Bmatrix}A_{\sigma\nu}-\begin{Bmatrix}\sigma\\ \nu\rho\end{Bmatrix}A_{\mu\sigma}\right]\lambda^\nu\lambda^\sigma\lambda^\mu \tag{2.148}$$

As $\lambda^\mu\lambda^\nu\lambda^\rho$ is a contravariant tensor of rank three, according to quotient law, the expression in the parenthesis is a covariant tensor of rank three since the expression on the l.h.s. is a scalar. Therefore the expression in the bracket is a covariant tensor of rank three, *i.e.*,

$$A_{\mu\nu;\rho}=\frac{\partial A_{\mu\nu}}{\partial x^\rho}-\begin{Bmatrix}\sigma\\ \mu\rho\end{Bmatrix}A_{\sigma\nu}-\begin{Bmatrix}\sigma\\ \nu\rho\end{Bmatrix}A_{\mu\sigma} \tag{2.149}$$

$A_{\mu\nu,\rho}$ is the covariant derivative of the second rank covariant tensor $A_{\mu\nu}$.

Analogously we can derive an expression for the covariant derivative of a contravariant tensor of rank two $A^{\mu\nu}$, as

$$A^{\mu\nu}{}_{;\rho}=\frac{\partial A^{\mu\nu}}{\partial x^\rho}+\begin{Bmatrix}\mu\\ \rho\sigma\end{Bmatrix}A^{\sigma\nu}+\begin{Bmatrix}\nu\\ \rho\sigma\end{Bmatrix}A^{\mu\sigma} \tag{2.150}$$

2.17.1 Rule of Covariant Differentiation

We consider the covariant derivative of a general mixed tensor and take the outer product $A_{\mu\nu}B^\gamma$ and find its covariant derivative

$$(A_{\mu\nu}B^\gamma);\rho=\frac{\partial}{\partial x^\rho}+(A_{\mu\nu}B^\gamma)+\begin{Bmatrix}\gamma\\ \rho\sigma\end{Bmatrix}A_{\mu\nu}B^\sigma-B^\gamma\left[\begin{Bmatrix}\sigma\\ \mu\rho\end{Bmatrix}A_{\sigma\nu}+\begin{Bmatrix}\sigma\\ \nu\rho\end{Bmatrix}A_{\mu\sigma}\right]$$
$$=B^\gamma\frac{\partial A_{\mu\nu}}{\partial x^\rho}+A_{\mu\nu}\frac{\partial B^\gamma}{\partial x^\rho}+A_{\mu\nu}\begin{Bmatrix}\gamma\\ \rho\sigma\end{Bmatrix}B^\sigma-B^\gamma\left[\begin{Bmatrix}\sigma\\ \mu\rho\end{Bmatrix}A_{\sigma\nu}+\begin{Bmatrix}\sigma\\ \nu\rho\end{Bmatrix}A_{\mu\sigma}\right]$$
$$=A_{\mu\nu;\rho}B^\gamma+A_{\mu\nu}B^\gamma_{;\rho} \tag{2.151}$$

Thus, the covariant derivative of a tensor consisting of an outer (or inner) product of two tensors is the sum of two terms obtained by outer (or inner) multiplication of each tensor with the covariant derivative of the other.

Furthermore, one can easily show the covariant differentiation for products, sums and differences obeys the same rule as in the case of ordinary differentiation.

Example 2.12

Prove that the covariant derivatives of the fundamental tensors $g_{\mu\nu}$, $g^{\mu\nu}$, δ^{μ}_{ν} are zero.

Solution

$$\begin{aligned} g_{\mu\nu;\rho} &= \frac{\partial g_{\mu\nu}}{\partial x^{\rho}} - \begin{Bmatrix} \sigma \\ \mu\rho \end{Bmatrix} g_{\sigma\nu} - \begin{Bmatrix} \sigma \\ \nu\rho \end{Bmatrix} g_{\mu\sigma} \\ &= \frac{\partial g_{\mu\nu}}{\partial x^{\rho}} - [\mu\rho,\nu] - [\nu\rho,\mu] \\ &= 0 \end{aligned}$$

Lastly since $g_{\mu\nu} g^{\rho\nu} = \delta^{\rho}_{\mu}$, it follows that the covariant derivative of $g^{\rho\nu}$ is also zero.

Problems

2.6 Calculate Christoffel's symbols corresponding to the metrics

(i) $ds^2 = (dr^1)^2 + r^2 d\theta^2 + r^2 \sin^2\theta d\phi^2$

(ii) $ds^2 = (dx^1)^2 + G(x^1, x^2)(dx^2)^2$

2.7 Compute all the Christoffel symbols for a space whose line element is given by

$$ds^2 = -e^{\lambda} dr^2 - r^2 d\theta^2 - r^2 \sin^2\theta d\phi^2 + e^{\mu} dt^2$$

2.8 The velocity vector-field of a fluid in motion in a plane is $v^{\mu} = (x, 2y)$ in Cartesian coordinates. Find its covariant derivative in polar coordinates.

2.9 Prove that the covariant derivative of an invariant is the same as the ordinary derivative.

2.10 Obtain the non-vanishing 3-index symbols for the metric

$$ds^2 = -dx^2 - dy^2 - dz^2 + f(x, y, z) dt^2$$

Hence obtain the equations of a geodesic.

PART III CURVATURE TENSOR

A space is Euclidean (or flat) if one could find a Cartesian coordinate system anywhere in it, however, if this is not possible, the space is non-Euclidean (or curved). For a given metric, $ds^2 = g_{\mu\nu} dx^\mu dx^\nu$, how one is to find whether the space is flat or curved. It is too cumbersome a procedure to try different coordinate transformations to determine whether it could be brought in the cartesian form $ds^2 = \sum_{\mu=1}^{3} (dx^\mu)^2$. For the sake of illustration, consider the metric. A simple glance at the metric $ds^2 = (dx^1)^2 + (x^1)^2 (dx^2)^2$ shows that it is a metric in terms of polar coordinates of a two-dimensional flat space with the coordinates $x^1 = r, x^2 = \theta$. Normally one desires a simple criterion for judging whether the space is flat or curved. The Riemannian–Christoffel curvature tensor furnishes such a criterion. It will turn out that for a flat space, the Riemannian–Christoffel curvature tensor will vanish at all points, irrespective of the coordinate system employed. The curvature tensor plays a central role in the general theory of relativity.

2.18 RIEMANNIAN COORDINATES

The introduction of a new system of coordinates called Riemannian coordinates, leads to convenience of simplification, since both kinds of Christoffel's 3-index symbols as well as the first partial derivatives of the fundamental tensors $g_{\mu\nu}$'s vanish at their origins. We proceed to verify these statements.

Let the point P_0 be the origin of the coordinates x_0^μ, in an n-dimensional Riemannian space, the suffix 0 indicating that the coordinates refer to the origin P_0. Next consider all the geodesics emanating from P_0. The unit vector at P_0 specifies the direction of each such geodesic. The components of the unit tangent vector at P_0 are

$$\lambda_0^\mu = \left(\frac{dx^\mu}{ds} \right)_0 \tag{2.152}$$

The equation of a geodesic is given by

$$\frac{d^2 x^\mu}{ds^2} + \begin{Bmatrix} \mu \\ \nu\sigma \end{Bmatrix} \frac{dx^\nu}{ds} \frac{dx^\sigma}{ds} = 0 \tag{2.153}$$

It follows that at the origin P_0, Eq. (2.153) gives

$$\left(\frac{d\lambda^\mu}{ds} \right)_0 = -\begin{Bmatrix} \mu \\ \nu\sigma \end{Bmatrix} \lambda_0^\nu \lambda_0^\sigma \tag{2.154}$$

Differentiating Eq. (2.154) with respect to the arc length s measured along the geodesic and adjusting the dummy suffixes, we get

$$\left(\frac{d^2\lambda^\mu}{ds^2} \right)_0 = A^\mu_{\alpha\beta\tau} \lambda_0^\alpha \lambda_0^\beta \lambda_0^\tau \tag{2.155}$$

where

$$A^{\mu}_{\alpha\beta\tau} = \left[-\frac{\partial}{\partial x^{\tau}} \begin{Bmatrix} \mu \\ \alpha\beta \end{Bmatrix} + 2 \begin{Bmatrix} \mu \\ \tau\sigma \end{Bmatrix} \begin{Bmatrix} \sigma \\ \alpha\beta \end{Bmatrix} \right]_0 \tag{2.156}$$

Let a geodesic originating at P_0 pass through a neighbouring point P and the arc length PP_0 be denoted by s. Expressing the coordinates x^{μ} of P in terms of those of P_0 by the Taylor series, we get

$$x^{\mu} = x_0^{\mu} + \lambda_0^{\mu}\, s + \frac{1}{2!}\left(\frac{d\lambda^{\mu}}{ds}\right)_0 s^2 + \frac{1}{3!}\left(\frac{d^2\lambda^{\mu}}{ds^2}\right)_0 s^3 + \cdots \quad (\mu = 1, 2, \ldots, n) \tag{2.157}$$

Rewriting Eq. (2.157) in the light of Eqs. (2.154) and (2.155), one gets

$$x^{\mu} = x_0^{\mu} + \lambda_0^{\mu}\, s - \frac{1}{2} \begin{Bmatrix} \mu \\ \alpha\beta \end{Bmatrix} \lambda_0^{\alpha} \lambda_0^{\beta} s^2 + \frac{1}{6} A^{\mu}_{\alpha\beta\tau} \lambda_0^{\alpha} \lambda_0^{\beta} \lambda_0^{\tau} s^3 + \cdots \tag{2.158}$$

Now define a new coordinate system with origin at λ_0^{μ} and the coordinates of point P in this system as

$$y^{\mu} = \lambda_0^{\mu} s \quad (\mu = 1, 2, \ldots, n) \tag{2.159}$$

Rewriting Eq. (2.158) in view of Eq. (2.159), we obtain

$$x^{\mu} = x_0^{\mu} + y^{\mu} - \frac{1}{2} \begin{Bmatrix} \mu \\ \alpha\beta \end{Bmatrix} y^{\alpha} y^{\beta} + \frac{1}{6} A^{\mu}_{\alpha\beta\tau} y^{\alpha} y^{\beta} y^{\tau} + \cdots \tag{2.160}$$

Obviously, Eq. (2.160) do not involve λ_0^{μ}'s and therefore are valid for all geodesics passing through P_0 and define the transformation from one coordinate system to the other. Since the Jacobian $|\partial x^{\mu}/\partial y^{\nu}| \neq 0$ at P_0, one can invert Eq. (2.160) and solve for y^{μ}'s. Thus,

$$y^{\mu} = (x^{\mu} - x_0^{\mu}) + F^{\mu}\left(x^1 - x_0^1, x^2 - x_0^2, \ldots, x^n - x_0^n\right) \tag{2.161}$$

This new coordinate system is the Riemannian coordinate system.

The equations of geodesics passing through the origin of Riemannian system, when expressed in terms of Reimannian coordinates assume a form as that of the straight lines in Euclidean geometry, as following.

Let the metric of the space in the x-coordinate system be given by

$$ds^2 = g_{\mu\nu}(x)dx^\mu dx^\nu \tag{2.162}$$

Transforming Eq. (2.162) through Eq. (2.160) to the Riemannian coordinate system, we get the equation for metric as

$$ds^2 = g'_{\mu\nu}(y)dy^\mu dy^\nu \tag{2.163}$$

The equation of a geodesic in Riemannian coordinates, is

$$\frac{d^2 y^\mu}{ds^2} + \begin{Bmatrix} \mu \\ \alpha\beta \end{Bmatrix}' \frac{dy^\alpha}{ds}\frac{dy^\beta}{ds} = 0 \tag{2.164}$$

The prime over the Christoffel's symbols implying that these are to be evaluated in terms of $g'_{\mu\nu}(y)$. Alternatively, the finite equations of the geodesics in the Riemannian coordinates are the ones given by Eq. (2.159), and thus these must be the solutions of the differential equations, Eqs. (2.164).

Substituting Eq. (2.159) into Eq. (2.164), gives at the point P_0

$$\begin{Bmatrix} \mu \\ \alpha\beta \end{Bmatrix}'_0 \lambda_0^\alpha \lambda_0^\beta = 0 \tag{2.165}$$

As λ_0^α may refer to any geodesic through P_0, we get the important result that all the Christoffel's symbols of the second kind at the origin of the Riemannian system are zero,

$$\begin{Bmatrix} \mu \\ \alpha\beta \end{Bmatrix}'_0 = 0 \quad (\mu,\ \alpha,\ \beta = 1, 2 \ldots, n) \tag{2.166}$$

All the Christoffel's symbols of the first kind $[\mu\alpha, \beta]$ as well are zero at P_0.

Further, since

$$[\mu\alpha, \beta]_0 + [\alpha\beta, \mu]_0 = 0 \tag{2.167}$$

one obtains

$$\left.\frac{\partial g'_{\alpha\beta}}{\partial y^\mu}\right|_0 = 0 \tag{2.168}$$

This completes the proof and justification of the statement made at the beginning.

2.19 RIEMANN—CHRISTOFFEL CURVATURE TENSOR

We know that ordinary differentiation with respect to coordinates is commutative. Thus, if $\phi(x)$ is a function of the coordinates x^μ $(\mu = 1, 2, \ldots, n)$, then

$$\frac{\partial^2 \phi(x)}{\partial x^\mu \partial x^\nu} = \frac{\partial^2 \phi(x)}{\partial x^\nu \partial x^\mu} \tag{2.169}$$

since x^μ and x^ν $(\mu \neq \nu)$ are independent of each other. Contrarily, the repeated process of covariant differentiation applied to a tensor will necessarily give rise to tensors of higher rank, but, in general, the covariant differentiation is not generally commutative.

Consider the covariant derivative of A_μ with respect to x^ν as

$$A_{\mu;\nu} = \frac{\partial A_\mu}{\partial x^\nu} - \begin{Bmatrix} \rho \\ \mu\nu \end{Bmatrix} A_\rho \tag{2.170}$$

A further differentiation of $A_{\mu;\nu}$ with respect to x^σ yields

$$\begin{aligned} A_{\mu;\nu\sigma} &= A_{\mu;\nu;\sigma} \\ &= \frac{\partial}{\partial x^\sigma}(A_{\mu;\nu}) - \begin{Bmatrix} \rho \\ \mu\sigma \end{Bmatrix} A_{\rho;\nu} - \begin{Bmatrix} \rho \\ \nu\sigma \end{Bmatrix} A_{\mu;\rho} \\ &= \frac{\partial^2 A_\mu}{\partial x^\sigma \partial x^\nu} - \frac{\partial}{\partial x^\sigma}\left[\begin{Bmatrix} \rho \\ \mu\nu \end{Bmatrix} A_\rho\right] - \begin{Bmatrix} \rho \\ \mu\sigma \end{Bmatrix}\left[\frac{\partial A_\rho}{\partial x^\nu} - \begin{Bmatrix} \alpha \\ \rho\nu \end{Bmatrix} A_\alpha\right] - \begin{Bmatrix} \rho \\ \nu\sigma \end{Bmatrix}\left[\frac{\partial A_\mu}{\partial x^\rho} - \begin{Bmatrix} \alpha \\ \mu\rho \end{Bmatrix} A_\alpha\right] \\ &= \frac{\partial^2 A_\mu}{\partial x^\sigma \partial x^\nu} - A_\rho \frac{\partial}{\partial x^\sigma}\begin{Bmatrix} \rho \\ \mu\nu \end{Bmatrix} - \begin{Bmatrix} \rho \\ \mu\nu \end{Bmatrix}\frac{\partial A_\rho}{\partial x^\sigma} + \begin{Bmatrix} \rho \\ \mu\sigma \end{Bmatrix}\begin{Bmatrix} \alpha \\ \rho\nu \end{Bmatrix} A_\alpha - \begin{Bmatrix} \rho \\ \mu\sigma \end{Bmatrix}\frac{\partial A_\rho}{\partial x^\nu} \\ &\quad + \begin{Bmatrix} \rho \\ \nu\sigma \end{Bmatrix}\begin{Bmatrix} \alpha \\ \mu\rho \end{Bmatrix} A_\alpha - \begin{Bmatrix} \rho \\ \nu\sigma \end{Bmatrix}\frac{\partial A_\mu}{\partial x^\rho} \end{aligned} \tag{2.171}$$

Rearranging Eq. (2.171), we get

$$\begin{aligned} A_{\mu;\nu\sigma} &= \frac{\partial^2 A_\mu}{\partial x^\sigma \partial x^\nu} - \begin{Bmatrix} \rho \\ \mu\nu \end{Bmatrix}\frac{\partial A_\rho}{\partial x^\sigma} - \begin{Bmatrix} \rho \\ \mu\sigma \end{Bmatrix}\frac{\partial A_\rho}{\partial x^\nu} - \begin{Bmatrix} \rho \\ \nu\sigma \end{Bmatrix}\frac{\partial A_\mu}{\partial x^\rho} \\ &\quad - A_\rho \frac{\partial}{\partial x^\sigma}\begin{Bmatrix} \rho \\ \mu\nu \end{Bmatrix} + A_\alpha\left[\begin{Bmatrix} \rho \\ \nu\sigma \end{Bmatrix}\begin{Bmatrix} \alpha \\ \mu\rho \end{Bmatrix} + \begin{Bmatrix} \rho \\ \mu\sigma \end{Bmatrix}\begin{Bmatrix} \alpha \\ \rho\nu \end{Bmatrix}\right] \end{aligned} \tag{2.172}$$

Next we reverse the order of covariant differentiation *i.e.*, differentiating, A_μ first w.r.t. σ and then w.r.t. ν, we get from Eq. (2.172), interchanging ν and σ. Thus,

$$\begin{aligned} A_{\mu;\sigma\nu} &= \frac{\partial^2 A_\mu}{\partial x^\nu \partial x^\sigma} - \begin{Bmatrix} \rho \\ \mu\sigma \end{Bmatrix}\frac{\partial A_\rho}{\partial x^\nu} - \begin{Bmatrix} \rho \\ \mu\nu \end{Bmatrix}\frac{\partial A_\rho}{\partial x^\sigma} - \begin{Bmatrix} \rho \\ \sigma\nu \end{Bmatrix}\frac{\partial A_\mu}{\partial x^\rho} - A_\rho \frac{\partial}{\partial x^\nu}\begin{Bmatrix} \rho \\ \mu\sigma \end{Bmatrix} \\ &\quad + A_\alpha\left[\begin{Bmatrix} \rho \\ \sigma\nu \end{Bmatrix}\begin{Bmatrix} \alpha \\ \mu\rho \end{Bmatrix} + \begin{Bmatrix} \rho \\ \mu\nu \end{Bmatrix}\begin{Bmatrix} \alpha \\ \rho\sigma \end{Bmatrix}\right] \end{aligned} \tag{2.173}$$

Subtracting Eq. (2.173) from Eq. (2.172), we get

$$A_{\mu;\nu\sigma} - A_{\mu;\sigma\nu} = A_\rho \left[\frac{\partial}{\partial x^\nu} \begin{Bmatrix} \rho \\ \mu\sigma \end{Bmatrix} - \frac{\partial}{\partial x^\sigma} \begin{Bmatrix} \rho \\ \mu\nu \end{Bmatrix} \right] + A_\alpha \left[\begin{Bmatrix} \rho \\ \mu\sigma \end{Bmatrix} \begin{Bmatrix} \alpha \\ \rho\nu \end{Bmatrix} - \begin{Bmatrix} \rho \\ \mu\nu \end{Bmatrix} \begin{Bmatrix} \alpha \\ \rho\sigma \end{Bmatrix} \right] \tag{2.174}$$

Further, we interchange the dummy indices ρ and α in the last term as

$$A_{\mu;\nu\sigma} - A_{\mu;\sigma\nu} = \left[\frac{\partial}{\partial x^\nu} \begin{Bmatrix} \rho \\ \mu\sigma \end{Bmatrix} - \frac{\partial}{\partial x^\sigma} \begin{Bmatrix} \rho \\ \mu\nu \end{Bmatrix} + \begin{Bmatrix} \alpha \\ \mu\sigma \end{Bmatrix} \begin{Bmatrix} \rho \\ \alpha\nu \end{Bmatrix} - \begin{Bmatrix} \alpha \\ \mu\nu \end{Bmatrix} \begin{Bmatrix} \rho \\ \alpha\sigma \end{Bmatrix} \right] A_\rho \tag{2.175}$$

where

$$\begin{aligned} R^\rho_{\mu\nu\sigma} &= \left[\frac{\partial}{\partial x^\nu} \begin{Bmatrix} \rho \\ \mu\sigma \end{Bmatrix} - \frac{\partial}{\partial x^\sigma} \begin{Bmatrix} \rho \\ \mu\nu \end{Bmatrix} + \begin{Bmatrix} \alpha \\ \mu\sigma \end{Bmatrix} \begin{Bmatrix} \rho \\ \sigma\nu \end{Bmatrix} - \begin{Bmatrix} \alpha \\ \mu\nu \end{Bmatrix} \begin{Bmatrix} \rho \\ \alpha\sigma \end{Bmatrix} \right] \\ &= \partial_\nu \begin{Bmatrix} \rho \\ \mu\sigma \end{Bmatrix} - \partial_\sigma \begin{Bmatrix} \rho \\ \mu\nu \end{Bmatrix} + \begin{Bmatrix} \alpha \\ \mu\sigma \end{Bmatrix} \begin{Bmatrix} \rho \\ \sigma\nu \end{Bmatrix} - \begin{Bmatrix} \alpha \\ \mu\nu \end{Bmatrix} \begin{Bmatrix} \rho \\ \alpha\sigma \end{Bmatrix} \end{aligned} \tag{2.176}$$

where $\partial_\nu = \partial / \partial x^\nu$.

The difference of two tensors is a tensor. As A_ρ is a covariant tensor in ρ, it follows from the quotient theorem that $R^\rho_{\mu\nu\sigma}$ is a tensor of fourth rank, covariant in $\mu\nu\sigma$ and contravariant in ρ. It is called the Riemannian–Christoffel tensor or the curvature tensor. It is only when this tensor vanishes, the covariant differentiation becomes permutable. This tensor was discovered independently first by Riemann and later on by Christoffel.

It may be remarked that $R^\rho_{\mu\nu\sigma}$ does not depend on vector A_ρ and is constituted totally from the components of the fundamental tensor $g_{\mu\nu}$ and their first and second partial derivatives with respect to the coordinates.

2.19.1 Covariant Curvature Tensor

We derive the completely covariant form of the curvature tensor, since this form facilitates the study of its symmetry properties, which is the subject matter of the next article.

For lowering the contravariant index ρ, we form the inner product of $R^\rho_{\mu\nu\sigma}$ with $g_{\lambda\rho}$ and get

$$\begin{aligned} R_{\lambda\mu\nu\sigma} &= g_{\lambda\rho} R^\rho_{\mu\nu\sigma} \\ &= g_{\lambda\rho}\partial_\nu \Gamma^\rho_{\mu\sigma} - g_{\lambda\rho}\partial_\sigma \Gamma^\rho_{\mu\nu} + g_{\lambda\rho}\Gamma^\alpha_{\mu\sigma}\Gamma^\rho_{\alpha\nu} - g_{\lambda\rho}\Gamma^\alpha_{\mu\nu}\Gamma^\rho_{\alpha\sigma} \end{aligned} \tag{2.177}$$

Expressing the first term on the r.h.s. as

$$\begin{aligned} g_{\lambda\rho} \frac{\partial \Gamma^\rho_{\mu\sigma}}{\partial x^\nu} &= \frac{\partial}{\partial x^\nu}\left(g_{\lambda\rho}\Gamma^\rho_{\mu\sigma}\right) - \frac{\partial g_{\lambda\rho}}{\partial x^\nu}\Gamma^\rho_{\mu\sigma} \\ &= \frac{\partial}{\partial x^\nu}(\Gamma_{\lambda,\mu\sigma}) - \frac{\partial g_{\lambda\rho}}{\partial x^\nu}\Gamma^\rho_{\mu\sigma} \end{aligned} \tag{2.178}$$

Rewriting Eq. (2.177) in the light of Eq. (2.178), one gets

$$R_{\lambda\mu\nu\sigma} = \frac{\partial\Gamma_{\lambda,\mu\sigma}}{\partial x^{\nu}} - \frac{\partial\Gamma_{\lambda,\mu\nu}}{\partial x^{\sigma}} - \frac{\partial g_{\lambda\rho}}{\partial x^{\nu}}\Gamma^{\rho}_{\mu\sigma} + \frac{\partial g_{\lambda\rho}}{\partial x^{\sigma}}\Gamma^{\rho}_{\mu\nu} + \Gamma^{\alpha}_{\mu\sigma}\Gamma_{\lambda,\alpha\nu} - \Gamma^{\alpha}_{\mu\nu}\Gamma_{\lambda,\alpha\sigma} \tag{2.179}$$

Exchanging the dummy index α in the last two terms in Eq. (2.179) to ρ, one gets

$$\begin{aligned} R_{\lambda\mu\nu\sigma} &= \frac{\partial\Gamma_{\lambda,\mu\sigma}}{\partial x^{\nu}} - \frac{\partial\Gamma_{\lambda,\mu\nu}}{\partial x^{\sigma}} - \frac{\partial g_{\lambda\rho}}{\partial x^{\nu}}\Gamma^{\rho}_{\mu\sigma} + \frac{\partial g_{\lambda\rho}}{\partial x^{\sigma}}\Gamma^{\rho}_{\mu\nu} + \Gamma^{\rho}_{\mu\sigma}\Gamma_{\lambda,\rho\nu} - \Gamma^{\rho}_{\mu\nu}\Gamma_{\lambda,\rho\sigma} \\ &= \frac{\partial\Gamma_{\lambda,\mu\sigma}}{\partial x^{\nu}} - \frac{\partial\Gamma_{\lambda,\mu\nu}}{\partial x^{\sigma}} + \Gamma^{\rho}_{\mu\sigma}\left(\Gamma_{\lambda,\rho\nu} - \frac{\partial g_{\lambda\rho}}{\partial x^{\nu}}\right) - \Gamma^{\rho}_{\mu\nu}\left(\Gamma^{\rho}_{\lambda,\rho\sigma} - \frac{\partial g_{\lambda\rho}}{\partial x^{\sigma}}\right) \end{aligned} \tag{2.180}$$

But since $\frac{\partial g_{\lambda\rho}}{\partial x^{\nu}} = \Gamma_{\rho,\lambda\nu} + \Gamma_{\lambda,\rho\nu}$, we rewrite Eq. (2.180) as

$$\begin{aligned} R_{\lambda\mu\nu\sigma} &= \frac{\partial\Gamma_{\lambda,\mu\sigma}}{\partial x^{\nu}} - \frac{\partial\Gamma_{\lambda,\mu\sigma}}{\partial x^{\sigma}} + \Gamma^{\rho}_{\mu\nu}\Gamma_{\rho,\lambda\sigma} - \Gamma^{\rho}_{\mu\sigma}\Gamma_{\rho,\lambda\nu} \\ &= \frac{\partial\Gamma_{\lambda,\mu\sigma}}{\partial x^{\nu}} - \frac{\partial\Gamma_{\lambda,\mu\nu}}{\partial x^{\sigma}} + g_{\rho\alpha}\left(\Gamma^{\rho}_{\mu\nu}\Gamma^{\alpha}_{\lambda\sigma} - \Gamma^{\rho}_{\mu\sigma}\Gamma^{\alpha}_{\lambda\nu}\right) \end{aligned} \tag{2.181}$$

Expressing the first two terms in terms of the partial derivatives of $g_{\mu\nu}$'s, we obtain

$$R_{\lambda\mu\nu\sigma} = \frac{1}{2}\left(\frac{\partial^2 g_{\lambda\sigma}}{\partial x^{\nu}\partial x^{\mu}} + \frac{\partial^2 g_{\mu\nu}}{\partial x^{\sigma}\partial x^{\lambda}} - \frac{\partial^2 g_{\mu\sigma}}{\partial x^{\nu}\partial x^{\lambda}} - \frac{\partial^2 g_{\lambda\nu}}{\partial x^{\sigma}\partial x^{\mu}}\right) + g_{\rho\alpha}\left(\Gamma^{\rho}_{\mu\nu}\Gamma^{\alpha}_{\lambda\sigma} - \Gamma^{\rho}_{\mu\sigma}\Gamma^{\alpha}_{\lambda\nu}\right) \tag{2.182}$$

2.20 SYMMETRIES AND ANTI-SYMMETRIES OF CURVATURE TENSOR

In writing the completely covariant form of the curvature tensor, the lowered index was written as the first of the four covariant indices. With this convention, the curvature tensor has the following symmetry and anti-symmetry properties.

1. $R_{\lambda\mu\nu\sigma}$ is anti-symmetrical or skew-symmetrical in the last two indices ν and σ, *i.e.*,

$$R_{\lambda\mu\nu\sigma} = -R_{\lambda\mu\sigma\nu} \tag{2.183}$$

This can be seen from Eq. (2.182)

$$\begin{aligned} R_{\lambda\mu\sigma\nu} &= \frac{1}{2}\left(\frac{\partial^2 g_{\lambda\nu}}{\partial x^{\sigma}\partial x^{\mu}} + \frac{\partial^2 g_{\mu\sigma}}{\partial x^{\nu}\partial x^{\lambda}} - \frac{\partial^2 g_{\mu\nu}}{\partial x^{\sigma}\partial x^{\lambda}} - \frac{\partial^2 g_{\lambda\sigma}}{\partial x^{\nu}\partial x^{\mu}}\right) + g_{\rho\alpha}\left(\Gamma^{\rho}_{\mu\sigma}\Gamma^{\alpha}_{\lambda\nu} - \Gamma^{\rho}_{\mu\nu}\Gamma^{\alpha}_{\lambda\sigma}\right) \\ &= -R_{\lambda\mu\nu\sigma} \end{aligned} \tag{2.184}$$

2. $R_{\lambda\mu\nu\sigma}$ is anti-symmetrical in the first two indices λ and μ, *i.e.*,

$$R_{\lambda\mu\nu\sigma} = -R_{\mu\lambda\nu\sigma} \tag{2.185}$$

   It may be similarly verified from Eq. (2.182).

3. $R_{\lambda\mu\nu\sigma}$ is symmetrical in interchange of the first pair of indices $(\lambda\mu)$ and the second pair of indices $(\nu\sigma)$; however, retaining the order of the indices in both the pairs, *i.e.*,

$$R_{\lambda\mu\nu\sigma} = R_{\nu\sigma\lambda\mu} \tag{2.186}$$

4. Lastly, the sum obtained by keeping any index fixed and permuting the last three indices in a cyclic manner vanishes, *i.e.*,

$$R_{\lambda\mu\nu\sigma} + R_{\lambda\nu\sigma\mu} + R_{\lambda\sigma\mu\nu} = 0 \tag{2.187}$$

In order to prove this result, we proceed as follows.

Writing the expressions for these tensors explicitly, we obtain

$$R_{\lambda\mu\nu\sigma} = \frac{1}{2}\left(\frac{\partial^2 g_{\lambda\sigma}}{\partial x^\nu \partial x^\mu} + \frac{\partial^2 g_{\mu\nu}}{\partial x^\sigma \partial x^\lambda} - \frac{\partial^2 g_{\mu\sigma}}{\partial x^\nu \partial x^\lambda} - \frac{\partial^2 g_{\lambda\nu}}{\partial x^\sigma \partial x^\mu}\right) + g_{\rho\alpha}\left(\Gamma^\rho_{\mu\nu}\Gamma^\alpha_{\lambda\sigma} - \Gamma^\rho_{\mu\sigma}\Gamma^\alpha_{\lambda\nu}\right) \tag{2.188}$$

$$R_{\lambda\nu\sigma\mu} = \frac{1}{2}\left(\frac{\partial^2 g_{\lambda\mu}}{\partial x^\sigma \partial x^\nu} + \frac{\partial^2 g_{\nu\sigma}}{\partial x^\mu \partial x^\lambda} - \frac{\partial^2 g_{\nu\mu}}{\partial x^\sigma \partial x^\lambda} - \frac{\partial^2 g_{\lambda\sigma}}{\partial x^\mu \partial x^\nu}\right) + g_{\rho\alpha}\left(\Gamma^\rho_{\nu\sigma}\Gamma^\alpha_{\lambda\mu} - \Gamma^\rho_{\nu\mu}\Gamma^\alpha_{\lambda\sigma}\right) \tag{2.189}$$

$$R_{\lambda\sigma\mu\nu} = \frac{1}{2}\left(\frac{\partial^2 g_{\lambda\nu}}{\partial x^\mu \partial x^\sigma} + \frac{\partial^2 g_{\sigma\mu}}{\partial x^\nu \partial x^\lambda} - \frac{\partial^2 g_{\sigma\nu}}{\partial x^\mu \partial x^\lambda} - \frac{\partial^2 g_{\lambda\mu}}{\partial x^\nu \partial x^\sigma}\right) + g_{\rho\alpha}\left(\Gamma^\rho_{\sigma\nu}\Gamma^\alpha_{\lambda\nu} - \Gamma^\rho_{\sigma\nu}\Gamma^\alpha_{\lambda\mu}\right) \tag{2.190}$$

Adding the above three Eqs.. (2.188), (2.189) and (2.190), the Eq. (2.187) follows.

2.21 NUMBER OF INDEPENDENT COMPONENTS OF THE CURVATURE TENSOR $R_{\lambda\mu\nu\sigma}$

The Riemann–Christoffel curvature tensor, being of rank four, has n^4 components in an n-dimensional space and for $n = 4$, the number of independent components is 256. However, due to symmetries of the curvature tensor, the number of independent components is reduced considerably.

In order to calculate the number of independent components, we divide the non-vanishing components into the following three categories for the present purpose.

1. Two distinct indices, $R_{\lambda\mu\lambda\mu}$, *i.e.*, those components where each index of the first pair has the same value as an index of the second pair, such as R_{1212}. Since only two of the indices are different in this category, due to the symmetry property, the two pairs must be identical and have the form $R_{\lambda\mu\lambda\mu}$ (no summation implied). The component $R_{\lambda\mu\mu\lambda} = -R_{\lambda\mu\lambda\mu}$. Obviously the number of independent components of $R_{\lambda\mu\lambda\mu}$ is equal to the number of index pairs $(\lambda\mu)$ with $\lambda \neq \mu$. For an n-dimensional space, the first index λ can be chosen in n ways and the second index μ can

be chosen in (n – 1) ways. Since $R_{\lambda\mu\lambda\mu} = R_{\mu\lambda\mu\lambda}$, the order of the two indices is inconsequential; we divide the total number of index pairs by 2. Thus, the resulting number of independent components in this category is

$$N_I = \frac{1}{2}n(n-1) \tag{2.191}$$

The cyclic identities, Eq. (2.187) do not reduce this number, since these identities are independent of the other algebraic identities only when the four indices are different. Explicitly stated, since in the present case only two of the four indices $\lambda\mu\nu\sigma$ are equal, the cyclic identities, Eq. (2.187) have either the form

$$R_{\lambda\mu\lambda\nu} + R_{\lambda\lambda\nu\mu} + R_{\lambda\nu\mu\lambda} = 0 \tag{2.192}$$

or

$$R_{\lambda\mu\nu\lambda} + R_{\lambda\nu\lambda\mu} + R_{\lambda\lambda\mu\nu} = 0 \tag{2.193}$$

Since $R_{\lambda\lambda\nu\mu} = 0 = R_{\lambda\lambda\mu\nu}$, these equations reduce to

$$R_{\lambda\mu\lambda\nu} + R_{\lambda\nu\mu\lambda} = 0 \tag{2.194}$$

$$R_{\lambda\mu\nu\lambda} + R_{\lambda\nu\lambda\mu} = 0 \tag{2.195}$$

which are satisfied due to the symmetry properties of Eqs. (2.185) and (2.186).

2. Three distinct indices, $R_{\lambda\mu\lambda\sigma}$, *i.e.*, those components where one of the indices in each pair is the same like R_{1213}. The first index pair can be chosen in $\frac{1}{2}n(n-1)$ ways as shown above. Having chosen the first pair, the third index can be chosen in $(n-2)$ ways. Thus, the total number of ways in which the indices $(\lambda\mu\lambda\sigma)$ can be chosen is

$$N_{II} = \frac{1}{2}n(n-1)(n-2) \tag{2.196}$$

3. Four distinct indices, $R_{\lambda\mu\nu\sigma}$ like R_{1234}.

   The first index pair can be chosen in $1/2n(n-1)$ ways. Having chosen the first pair, the second pair can be chosen in $\frac{1}{2}(n-2)(n-3)$ ways. As the index pairs can be interchanged, Eq. (2.186), one has to divide this number by a factor of 2. The number of independent ways of choosing the four indices is $\frac{1}{2}n(n-1)\times\frac{1}{2}(n-2)(n-3)\times\frac{1}{2}$. Further, as the four indices are distinct, the cyclic rotation, Eq. (2.187) shows that for a set of four different indices, only two of these components are independent, therefore only two-thirds of these components are independent. The number of independent components of this category, becomes

$$N_{III} = \frac{1}{12}n(n-1)(n-2)(n-3) \tag{2.197}$$

The total number of independent components of curvature tensor $R_{\lambda\mu\nu\sigma}$ becomes

$$\begin{aligned} N &= N_I + N_{II} + N_{III} \\ &= \frac{1}{12}n^2(n^2-1) \end{aligned} \tag{2.198}$$

For a few small values of n, the number of independent components of the curvature tensor is, as follows:

n, dimensionality of space	1	2	3	4	5
Number of independent components	0	1	6	20	50

One-dimensional space has no independent component of the curvature tensor $R_{\lambda\mu\nu\sigma}$, implying thereby that it cannot be curved; *i.e.*, it is not a non-Euclidean space, in the sense of general theory of relativity.

In a two-dimensional space, the indices $\lambda, \mu, \nu, \sigma$ can take only the values 1 and 2. Out of total 16 components, the only non-zero components of the curvature tensor are $R_{1212}, R_{1221}, R_{2112}, R_{2121}$; all of which are related to each other through the Eqs. (2.183), (2.185) and (2.186). As such, the only independent component is one and it is taken as R_{1212} conventionally.

For a three-dimensional space, there are 6 independent components. Due to three independent rotations in such a space, three of the components can be made zero, and thus there remain only 3 independent components.

In a four-dimensional space, the Riemann–Christoffel tensor $R_{\lambda\mu\nu\sigma}$ has 20 independent components. These are obtained as follows, by writing the indices in the order $(\lambda\mu\nu\sigma)$. There result 21 components, *i.e.*,

$$\begin{matrix} R_{1212} & R_{1223} & R_{1313} & R_{1324} & R_{1423} & R_{2323} & R_{2424} \\ R_{1213} & R_{1224} & R_{1314} & R_{1334} & R_{1424} & R_{2324} & R_{2434} \\ R_{1214} & R_{1234} & R_{1323} & R_{1414} & R_{1434} & R_{2334} & R_{3434} \end{matrix} \tag{2.199}$$

The three components with all the indices different are related through the cyclic identity, *i.e.*,

$$R_{1234} + R_{1342} + R_{1423} = 0 \tag{2.200}$$

thus reducing the number to 20.

Now choosing a coordinate system which is cartesian at a given point and carrying out a general rotation of the axes without altering the values of $g_{\mu\nu}$'s at that point; one can arrange the components of the curvature tensor so that six of these become zero; since in a four-dimensional space, one can perform six independent rotations. Thus the curvature of a four-dimensional space at a point is specified by 14 components.

2.22 THE BIANCHI IDENTITIES

In Riemannian coordinates, at the origin of the coordinate system, both the Christoffel's 3-index symbols of the first and second kind $\Gamma^{\alpha}_{\mu\nu}$ and $\Gamma_{\alpha;\mu\nu}$ and the first order derivatives of the fundamental tensors $g_{\mu\nu}$, vanish. In the framework of Riemannian coordinate system, we establish an identity, that is known as Bianchi identities.

The Riemannian–Christoffel curvature tensor in any coordinate system is

$$R^{\rho}_{\mu\nu\sigma} = \partial_{\nu}\Gamma^{\rho}_{\mu\sigma} - \partial_{\sigma}\Gamma^{\rho}_{\mu\nu} + \Gamma^{\alpha}_{\mu\sigma}\Gamma^{\rho}_{\alpha\nu} - \Gamma^{\alpha}_{\mu\nu}\Gamma^{\rho}_{\alpha\sigma} \tag{2.201}$$

At the origin 0, of the Riemannian coordinate system (y), it becomes

$$R^{\rho}_{\mu\nu\sigma}(y) = \frac{\partial}{\partial y^{\nu}}\Gamma^{\rho}_{\mu\sigma} - \frac{\partial}{\partial y^{\sigma}}\Gamma^{\rho}_{\mu\nu} \tag{2.202}$$

since the $\Gamma^{\alpha}_{\mu\sigma}$'s vanish at the origin. The Christoffel symbols $\Gamma^{\rho}_{\mu\sigma}$ and $\Gamma^{\rho}_{\mu\nu}$ are to be evaluated at the origin.

At the origin, the covariant derivative reduces to the ordinary partial derivative. The covariant derivative of $R^{\rho}_{\mu\nu\sigma}$ is

$$R^{\rho}_{\mu\nu\sigma,\alpha} = \frac{\partial^2}{\partial y^{\alpha}\partial y^{\nu}}\Gamma^{\rho}_{\mu\sigma} - \frac{\partial^2}{\partial y^{\alpha}\partial y^{\sigma}}\Gamma^{\rho}_{\mu\nu} \tag{2.203}$$

In this make cyclic permutations of ν, σ, α thus

$$R^{\rho}_{\mu\sigma\alpha;\nu} = \frac{\partial^2}{\partial y^{\nu}\partial y^{\sigma}}\Gamma^{\rho}_{\mu\alpha} - \frac{\partial^2}{\partial y^{\nu}\partial y^{\alpha}}\Gamma^{\rho}_{\mu\sigma} \tag{2.204}$$

$$R^{\rho}_{\mu\alpha\nu,\sigma} = \frac{\partial^2}{\partial y^{\sigma}\partial y^{\alpha}}\Gamma^{\rho}_{\mu\nu} - \frac{\partial^2}{\partial y^{\sigma}\partial y^{\nu}}\Gamma^{\rho}_{\mu\alpha} \tag{2.205}$$

Adding Eqs. (2.203), (2.204) and (2.205), we get

$$R^{\rho}_{\mu\nu\sigma,\alpha} + R^{\rho}_{\mu\sigma\alpha,\nu} + R^{\rho}_{\mu\alpha\nu,\sigma} = 0 \tag{2.206}$$

This is the Bianchi Identities. As Eq. (2.206) is a tensor equation and holds at the origin of the Riemannian system, this will hold for any coordinate system at the origin. Further, since any point can be selected as origin, these will hold at all points of space.

The covariant form of the Bianchi Identities, is obtained by multiplying Eq. (2.206) by $g_{\lambda\rho}$, therefore

$$R_{\lambda\mu\nu\sigma;\alpha} + R_{\lambda\mu\sigma\alpha;\nu} + R_{\lambda\mu\alpha\nu;\sigma} = 0 \tag{2.207}$$

2.23 THE RICCI TENSOR

On contracting the Riemannian–Christoffel tensor $R^{\rho}_{\mu\nu\sigma}$, one can obtain tensors of lower rank. One may imagine that this contraction may be effected in three different ways, but this is not the case. Let us contract ρ with μ so $R^{\rho}_{\rho\nu\sigma} = g^{\rho\mu}R_{\rho\mu\nu\sigma}$. This is obviously zero since $R_{\rho\mu\nu\sigma}$ is anti-symmetric with respect to ρ, μ whereas $g^{\rho\mu}$ is symmetric. In addition, $R^{\rho}_{\mu\nu\sigma} = -R^{\rho}_{\mu\sigma\nu}$; there is only one way of contracting it with respect to indices ρ, σ. This contraction gives the Ricci tensor $R_{\mu\nu}$ as

$$\begin{aligned} R_{\mu\nu} &= R^{\rho}_{\mu\nu\rho} \\ &= \frac{\partial}{\partial x^{\nu}}\Gamma^{\rho}_{\mu\rho} - \frac{\partial}{\partial x^{\rho}}\Gamma^{\rho}_{\mu\nu} + \Gamma^{\alpha}_{\mu\rho}\Gamma^{\rho}_{\alpha\nu} - \Gamma^{\alpha}_{\mu\nu}\Gamma^{\rho}_{\alpha\rho} \end{aligned} \tag{2.208}$$

Further, let us show that $R_{\mu\nu}$ is symmetric with respect to μ, ν. The last three terms are obviously symmetric and the first term is also symmetric as shown below.

Recalling that
$$\Gamma^{\rho}_{\mu\rho} = \frac{\partial}{\partial x^{\mu}}\left[\log\sqrt{(-g)}\right] \tag{2.209}$$

Therefore,
$$\begin{aligned} \frac{\partial}{\partial x^{\nu}}\Gamma^{\rho}_{\mu\rho} &= \frac{\partial^2}{\partial x^{\nu}\partial x^{\mu}}\left[\log\sqrt{(-g)}\right] \\ &= \frac{\partial^2}{\partial x^{\mu}\partial x^{\nu}}\left[\log\sqrt{(-g)}\right] \end{aligned} \tag{2.210}$$

Thus,
$$R_{\mu\nu} = R_{\nu\mu} \tag{2.211}$$

Contracting the Ricci tensor $R_{\mu\nu}$, we obtain

$$R = g^{\mu\nu}R_{\mu\nu} = g^{\mu\sigma}g^{\rho\nu}R_{\rho\mu\nu\sigma} \tag{2.212}$$

R is called the **scalar curvature** or the **curvature invariant**. It is defined in such a way that it is positive for the surface of a sphere in three dimensions.

2.24 THE CONTRACTED BIANCHI IDENTITIES OR THE EINSTEIN TENSOR

The Bianchi relation, Eq. (2.206), involves five suffixes. Let us contract it twice and get a relation with one free index. Rewriting the Eq. (2.206), we get

$$R^{\rho}_{\sigma\mu\nu;\alpha} + R^{\rho}_{\sigma\nu\alpha;\mu} + R^{\rho}_{\sigma\alpha\mu;\nu} = 0 \tag{2.213}$$

Contracting the indices ρ and ν, we get

$$R_{\sigma\mu;\alpha} + R^{\rho}_{\sigma\rho\alpha;\mu} + R^{\rho}_{\sigma\alpha\mu;\rho} = 0 \tag{2.214}$$

or
$$R_{\sigma\mu;\alpha} - R^{\rho}_{\sigma\alpha\rho;\mu} + R^{\rho}_{\sigma\alpha\mu;\rho} = 0 \tag{2.215}$$

where the symmetry property of the curvature tensor was used.

Now
$$R^{\rho}_{\sigma\alpha\rho;\mu} = R_{\sigma\alpha;\mu}$$

So Eq. (2.215) becomes

$$R_{\sigma\mu;\alpha} - R_{\sigma\alpha;\mu} + R^{\rho}_{\sigma\alpha\mu;\rho} = 0 \tag{2.216}$$

Recalling that the covariant derivatives of $g_{\mu\nu}$'s are zero, and multiplying Eq. (8.216) by $g^{\sigma\lambda}$, we get

$$\left(g^{\sigma\lambda} R_{\sigma\mu}\right)_{;\alpha} - \left(g^{\sigma\lambda} R_{\sigma\alpha}\right)_{;\mu} + \left(g^{\sigma\lambda} R^{\rho}_{\sigma\alpha\mu}\right)_{;\rho} = 0 \tag{2.217}$$

or
$$R^{\lambda}_{\mu;\alpha} - R^{\lambda}_{\alpha;\mu} + \left(g^{\sigma\lambda} R^{\rho}_{\sigma\alpha\mu}\right)_{;\rho} = 0 \tag{2.218}$$

Rewriting the last term,

$$\begin{aligned} g^{\sigma\lambda} R^{\rho}_{\sigma\alpha\mu} &= g^{\sigma\lambda} g^{\rho\beta} R_{\beta\sigma\alpha\mu} \\ &= g^{\sigma\lambda} g^{\rho\beta} R_{\sigma\beta\mu\alpha} = g^{\rho\beta} g^{\sigma\lambda} R_{\sigma\beta\mu\alpha} \\ &= g^{\rho\beta} R^{\lambda}_{\beta\mu\alpha} \end{aligned} \tag{2.219}$$

So rewriting Eq. (2.218) in view of Eq. (2.219)

$$R^{\lambda}_{\mu;\lambda} - R^{\lambda}_{\alpha;\mu} + \left(g^{\rho\beta} R^{\lambda}_{\beta\mu\alpha}\right)_{;\rho} = 0 \tag{2.220}$$

On contracting λ and α, one gets

$$R^{\lambda}_{\mu;\lambda} - R^{\lambda}_{\lambda;\mu} + \left(g^{\rho\beta} R_{\beta\mu}\right)_{;\rho} = 0 \tag{2.221}$$

In the last term, changing the dummy index ρ to λ and using the result $g^{\lambda\beta} R_{\beta\mu} = R^{\lambda}_{\mu}$, we get

$$R^{\lambda}_{\mu;\lambda} - R^{\lambda}_{\lambda;\mu} + R^{\lambda}_{\mu;\lambda} = 0 \tag{2.222}$$

or
$$R^{\lambda}_{\mu;\lambda} - \frac{1}{2} R^{\lambda}_{\lambda;\mu} = 0 \tag{2.223}$$

Since
$$R^{\lambda}_{\lambda;\mu} = \frac{\partial}{\partial x^{\mu}} R^{\lambda}_{\lambda} = \frac{\partial}{\partial x^{\nu}}\left(g^{\nu}_{\mu} R^{\lambda}_{\lambda}\right) \tag{2.224}$$

Changing the dummy index λ in the first term of Eq. (2.223) to ν and employing the Eq. (2.224), we have

$$\left(R^{\nu}_{\mu} - \frac{1}{2} g^{\nu}_{\mu} R\right)_{;\nu} = 0 \tag{2.225}$$

The tensor $$G_{\mu}^{\nu} = R_{\mu}^{\nu} - \frac{1}{2} g_{\mu}^{\nu} R \tag{2.226}$$

is the Einstein tensor. It has a central role in the general theory of relativity. It is also called contracted Binachi identities as it was obtained from Binachi identities through contraction.

Example 2.13

Find the Christoffel symbols and the curvature of the two-dimensional space of the surface of a sphere.

Solution

The line element on the surface of a sphere of radius a is

$$ds^2 = a^2 (d\theta^2 + \sin^2 \theta d\varphi^2)$$

Let $x^1 = \theta, x^2 = \phi$, then

$$ds^2 = a^2 [(dx^1)^2 + \sin^2 \theta (dx^2)^2]$$

Thus, $g_{11} = a^2$, $g_{22} = a^2 \sin^2 \theta$, $g_{12} = g_{21} = 0$

$$|g| = \begin{vmatrix} g_{11} & g_{12} \\ g_{21} & g_{22} \end{vmatrix} = \begin{vmatrix} a^a & 0 \\ 0 & a^2 \sin^2 \theta \end{vmatrix} = a^4 \sin^2 \theta$$

$$g^{11} = \frac{\text{cofactor of } g_{11}}{|g|} = \frac{a^2 \sin^2 \theta}{a^4 \sin^2 \theta} = \frac{1}{a^2}$$

$$g^{12} = g^{21} = 0$$

$$g^{22} = \frac{1}{a^2 \sin^2 \theta}$$

The non-vanishing Christoffel symbols are

$$\Gamma_{22}^{1} = -\sin\theta\cos\theta; \quad \Gamma_{21}^{2} = \Gamma_{12}^{2} = \cot = \theta$$

There is only one independent covariant component of the Riemann–Christoffel tensor, *e.g.*,

$$R_{1221} = -a^2 \sin^2 \theta$$

The Ricci tensor $R_{\mu\nu}$ has two non-zero components

$$R_{11} = -1; \quad R_{22} = -\sin^2 \theta$$

The scalar curvature R is given by

$$\begin{aligned} R &= R_{\mu}^{\mu} = g^{11} R_{11} + g^{22} R_{22} \\ &= -2/a^2 \end{aligned}$$

2.25 UNIFORM VECTOR FIELD

We will like to show that the construction of a uniform vector field is only possible when the curvature tensor vanishes. Stated alternatively, when the Riemann–Christoffel tensor vanishes, the differential equations

$$A_{\mu;\nu} = \frac{\partial A_\mu}{\partial x^\nu} - \Gamma^\rho_{\mu\nu} A_\rho = 0 \tag{2.227}$$

are integrable.

Let A_μ be a covariant vector and its covariant derivative, by definition, is

$$A_{\mu;\nu} = \frac{\partial A_\mu}{\partial x^\nu} - \Gamma^\rho_{\mu\nu} A_\rho \tag{2.228}$$

If $A_{\mu;\nu} = 0$, then

$$\frac{\partial A_\mu}{\partial x^\nu} - \Gamma^\rho_{\mu\nu} A_\rho = 0 \tag{2.229}$$

We intend to show that Eq. (2.227) is integrable only when the curvature tensor $R^\rho_{\mu\nu\sigma} = 0$.

From Eq. (2.229), we have

$$\frac{\partial A_\mu}{\partial x^\nu} = \Gamma^\rho_{\mu\nu} A_\rho$$

or

$$\int \frac{\partial A_\mu}{\partial x^\nu} dx^\nu = \int A_\rho dx^\nu \Gamma^\rho_{\mu\nu}$$

or

$$\int dA_\mu = \int A_\rho dx^\nu \Gamma^\rho_{\mu\nu}$$

or

$$A_\mu = \int A_\rho dx^\nu \Gamma^\rho_{\mu\nu} \tag{2.230}$$

Thus Eq. (2.228) is integrable only when the r.h.s. of Eq. (2.229) is integrable, which will be the case if $A_\rho dx^\nu \Gamma^\rho_{\mu\nu}$ is a perfect differential. Thus writing it as

$$A_\rho \Gamma^\rho_{\mu\nu} dx^\nu = dB_\mu$$

But

$$dB_\mu = \frac{\partial B_\mu}{\partial x^\nu} dx^\nu$$

or

$$\left(A_\rho \Gamma^\rho_{\mu\nu} - \frac{\partial B_\mu}{\partial x^\nu} \right) dx^\nu = 0$$

or

$$A_\rho \Gamma^\rho_{\mu\nu} = \frac{\partial B_\mu}{\partial x^\nu} \tag{2.231}$$

Differentiating it w.r.t. x^σ, we get

$$\frac{\partial A_\rho}{\partial x^\sigma}\Gamma^\rho_{\mu\nu} + A_\rho \frac{\partial \Gamma^\rho_{\mu\nu}}{\partial x^\sigma} = \frac{\partial^2 B_\mu}{\partial x^\sigma \partial x^\nu} \tag{2.232}$$

Interchanging ν and σ, one gets

$$\frac{\partial A_\rho}{\partial x^\nu}\Gamma^\rho_{\mu\sigma} + A_\rho \frac{\partial \Gamma^\rho_{\mu\sigma}}{\partial x^\nu} = \frac{\partial^2 B_\mu}{\partial x^\nu \partial x^\sigma} \tag{2.233}$$

Subtracting Eq. (2.233) from Eq. (2.232), we get

$$\frac{\partial A_\rho}{\partial x^\sigma}\Gamma^\rho_{\mu\nu} + A_\rho \frac{\partial \Gamma^\rho_{\mu\nu}}{\partial x^\sigma} - \frac{\partial A_\rho}{\partial x^\nu}\Gamma^\rho_{\mu\sigma} - A_\rho \frac{\partial \Gamma^\rho_{\mu\sigma}}{\partial x^\nu} = 0 \tag{2.234}$$

Employing Eq. (2.229), one writes Eq. (2.234) as

$$\Gamma^\rho_{\mu\nu}\Gamma^\beta_{\rho\sigma}A_\beta + A_\rho \frac{\partial \Gamma^\rho_{\mu\nu}}{\partial x^\sigma} - \Gamma^\beta_{\rho\nu}\Gamma^\rho_{\mu\sigma}A_\beta - A_\rho \frac{\partial \Gamma^\rho_{\mu\sigma}}{\partial x^\nu} = 0$$

or

$$\Gamma^\rho_{\mu\nu}A_\beta\Gamma^\beta_{\rho\sigma} + A_\beta \frac{\partial \Gamma^\beta_{\mu\nu}}{\partial x^\sigma} - A_\beta\Gamma^\beta_{\rho\nu}\Gamma^\rho_{\mu\sigma} - A_\beta \frac{\partial \Gamma^\beta_{\mu\sigma}}{\partial x^\nu} = 0$$

or

$$A_\beta\left(\frac{\partial \Gamma^\beta_{\mu\nu}}{\partial x^\sigma} - \frac{\partial \Gamma^\beta_{\mu\sigma}}{\partial x^\nu} + \Gamma^\rho_{\mu\nu}\Gamma^\beta_{\rho\sigma} - \Gamma^\rho_{\mu\sigma}\Gamma^\beta_{\rho\nu}\right) = 0$$

or

$$A_\beta R^\beta_{\mu\sigma\nu} = 0$$

Since A_β is arbitrary, $R^\beta_{\mu\sigma\nu} = 0$.

Thus, the r.h.s. Eq. (2.229) is integrable only when the curvature tensor vanishes. In conclusion, Eq. (2.228) is integrable only when

$$R^\beta_{\mu\sigma\nu} = 0 \tag{2.235}$$

We can carry the vector A_μ to any point by parallel displacement and the result will be independent of the path of integration and one can obtain a uniform vector field. The construction of a uniform vector field is possible only when the curvature tensor $R^\beta_{\mu\sigma\nu} = 0$.

2.26 THE CONDITION FOR FLAT SPACE–TIME

A region of the world in which it is possible to construct a Galilean frame of reference, is said to be flat. It is possible only when $g_{\mu\nu}$ are constants. Therefore, when $g_{\mu\nu}$ are constants, the space–time is flat.

When $g_{\mu\nu}$ are constants, the 3-index symbols all vanish. However, the 3-index symbols not being tensors, will not vanish in all coordinate systems. The Riemann–Christoffel curvature tensor, being constituted of products and derivatives of the 3-index symbols, will vanish. Furthermore, this being a tensor, will continue to vanish in all oordinate systems. Therefore, the vanishing of curvature tensor, is a necessary condition for the space-tme to be flat. This condition is also sufficient as we will show below.

We showed in Section 2.25, that when $R^{\rho}_{\mu\nu\sigma}=0$, it is possible to construct a uniform vector field by parallel displacement of a vector over the region of space.

If $A^{\mu}_{(\alpha)}$ are four uniform vector fields for $\mu = 0,1,2,3$ (α is not a tensor suffix); then

$$A^{\mu}_{(\alpha);\rho}=0$$

i.e.,
$$\frac{\partial A^{\mu}_{(\alpha)}}{\partial x^{\rho}}=-\Gamma^{\mu}_{\sigma\rho}A^{\sigma}_{(\alpha)} \tag{2.236}$$

Consider a unit mesh constituted by four-vectors at any point. One edge corresponding to the vector $A^{\mu}_{(\alpha)}=dx^{\mu}$ be (1, 0, 0, 0) dx'^{α}. The second edge be $A^{\mu}_{(\alpha)}=dx^{\mu}$ be (0, 1, 0, 0) dx'^{β} and so on and so forth.

Now ds^2 being invariant, we have

$$\begin{aligned} g'_{\alpha\beta}dx'^{\alpha}dx'^{\beta} &= g_{\mu\nu}dx^{\mu}dx^{\nu} \\ &= g_{\mu\nu}A^{\mu}_{(\alpha)}A^{\nu}_{(\beta)}dx'^{\alpha}dx'^{\beta} \end{aligned} \tag{2.237}$$

or
$$g'_{\alpha\beta}=g_{\mu\nu}A^{\mu}_{(\alpha)}A^{\nu}_{(\beta)} \tag{2.238}$$

Differentiating Eq. (2.238), we obtain

$$\begin{aligned} \frac{\partial g'_{\alpha\beta}}{\partial x^{\rho}} &= g_{\mu\nu}A^{\mu}_{(\alpha)}\frac{\partial A^{\nu}_{(\beta)}}{\partial x^{\rho}}+g_{\mu\nu}A^{\nu}_{(\beta)}\frac{\partial A^{\mu}_{(\alpha)}}{\partial x^{\rho}}+A^{\mu}_{(\alpha)}A^{\nu}_{(\beta)}\frac{\partial g_{\mu\nu}}{\partial x^{\rho}} \\ &= -g_{\mu\nu}A^{\mu}_{(\alpha)}A^{\sigma}_{(\beta)}\Gamma^{\nu}_{\sigma\rho}-g_{\mu\nu}A^{\nu}_{(\beta)}A^{\sigma}_{(\alpha)}\Gamma^{\mu}_{\sigma\rho}+A^{\mu}_{(\alpha)}A^{\nu}_{(\beta)}\frac{\partial g_{\mu\nu}}{\partial x^{\rho}} \end{aligned} \tag{2.239}$$

where use was made of Eq. (2.236)

Changing the dummy indices, σ with ν in the first term, σ with μ in the second; we get

$$\begin{aligned} \frac{\partial g'_{\alpha\beta}}{\partial x^{\rho}} &= A^{\mu}_{(\alpha)}A^{\nu}_{(\beta)}\left(-g_{\mu\sigma}\Gamma^{\sigma}_{\nu\rho}-g_{\sigma\nu}\Gamma^{\sigma}_{\mu\rho}+\frac{\partial g_{\mu\nu}}{\partial x^{\rho}}\right) \\ &= A^{\mu}_{(\alpha)}A^{\nu}_{(\beta)}\left(-\Gamma_{\mu;\nu\rho}-\Gamma_{\nu;\mu\rho}+\frac{\partial g_{\mu\nu}}{\partial x^{\rho}}\right)=0 \end{aligned} \tag{2.240}$$

Therefore, when $g'_{\alpha\beta}$ is constant, the space–time is flat.

2.27 PARALLEL DISPLACEMENT AND AFFINE CONNECTIONS

Parallelism in Euclidean space implies that one gets a parallel field of vectors A^μ if the components of A^μ are constants under a parallel translation of a vector. Consider $x^\mu(p)$, a family of curves defined by a parameter p, then parallelism means that

$$\frac{dA^\mu}{dp} = 0 \tag{2.241}$$

where $\mu = 1, 2, 3, 4, \ldots, n$.

Alternatively stated

$$\frac{dA^\mu}{dp} = \frac{\partial A^\mu}{\partial x^\nu}\frac{dx^\nu}{dp} = 0 \tag{2.242}$$

The change in the value of A^μ for an infinitesimal displacement δx^ν along the curve is

$$\delta A^\mu = \frac{\partial A^\mu}{\partial x^\nu}\delta x^\nu = 0 \tag{2.243}$$

The concept of parallelism in Euclidean space is described by any of the Eqs. (2.241)–(2.243).

Next we generalize this concept of parallelism for a general Riemannian space, under a coordinate transformation $x \to x'$, the vector $A^\mu(x) \to A'^\mu(x')$, as

$$A'^\mu(x') = \frac{\partial x'^\mu}{\partial x^\nu}A^\nu(x) \tag{2.244}$$

The condition of parallelism in Euclidean space, $dA^\mu/dp = 0$, *i.e.*, does not hold, in general, for the vector $A'^\mu(x')$ in Reimannian space, as shown below.

Thus

$$\begin{aligned}\frac{dA'^\mu(x')}{dp} &= \frac{d}{dp}\left(\frac{\partial x'^\mu}{\partial x^\nu}A^\nu(x)\right)\\ &= \frac{d}{dp}\left(\frac{\partial x'^\mu}{\partial x^\nu}\right)A^\nu(x) + \frac{dA^\nu(x)}{dp}\frac{\partial x'^\mu}{\partial x^\nu}\\ &= \frac{\partial^2 x'^\mu}{\partial x^\rho \partial x^\nu}\frac{dx^\rho}{dp}A^\nu(x)\end{aligned} \tag{2.245}$$

where we put $\frac{dA^\mu}{dp} = 0$. The factor $\frac{\partial^2 x'^\mu}{\partial x^\rho \partial x^\nu} = 0$, only when the coordinate transformation is a linear transformation. Thus in the case of a general coordinate transformation, $\frac{\partial^2 x'^\mu}{\partial x^\rho \partial x^\nu} \neq 0$.

Next we express the terms in Eq. (2.245) in terms of quantities in the new coordinate system. Thus

$$A^{\nu}(x) = \frac{\partial x^{\nu}}{\partial x'^{\lambda}} A'^{\lambda}(x') \tag{2.246}$$

and

$$\frac{dx^{\rho}}{dp} = \frac{\partial x^{\rho}}{\partial x'^{\sigma}} \frac{dx'^{\sigma}}{dp} \tag{2.247}$$

Writing Eq. (2.245) in view of Eqs. (2.246) and (2.247), we have

$$\frac{dA'^{\mu}(x')}{dp} = \left(\frac{\partial^2 x'^{\mu}}{\partial x^{\rho} \partial x^{\nu}} \frac{\partial x^{\rho}}{\partial x'^{\sigma}} \frac{\partial x^{\nu}}{\partial x'^{\lambda}} \right) \frac{dx'^{\sigma}}{dp} A'^{\lambda}(x') \tag{2.248}$$

Putting

$$\frac{\partial^2 x'^{\mu}}{\partial x^{\rho} \partial x^{\nu}} \frac{\partial x^{\rho}}{\partial x'^{\sigma}} \frac{\partial x^{\nu}}{\partial x'^{\lambda}} = -\Gamma^{\mu}_{\sigma\lambda} \tag{2.249}$$

Eq. (2.248) becomes

$$\frac{dA'^{\mu}(x')}{dp} = -\Gamma^{\mu}_{\sigma\lambda} \frac{dx'^{\sigma}}{dp} A'^{\lambda}(x') \tag{2.250}$$

The coefficients are not to be understood as Christoffel's 3-index symbols. It is clear that the increment of the vector component $A'^{\mu}(x')$ is a linear function of the components $A'^{\lambda}(x')$ and the displacement $\delta x'^{\sigma}$ tangent to the curve along which the displacement is made. In addition, if the components of $\delta x'^{\sigma} A'^{\mu}(x)$ are constant in a particular coordinate system, then a more general law like Eq. (2.250) will be required to hold in an arbitrary frame.

Let us now seek the generalization of Eq. (2.250), confining to the increments of differentials, we rewrite Eq. (2.250) axiomatically as

$$\delta A^{\mu}(x) = -\Gamma^{\mu}_{\alpha\beta} A^{\alpha} \delta x^{\beta} \tag{2.251}$$

where $\Gamma^{\mu}_{\alpha\beta}$ are any set of functions of coordinates and no longer of the form, Eq. (2.249). $\Gamma^{\mu}_{\alpha\beta}$ are called the affine connections of a contravariant vector. Eq. (2.251) implies a general law for parallel displacement of a vector A^{μ} at x into quantities $A^{\mu} + \delta A^{\mu}$ at the point $x + \delta x$. If $A^{\mu} + \delta A^{\mu}$ has to be vector, some restrictions have to be imposed on the $\Gamma^{\mu}_{\alpha\beta}$ coefficients, *i.e.*, $\Gamma^{\mu}_{\alpha\beta}$ have to transform in a particular manner. We set ourselves to find that.

In the unprimed system, we have

$$A^{\mu}(x + dx) = A^{\mu} + \delta A^{\mu} = A^{\mu} - \Gamma^{\mu}_{\alpha\beta} A^{\alpha} \delta x^{\beta} \tag{2.252}$$

and in the primed frame, $x \to x'$, $A^{\mu}(x) \to A'^{\mu}(x')$, we get

$$A'^{\mu}(x'+dx') = A'^{\mu}(x') + \delta A'(x')$$
$$= A'^{\mu}(x') - \Gamma^{\mu}_{\rho\sigma} A'^{\rho} \delta x'^{\sigma} \tag{2.253}$$

If $A^{\mu}(x+dx)$ is a vector provided under coordinate transformation, it behaves like a contravariant vector, *i.e.*,

$$A'^{\mu}(x'+dx') = A^{\alpha}(x+dx)\left(\frac{\partial x'^{\mu}}{\partial x^{\alpha}}\right)_{x+dx} \tag{2.254}$$

Substituting for $A^{\alpha}(x+dx)$ and $A'^{\mu}(x'+dx')$ from Eqs. (2.252) and (2.253), we obtain

$$A'^{\mu}(x') + \delta A'^{\mu}(x') = \left(\frac{\partial x'^{\mu}}{\partial x^{\alpha}}\right)_{x+dx} (A^{\alpha}(x) + \delta A^{\alpha}(x)) \tag{2.255}$$

Using Eq. (2.251) for $A^{\mu}(x)$ and $\delta A'^{\mu}(x')$, we get

$$A'^{\mu}(x') - \Gamma'^{\mu}_{\rho\sigma} A'^{\sigma} \delta x'^{\rho} = \left(\frac{\partial x'^{\mu}}{\partial x^{\alpha}}\right)_{x+dx} \left(A^{\alpha}(x) - \Gamma^{\alpha}_{\beta\delta} A^{\delta} \delta x^{\beta}\right) \tag{2.256}$$

Expanding $(\frac{\partial x'^{\mu}}{\partial x^{\alpha}})_{x+dx}$ around the point x into a Taylor series and retaining only the first order terms, we obtain

$$\left(\frac{\partial x'^{\mu}}{\partial x^{\alpha}}\right)_{x+dx} = \left(\frac{\partial x'^{\mu}}{\partial x^{\alpha}}\right)_{x} + \frac{\partial}{\partial x^{\rho}}\left(\frac{\partial x'^{\mu}}{\partial x^{\alpha}}\right)\delta x^{\rho} + \cdots \tag{2.257}$$

Therefore, Eq. (2.256), in view of Eq. (2.257) becomes

$$A'^{\mu}(x') - \Gamma'^{\mu}_{\rho\sigma} A'^{\sigma} \delta x'^{\rho} = \frac{\partial x'^{\mu}}{\partial x^{\alpha}} A^{\alpha}(x) + \frac{\partial^2 x'^{\mu}}{\partial x^{\rho} \partial x^{\alpha}} A^{\alpha}(x) \delta x^{\rho} - \Gamma^{\alpha}_{\beta\delta} \frac{\partial x'^{\mu}}{\partial x^{\alpha}} A^{\delta} \delta x^{\beta} + \cdots \tag{2.258}$$

As the first terms on both the sides of Eq. (2.258) are the same, we get

$$\Gamma'^{\mu}_{\rho\sigma} A'^{\sigma} \delta x'^{\rho} = \Gamma^{\alpha}_{\beta\delta} \frac{\partial x'^{\mu}}{\partial x^{\alpha}} A^{\delta} \delta x^{\beta} - \frac{\partial^2 x'^{\mu}}{\partial x^{\rho} \partial x^{\alpha}} A^{\alpha}(x) \delta x^{\rho}$$
$$= \left[\Gamma^{\alpha}_{\beta\delta} \frac{\partial x'^{\mu}}{\partial x^{\alpha}} - \frac{\partial^2 x'^{\mu}}{\partial x^{\beta} \partial x^{\delta}}\right] A^{\delta} \delta x^{\beta} \tag{2.259}$$

Now $$A^{\delta}(x) = \frac{\partial x^{\delta}}{\partial x'^{\sigma}} A'^{\delta}(x')$$

and $\delta x^{\beta} = \dfrac{\partial x^{\beta}}{\partial x'^{\rho}} \delta x'^{\rho}$, so Eq. (2.259) becomes

$$\Gamma'^{\mu}_{\rho\sigma} A'^{\sigma} \delta x'^{\rho} = \left[\Gamma^{\alpha}_{\beta\delta} \frac{\partial x'^{\mu}}{\partial x^{\alpha}} - \frac{\partial^2 x'^{\mu}}{\partial x^{\beta} \partial x^{\delta}} \right] \frac{\partial x^{\beta}}{\partial x'^{\rho}} \frac{\partial x^{\delta}}{\partial x'^{\sigma}} A'^{\sigma}(x') \delta x'^{\rho} \tag{2.260}$$

As $A'^{\sigma}(x')\delta x'^{\rho}$ are arbitrary, the law of transformation for $\Gamma^{\alpha}_{\beta\delta}$ is

$$\Gamma'^{\mu}_{\rho\sigma} = \frac{\partial x'^{\mu}}{\partial x^{\alpha}} \frac{\partial x^{\beta}}{\partial x'^{\rho}} \frac{\partial x^{\delta}}{\partial x'^{\sigma}} \Gamma^{\alpha}_{\beta\delta} - \frac{\partial^2 x'^{\mu}}{\partial x^{\beta} \partial x^{\delta}} \frac{\partial x^{\beta}}{\partial x'^{\rho}} \frac{\partial x^{\delta}}{\partial x'^{\sigma}} \tag{2.261}$$

The second term on the r.h.s. can be arranged as follows:

$$\begin{aligned}
\frac{\partial x^{\beta}}{\partial x'^{\rho}} \frac{\partial x^{\delta}}{\partial x'^{\sigma}} \frac{\partial^2 x'^{\mu}}{\partial x^{\beta} \partial x^{\delta}} &= \frac{\partial x^{\beta}}{\partial x'^{\rho}} \frac{\partial}{\partial x^{\beta}} \left(\frac{\partial x^{\delta}}{\partial x'^{\sigma}} \frac{\partial x'^{\mu}}{\partial x^{\delta}} \right) - \frac{\partial x^{\beta}}{\partial x'^{\rho}} \frac{\partial x'^{\mu}}{\partial x^{\delta}} \frac{\partial^2 x^{\delta}}{\partial x^{\beta} \partial x'^{\sigma}} \\
&= \frac{\partial x^{\beta}}{\partial x'^{\rho}} \frac{\partial}{\partial x^{\beta}} (\delta^{\mu}_{\sigma}) - \frac{\partial x'^{\mu}}{\partial x^{\delta}} \frac{\partial}{\partial x^{\beta}} \frac{\partial x^{\delta}}{\partial x'^{\sigma}} \frac{\partial x^{\beta}}{\partial x'^{\rho}} \\
&= \frac{\partial x^{\beta}}{\partial x'^{\rho}} \frac{\partial}{\partial x^{\beta}} (\delta^{\mu}_{\sigma}) - \frac{\partial x'^{\mu}}{\partial x^{\delta}} \frac{\partial^2 x^{\delta}}{\partial x'^{\rho} \partial x'^{\sigma}}
\end{aligned} \tag{2.262}$$

Since δ^{μ}_{σ} is a constant, the first term vanishes and the required law of transformation for, the set of coefficients $\Gamma^{\alpha}_{\beta\gamma}$ from one coordinate system (x) to $\Gamma'^{\mu}_{\rho\sigma}$ in another coordinate system (x') is given by

$$\Gamma'^{\mu}_{\rho\sigma} = \frac{\partial x'^{\mu}}{\partial x^{\alpha}} \left(\frac{\partial x^{\beta}}{\partial x'^{\rho}} \frac{\partial x^{\gamma}}{\partial x'^{\sigma}} \Gamma^{\alpha}_{\beta\gamma} - \frac{\partial^2 x^{\alpha}}{\partial x'^{\rho} \partial x'^{\sigma}} \right) \tag{2.263}$$

The transformation law for $\Gamma'^{\mu}_{\rho\sigma}$ consists of two terms; the first term depends on $\Gamma^{\alpha}_{\beta\gamma}$ in the old coordinate system (x) and has the similar form of the transformation law for a tensor. The second term does not depend on $\Gamma^{\alpha}_{\beta\gamma}$ but is symmetric in the subscripts ρ and σ. Therefore, we infer that the law of transformation is not homogeneous in the coefficients $\Gamma^{\alpha}_{\beta\gamma}$ and is different from the law of transformation of tensors. If $\Gamma^{\alpha}_{\beta\gamma}$ are zero in one coordinate system, they will not be zero in another coordinate system. However, if $\Gamma^{\alpha}_{\beta\gamma}$ are symmetric in one coordinate system, they will continue to be symmetric in another coordinate system.

The following conclusions can be drawn from the transformation law, Eq. (2.263).

1. If the two coordinate systems are connected with each other by a linear transformation, then $\frac{\partial^2 x'^{\alpha}}{\partial x'^{\rho} \partial x'^{\sigma}} = 0$ and $\Gamma^{\alpha}_{\beta\gamma}$ does transform like a tensor. If a vector has constant components in one system, it will have constant components in all other coordinate systems as well, provided these are connected with each other through a linear transformation.

2. In the case of two vector fields having connections, ${}_{(1)}\Gamma^{\alpha}_{\beta\gamma}$ and ${}_{(2)}\Gamma^{\alpha}_{\beta\gamma}$ where ${}_{(1)}\Gamma^{\alpha}_{\beta\gamma} \neq {}_{(2)}\Gamma^{\alpha}_{\beta\gamma}$, then their difference is a tensor, since the inhomogeneous term in Eq. (2.263) depends only on coordinates and we have

$$ {}_{(1)}\Gamma'^{\mu}_{\rho\sigma} = \frac{\partial x'^{\mu}}{\partial x^{\alpha}}\left(\frac{\partial x^{\beta}}{\partial x'^{\rho}}\frac{\partial x^{\gamma}}{\partial x'^{\sigma}}\,{}_{(1)}\Gamma^{\alpha}_{\beta\gamma} - \frac{\partial^2 x'^{\alpha}}{\partial x'^{\rho}\partial x'^{\sigma}}\right) \tag{2.264} $$

and

$$ {}_{(2)}\Gamma'^{\mu}_{\rho\sigma} = \frac{\partial x'^{\mu}}{\partial x^{\alpha}}\left(\frac{\partial x^{\beta}}{\partial x'^{\rho}}\frac{\partial x^{\gamma}}{\partial x'^{\sigma}}\,{}_{(2)}\Gamma^{\alpha}_{\beta\gamma} - \frac{\partial^2 x'^{\alpha}}{\partial x'^{\rho}\partial x'^{\sigma}}\right) \tag{2.265} $$

and on subtracting, one get

$$ {}_{(1)}\Gamma'^{\mu}_{\rho\sigma} - {}_{(2)}\Gamma'^{\mu}_{\rho\sigma} = \frac{\partial x'^{\mu}}{\partial x^{\alpha}}\left(\frac{\partial x^{\beta}}{\partial x'^{\rho}}\frac{\partial x^{\gamma}}{\partial x'^{\sigma}}\left\{{}_{(1)}\Gamma^{\alpha}_{\beta\gamma} - {}_{(2)}\Gamma^{\alpha}_{\beta\gamma}\right\}\right) \tag{2.266} $$

which shows that the difference transforms like a tensor.

2.28 AFFINE CONNECTIONS FOR COVARIANT VECTOR

We have seen that the increment δA^{μ} in the components of a contravariant vector A^{μ} on parallel displacement is given by Eq. (2.251), as

$$ \delta A^{\mu}(x) = -\Gamma^{\mu}_{\alpha\beta} A^{\alpha}\delta x^{\beta} \tag{2.267} $$

One can derive a similar law for covariant vector A_{μ}, but the affine connections $\Gamma^{\mu}_{\alpha\beta}$ need not be the same for the contravariant and covariant vectors. Let these laws be

$$ \delta A^{\mu} = -{}_{I}\Gamma^{\mu}_{\rho\sigma} A^{\rho}\delta x^{\sigma} \quad \text{for contravariant} \tag{2.268} $$

$$ \delta A_{\rho} = {}_{II}\Gamma^{\mu}_{\rho\sigma} A_{\mu}\delta x^{\sigma} \quad \text{for covariant} \tag{2.269} $$

We used the prefix I and II for the affine connections.

By following the similar procedure as in the case of contravariant vector for the law of transformation of $\Gamma^{\mu}_{\rho\sigma}$, Eq. (2.263), one can show especially that the same law of transformation of $\Gamma^{\mu}_{\rho\sigma}$ will hold for the covariant vector. Alternately, we can show that ${}_{(1)}\Gamma^{\mu}_{\rho\sigma} = {}_{(2)}\Gamma^{\mu}_{\rho\sigma} \equiv \Gamma^{\mu}_{\rho\sigma}$, say by displacing two vectors A^{μ} and B_{μ} parallel to themselves along an infinitesimal path δx^{μ} and considering the change in their scalar product $(A^{\mu}B_{\mu})$. It is given as

$$ \delta(A^{\mu}B_{\mu}) = A^{\mu}\delta B_{\mu} + B_{\mu}\delta A^{\mu} \tag{8.270} $$

Substituting the values of δB_μ and δA^μ from Eqs. (2.269) and (2.268), we obtain

$$\delta(A^\mu B_\mu) = A_\mu B^\rho \left({}_{II}\Gamma^\mu_{\rho\sigma} - {}_{I}\Gamma^\mu_{\rho\sigma} \right) \delta x^\sigma \tag{2.271}$$

Geometrically, when two vectors are displaced parallel to themselves, their scalar product remains constant. This will be so provided

$${}_{I}\Gamma^\mu_{\rho\sigma} = {}_{II}\Gamma^\mu_{\rho\sigma} \tag{2.272}$$

which establishes the fact that the connections for the contravariant and covariant vectors are equal.

2.29 AFFINE CONNECTIONS AND THE METRIC TENSOR

We set ourselves the task of determining the relationship of the affine connections $\Gamma^\mu_{\alpha\beta}$ with the metric properties of the space. Consider the Riemannian space characterised by a quadratic metric $ds^2 = g_{\mu\nu} dx^\mu dx^\nu$, where $g_{\mu\nu}$ is the metric tensor. It will be found that a unique relationship is found between the connections $\Gamma^\mu_{\alpha\beta}$ and the first partial derivatives of the components of the metric tensor $g_{\mu\nu}$, provided we impose the following condition.

On parallel displacement of the vector A^μ along the curve, the length of the vector $A^\mu = g_{\mu\nu} A^\mu A^\nu$ remains constant. This condition is expressed as

$$\frac{d}{ds}(g_{\mu\nu} A^\mu A^\nu) = 0 \tag{2.273}$$

or

$$\frac{\partial g_{\mu\nu}}{\partial x^\rho}\frac{dx^\rho}{ds} A^\mu A^\nu + g_{\mu\nu}\frac{dA^\mu}{ds} A^\nu + g_{\mu\nu} A^\mu \frac{dA^\nu}{ds} = 0 \tag{2.274}$$

Substituting for $\frac{\partial A^\mu}{\partial s}$ and $\frac{\partial A^\nu}{\partial s}$ from the law of parallel displacement, Eq. (2.250) and replacing the parameter p by s, we obtain

$$\frac{\partial g_{\mu\nu}}{\partial x^\rho}\frac{dx^\rho}{ds} A^\mu A^\nu - g_{\mu\nu}\Gamma^\mu_{\alpha\rho} A^\alpha \frac{dx^\rho}{ds} A^\nu + g_{\mu\nu} A^\mu \Gamma^\nu_{\alpha\rho} A^\alpha \frac{dx^\rho}{ds} = 0 \tag{2.275}$$

Changing the dummy suffixes $(\mu \leftrightarrow \alpha)$ and $(\nu \leftrightarrow \alpha)$ in the second and third terms in Eq. (2.275), we obtain

$$\left(\frac{\partial g_{\mu\nu}}{\partial x^\rho} - g_{\nu\alpha}\Gamma^\alpha_{\mu\rho} - g_{\mu\alpha}\Gamma^\alpha_{\nu\rho} \right) A^\mu A^\nu \frac{dx^\rho}{ds} = 0 \tag{2.276}$$

As the vector A^μ and displacement δx^ρ are arbitrary, the above equation will be valid only when the expression in the parenthesis vanishes for all combinations of $\mu\nu\rho$. Since the expression in the parenthesis is symmetric in μ and ν and, we get

$$\frac{\partial g_{\mu\nu}}{\partial x^\rho} - g_{\nu\alpha}\Gamma^\alpha_{\mu\rho} - g_{\mu\alpha}\Gamma^\alpha_{\nu\rho} = 0 \tag{2.277}$$

The cyclic interchange of the suffixes $\mu\nu\rho$, gives the two more equations

$$\frac{\partial g_{\nu\rho}}{\partial x^{\mu}} - g_{\rho\alpha}\Gamma^{\alpha}_{\nu\mu} - g_{\nu\alpha}\Gamma^{\alpha}_{\rho\mu} = 0 \tag{2.278}$$

$$\frac{\partial g_{\rho\mu}}{\partial x^{\nu}} - g_{\mu\alpha}\Gamma^{\alpha}_{\rho\nu} - g_{\rho\alpha}\Gamma^{\alpha}_{\mu\nu} = 0 \tag{2.279}$$

Subtracting Eqs. (2.278) and (2.279) and adding to Eq. (2.277), one gets

$$g_{\nu\alpha}\Gamma^{\alpha}_{\rho\mu} = \frac{1}{2}\left(\frac{\partial g_{\mu\nu}}{\partial x^{\rho}} + \frac{\partial g_{\nu\rho}}{\partial x^{\mu}} - \frac{\partial g_{\rho\mu}}{\partial x^{\nu}}\right) \tag{2.280}$$

Multiplying Eq. (2.280) by $g^{\sigma\nu}$, we obtain the expression for $\Gamma^{\sigma}_{\rho\mu}$

$$\Gamma^{\sigma}_{\rho\mu} = \frac{1}{2}g^{\sigma\nu}\left(\frac{\partial g_{\mu\nu}}{\partial x^{\rho}} + \frac{\partial g_{\nu\rho}}{\partial x^{\mu}} - \frac{\partial g_{\rho\mu}}{\partial x^{\nu}}\right) \tag{2.281}$$

Thus, the affine connections are related to the metric tensor $g_{\mu\nu}$. This conclusion is quite in order since just as the metric tensor $g_{\mu\nu}$ characterizes the property of the space, so does the affine connections.

The l.h.s. of Eq. (2.281), is the Christoffel symbol of the second kind, so

$$\Gamma^{\sigma}_{\rho\mu} = \left\{\begin{matrix}\sigma \\ \rho\mu\end{matrix}\right\} \tag{2.282}$$

and the r.h.s. of Eq. (2.280) is the Christoffel symbol of the first kind

$$g_{\nu\alpha}\Gamma^{\alpha}_{\rho\mu} = [\rho\mu, \nu] \tag{2.283}$$

The concept of parallel displacement is a geometrical operation and is independent of the existence of a metric tensor. A space in which the law of parallel displacement holds, is called an affinely connected space and the quantities $\Gamma^{\mu}_{\rho\nu}$ are the components of the affine connection. These components do not constitute a tensor and their transformation is governed by Eq. (2.263) and are called pseudotensors.

2.30 PARALLEL DISPLACEMENT AND COVARIANT DIFFERENTIATION

It is to illustrate how the concept of parallel displacement is helpful in getting tensors by covariant differentiation. Consider a vector $A^{\mu}(x)$ at the point x, *i.e.*, the components of this vector are functions of the coordinates. Let us take the value of $A^{\mu}(x)$ at the point x^{ν} and then displace the vector to a point

$x^\nu + \delta x^\nu$. The value of the vector field at the point $x+\delta x$ minus the value of the vector atarrived at $x^\nu + \delta x^\nu$ through the process of parallel displacement is again a vector and is given by

$$A^\mu(x+\delta x) - [A^\mu(x) + \delta A^\mu(x)] \tag{2.284}$$

Now
$$A^\mu(x+\delta x) = A^\mu + dA^\mu = A^\mu(x) + \frac{\partial A^\mu(x)}{\partial x^\nu}\delta x^\nu \tag{2.285}$$

The difference of Eq. (2.284) is given by

$$\begin{aligned} dA^\mu - \delta A^\mu(x) &= \frac{\partial A^\mu(x)}{\partial x^\nu}\delta x^\nu + \Gamma^\mu_{\rho\nu} A^\rho \delta x^\nu \\ &= \left(\frac{\partial A^\mu}{\partial x^\nu} + \Gamma^\mu_{\rho\nu} A^\rho\right)\delta x^\nu \end{aligned} \tag{2.286}$$

where we made use of Eq. (2.267).

As the l.h.s. is a vector, the expression on the r.h.s. is also a vector. Further, since δx^ν is a contravariant vector, the expression in the parenthesis is a mixed vector, contravariant in μ and covariant in ν. Thus,

$$A^\mu{}_{;\nu} = \frac{\partial A^\mu}{\partial x^\nu} + \Gamma^\mu_{\rho\nu} A^\rho \tag{2.287}$$

$A^\mu{}_{;\nu}$ is the covariant derivative of the vector A^μ.

Analogously, we define the covariant derivative of a tensor field, which are themselves tensors. Take a mixed tensor A^μ_ν of rank two. We consider the difference of two tensor fields at $(x+\delta x)$ of the field $A^\mu_\nu(x+\delta x)$ and $[A^\mu_\nu(x) + \delta A^\mu_\nu(x)]$ obtained by parallel displacement.

Now
$$A^\mu_\nu(x+\delta x) = A^\mu_\nu(x) + \frac{\partial A^\mu_\nu}{\partial x^\rho}\delta x^\rho \tag{2.288}$$

and
$$A^\mu_\nu(x) + \delta A^\mu_\nu(x) = A^\mu_\nu(x) - \Gamma^\mu_{\rho\sigma} A^\sigma_\nu \delta x^\rho + \Gamma^\sigma_{\nu\rho} A^\mu_\sigma \delta x^\rho \tag{2.289}$$

The difference of these equations is

$$A^\mu_\nu(x+\delta x) - \left[A^\mu_\nu(x) + \delta A^\mu_\nu(x)\right] = \left(\frac{\partial A^\mu_\nu}{\partial x^\rho} + \Gamma^\mu_{\rho\sigma} A^\sigma_\nu - \Gamma^\mu_{\rho\sigma} A^\sigma_\nu\right)\delta x^\rho \tag{2.290}$$

The quantity in the parenthesis on the r.h.s. of Eq. (2.290) is a tensor that we call the covariant derivative of A^μ_ν, so

$$A^\mu_{\nu\,;\rho} = \frac{\partial A^\mu_\nu}{\partial x^\rho} + \Gamma^\mu_{\rho\sigma} A^\sigma_\nu - \Gamma^\mu_{\rho\sigma} A^\sigma_\nu \tag{2.291}$$

2.31 ENERGY-MOMENTUM TENSOR FOR A PERFECT FLUID

A perfect fluid is visualized as a mechanical medium that has no heat conduction or viscosity and does not sustain any shearing stress and in which the normal stresses are isotropic. Therefore, for a local observer, the normal components of stress, are equal to the isotropic hydrostatic pressure. Let $T_0^{\mu\nu}$ denote the energy-momentum tensor in proper coordinate system in which the fluid is supposed to be at rest at the origin, so

$$T_0^{ij} = p_0\delta_{ij}, \qquad T_0^{00} = \rho_0 c^2 \qquad T_0^{i0} = 0 \quad i, j = 1,2,3. \tag{2.292}$$

where $x^\mu = (x^0 = ct, x^1, x^2, x^3)$. p and ρ_0 respectively denote pressure and density of the perfect fluid in the proper coordinate system. In proper coordinate system, Galilean coordinate system holds, in which the line-element is given by

$$ds^2 = c^2dt^2 - dx^2 - dy^2 - dz^2 \tag{2.293}$$

The fundamental tensor $g_0^{\mu\nu}$ in the proper coordinate system is given by

$$g_{0\mu\nu} = g_0^{\mu\nu} = \begin{pmatrix} 1 & 0 & 0 & 0 \\ 0 & -1 & 0 & 0 \\ 0 & 0 & -1 & 0 \\ 0 & 0 & 0 & -1 \end{pmatrix} \tag{2.294}$$

Thus, the energy momentum tensor that describes a perfect fluid in a local frame is

$$T_0^{\mu\nu} = \begin{pmatrix} \rho_0 c^2 & 0 & 0 & 0 \\ 0 & p_0 & 0 & 0 \\ 0 & 0 & p_0 & 0 \\ 0 & 0 & 0 & p_0 \end{pmatrix} \tag{2.295}$$

Let, $T^{\mu\nu}$ and $g^{\mu\nu}$ respectively be the energy-momentum tensor and fundamental tensor in an arbitrary coordinate system. Tensor law of transformation gives

$$\begin{aligned} T^{\mu\nu} &= T_0^{\alpha\beta} \frac{\partial x^\mu}{\partial x_0^\alpha} \frac{\partial x^\nu}{\partial x_0^\beta} \\ &= \sum_{\alpha=1}^{4} T_0^{\alpha\alpha} \frac{\partial x^\mu}{\partial x_0^\alpha} \frac{\partial x^\nu}{\partial x_0^\alpha} \end{aligned} \tag{2.296}$$

Rewriting it, we get

$$T^{\mu\nu} = \rho_0 c^2 \frac{\partial x^\mu}{\partial x_0^0 \partial x_0^0} + p_0 \sum_{\alpha=1}^{3} \frac{\partial x^\mu}{\partial x_0^\alpha} \frac{\partial x^\mu}{\partial x_0^\alpha} \tag{2.297}$$

Also

$$g^{\mu\nu} = g_0^{\alpha\beta} \frac{\partial x^\mu}{\partial x_0^\alpha} \frac{\partial x^\nu}{\partial x_0^\beta}$$

$$= \sum_{\alpha=1}^{4} g_0^{\alpha\alpha} \frac{\partial x^\mu}{\partial x_0^\alpha} \frac{\partial x^\nu}{\partial x_0^\alpha}$$

$$= \frac{\partial x^\mu}{\partial x_0^0} \frac{\partial x^\nu}{\partial x_0^0} - \sum_{\alpha=1}^{3} \frac{\partial x^\mu}{\partial x_0^\alpha} \frac{\partial x^\nu}{\partial x_0^\alpha}$$

or

$$\sum_{\alpha=1}^{3} \frac{\partial x^\mu}{\partial x_0^\alpha} \frac{\partial x^\nu}{\partial x_0^\alpha} = -g^{\mu\nu} + \frac{\partial x^\mu}{\partial x_0^0} \frac{\partial x^\nu}{\partial x_0^0} \tag{2.298}$$

Making use of Eq. (2.298) in Eq. (2.297)

$$T^{\mu\nu} = \rho_0 c^2 \frac{\partial x^\mu}{\partial x_0^0} \frac{\partial x^\nu}{\partial x_0^0} + p_0 \left(-g^{\mu\nu} + \frac{\partial x^\mu}{\partial x_0^0} \frac{\partial x^\nu}{\partial x_0^0} \right)$$

$$= (\rho_0 c^2 + p_0) \frac{\partial x^\mu}{\partial x_0^0} \frac{\partial x^\nu}{\partial x_0^0} - p_0 g^{\mu\nu} \tag{2.299}$$

As the fluid is at rest in the proper coordinate system, its velocity components are

$$\frac{dx_0^1}{ds} = \frac{dx_0^2}{ds} = \frac{dx_0^3}{ds} = 0, \frac{dx_0^0}{ds} = 1 \tag{2.300}$$

Now

$$\frac{\partial x^\mu}{ds} = \frac{\partial x^\mu}{\partial x_0^\nu} \frac{\partial x_0^\nu}{ds} = \sum_{i=1}^{3} \frac{\partial x^\mu}{\partial x_0^i} \frac{\partial x_0^i}{ds} + \frac{\partial x^\mu}{\partial x_0^0} \frac{\partial x_0^0}{ds}$$

$$= \frac{\partial x^\mu}{dx_0^0}$$

or

$$\frac{\partial x^\mu}{ds} = \frac{\partial x^\mu}{\partial x_0^0} \tag{2.301}$$

Making use of Eq. (2.301) in Eq. (2.299), we get

$$T^{\mu\nu} = (\rho_0 c^2 + p_0) \frac{dx^\mu}{ds} \frac{dx^\nu}{ds} - p_0 g^{\mu\nu} \tag{2.302}$$

The covariant and mixed forms of $T^{\mu\nu}$ tensor are

$$T_{\mu\nu} = (\rho_0 c^2 + p_0)\frac{dx_\mu}{ds}\frac{dx_\nu}{ds} - p_0 g_{\mu\nu} \tag{2.303}$$

$$T^\mu_\nu = (\rho_0 c^2 + p_0)\frac{dx^\mu}{ds}\frac{dx_\nu}{ds} - p_0 g^\mu_\nu \tag{2.304}$$

The invariant associated with $T^{\mu\nu}$ is $T = T^\mu_\mu$ as

$$\begin{aligned} T = T^\mu_\mu &= (\rho_0 c^2 + p_0)\frac{dx^\mu}{ds}\frac{dx_\mu}{ds} - p_0 g^\mu_\mu \\ &= (\rho_0 c^2 + p_0) - 4p_0 \\ &= \rho_0 c^2 - 3p_0 \end{aligned} \tag{2.305}$$

Another invariant is

$$T^{\mu\nu}T_{\mu\nu} = \rho_0^2 c^4 + 3p_0^2 \tag{2.306}$$

An incoherent matter, like a collection of dust particles, is not able to sustain any stress, and the energy-momentum tensor for it is given by

$$T^\mu_\nu = \rho_0 c^2 \frac{dx^\mu}{ds}\frac{dx^\nu}{ds} \tag{2.307}$$

The energy-momentum tensor $T^{\mu\nu}$, Eq. (2.307) refers to only the material properties of the medium. However, if the fluid is composed of electrically charged particles, an additional contribution $S^{\mu\nu}$ to the energy-momentum tensor arises due to the electromagnetic field and the energy-momentum tensor becomes

$$\begin{aligned} T^{\mu\nu} &= (T^{\mu\nu})_{mat} + (T^{\mu\nu})_{em} \\ &= (\rho_0 c^2 + p)\frac{dx^\mu}{ds}\frac{dx^\nu}{ds} - pg^{\mu\nu} + \frac{1}{4}(F^{\alpha\beta}F_{\alpha\beta}g^{\mu\nu} - F^{\mu\alpha}F^\nu_\alpha) \end{aligned} \tag{2.308}$$

Example 2.14

In the rest frame of a perfect fluid, its stress-energy tensor in terms of mass-energy density ρ and pressure p, is the diagonal tensor

$$T^{\mu\nu} = \begin{pmatrix} \rho c^2 & 0 & 0 & 0 \\ 0 & p & 0 & 0 \\ 0 & 0 & p & 0 \\ 0 & 0 & 0 & p \end{pmatrix}$$

If a fluid element of proper density ρ and pressure p, is moving with 4-velocity u^{μ}, what is its stress-energy?

Solution

In the rest frame of the fluid element

$$u^0 = 1, u^1 = u^2 = u^3 = 0$$

so that

$$T^{\mu\nu} = (\rho c^2 + p)u^{\mu}u^{\nu} - pg^{\mu\nu}$$

as this tensor equation is true in the rest frame, it must hold in any frame.

Example 2.15

Obtain the energy-momentum tensor $T^{\mu\nu}$ for a material medium composed of moving particles and show that the conservation of energy and momentum are contained in the equation $(T_{\mu\nu})_{,\nu} = 0$.

Solution

The energy-momentum in an arbitrary frame is given by

$$T^{\mu\nu} = (\rho_0 c^2 + p_0)\frac{dx^{\mu}}{ds}\frac{dx^{\nu}}{ds} - p_0 g^{\mu\nu} \tag{i}$$

Now the energy-momentum tensor in the proper coordinate system is

$$T_0^{\mu\nu} = \begin{pmatrix} \rho_0 c^2 & 0 & 0 & 0 \\ 0 & p_{0xx} & p_{0xy} & p_{0xz} \\ 0 & p_{0yx} & p_{0yy} & p_{0yz} \\ 0 & p_{0zx} & p_{0zy} & p_{0zz} \end{pmatrix} \tag{ii}$$

where p_{xx}, p_{xy}, etc., are the internal stresses.

Let the coordinate density and velocity of moving particles, relative to Galilean coordinate system, be ρ and $\mathbf{v}(\mu, \nu, w)$, respectively. Then

$$T^{\mu\nu} = \begin{pmatrix} \rho c^2 & c\rho u & c\rho v & c\rho w \\ c\rho u & p_{xx} + \rho u^2 & p_{xy} + \rho uv & p_{xz} + \rho uw \\ c\rho v & p_{yx} + \rho vu & p_{yy} + \rho v^2 & p_{yz} + \rho vw \\ c\rho w & p_{zx} + \rho wu & p_{zy} + \rho wv & p_{zz} + \rho w^2 \end{pmatrix} \tag{iii}$$

Let us take the equation

$$\frac{\partial T^{\mu\nu}}{\partial x^{\nu}} = 0 \tag{iv}$$

We write it out explicitly. For $\mu = 0$, it gives

$$\frac{\partial T^{0\nu}}{\partial x^\nu} = 0$$

$$\frac{\partial T^{00}}{\partial x^0} + \frac{\partial T^{01}}{\partial x^1} + \frac{\partial T^{02}}{\partial x^2} + \frac{\partial T^{03}}{\partial x^3} = 0$$

or

$$\frac{d\rho}{dt} + \frac{\partial(\rho u)}{\partial x} + \frac{\partial(\rho v)}{\partial y} + \frac{\partial(\rho w)}{\partial z} = 0 \qquad \text{(v)}$$

For $\mu = 1$, Eq. (iv) gives

$$\frac{\partial T^{1\nu}}{\partial x^\nu} = 0 \qquad \text{(vi)}$$

or

$$\frac{\partial T^{10}}{\partial x^0} + \frac{\partial T^{11}}{\partial x^1} + \frac{\partial T^{12}}{\partial x^2} + \frac{\partial T^{13}}{\partial x^3} = 0$$

$$\begin{aligned}\frac{\partial p_{xx}}{\partial x} + \frac{\partial p_{xy}}{\partial y} + \frac{\partial p_{xz}}{\partial z} &= -\left[\frac{\partial}{\partial x}(\rho u^2) + \frac{\partial}{\partial y}(\rho uv) + \frac{\partial}{\partial z}(\rho uw) + \frac{\partial}{\partial t}(\rho u)\right] \\ &= -u\left[\frac{\partial}{\partial x}(\rho u) + \frac{\partial}{\partial y}(\rho v) + \frac{\partial}{\partial z}(\rho w) + \frac{\partial}{\partial t}(\rho)\right] \\ &\quad -\rho\left[u\frac{\partial u}{\partial x} + v\frac{\partial u}{\partial y} + w\frac{\partial u}{\partial z} + \frac{\partial u}{\partial t}\right]\end{aligned}$$

or

$$= -u \cdot 0 - \rho\frac{du}{dt}$$

$$= -\rho\frac{du}{dt}$$

or

$$\frac{\partial p_{xx}}{\partial x} + \frac{\partial p_{xy}}{\partial y} + \frac{\partial p_{xz}}{\partial z} = -\rho\frac{du}{dt} \qquad \text{(vii)}$$

where we made use of the result

$$\frac{d}{dt} = u\frac{\partial}{\partial x} + v\frac{\partial}{\partial y} + w\frac{\partial}{\partial z} + \frac{\partial}{\partial t} \qquad \text{(viii)}$$

Analogously, for $\mu = 2, 3$; the Eq. (iv) gives

$$\frac{\partial p_{yx}}{\partial x} + \frac{\partial p_{yy}}{\partial y} + \frac{\partial p_{yz}}{\partial z} + = -\rho\frac{dv}{dt} \qquad \text{(ix)}$$

$$\frac{\partial p_{zx}}{\partial x} + \frac{\partial p_{zy}}{\partial y} + \frac{\partial p_{zz}}{\partial z} = -\rho\frac{dw}{dt} \qquad \text{(x)}$$

The Eq. (v) is the equation of continuity and Eqs. (viii), (ix), (x) are the x-, y- and z- components of the equation of motion without impressed force in hydrodynamics. These equations express the conservation of mass and momentum. Thus, the Eq. (iv) expresses the law of conservation of mass and momentum, relative to Galilean coordinates.

Problems

2.11 Show that the component R_{1212} of the curvature tensor for a two dimensional space with the metric $ds^2 = dx^2 + f(x,y)dy^2$ equals $-1/2\, \partial^2 f / \partial x^2 + 1/4f(\frac{\partial f}{\partial x})^2$.

2.12 In the space–time whose metric is given

$$ds^2 = e^{2\phi}(dx^0)^2 + e^{2\theta}(dx^1)^2 + (dx^2)^2 + (dx^3)^2$$

where ϕ and θ are functions of x^1 only, prove that the Riemann–Christoffel tensor vanishes if and only if $\phi'' - \theta'\phi' + \phi'^2 = 0$, where the dashes denote differentiation with respect to x^1.

2.13 Show that when the Riemann–Christoffel tensor vanishes, the differential equations

$$A_{\mu,\nu} = \frac{\partial A_\mu}{\partial x^\nu} - \Gamma^\alpha_{\mu\nu} A_\alpha = 0$$

are integrable.

2.14 The four independent relations among the components of the curvature tensor are

$$R_{\rho\mu\nu\lambda} = -R_{\rho\mu\lambda\nu} = -R_{\mu\rho\nu\lambda} = R_{\nu\lambda\rho\mu}$$

$$R_{\rho\mu\nu\lambda} + R_{\rho\nu\lambda\mu} + R_{\rho\lambda\mu\nu} = 0$$

Show that combining these with each other, the following relations can be obtained.

(i) $R_{\rho\mu\nu\lambda} = -R_{\lambda\nu\rho\mu} = -R_{\nu\lambda\mu\rho} = R_{\lambda\nu\mu\rho}$

(ii) $R_{\rho\mu\nu\lambda} + R_{\nu\mu\lambda\rho} + R_{\lambda\mu\rho\nu} = 0$ (second index fixed)

(iii) $R_{\rho\mu\nu\lambda} + R_{\mu\lambda\nu\rho} + R_{\lambda\rho\nu\mu} = 0$ (third index fixed)

(iv) $R_{\rho\mu\nu\lambda} + R_{\mu\nu\rho\lambda} + R_{\nu\rho\mu\lambda} = 0$

2.15 Show that in Galilean coordinates, energy-momentum tensor is given by

$$T^{\mu\nu} = \rho \frac{dx^\mu}{ds}\frac{dx^\nu}{ds}$$

where ρ is the coordinate density of matter.

3

Einstein's Field Equations

3.1 INTRODUCTION

Since special relativity requires all objects and particles to be limited by the speed of light, all forces and interactions must travel at or below the speed of light. Newton's gravitational theory is in contradiction with this principle because it states that the gravitational force acts instantaneously. Einstein spent many years attempting to create a gravitational theory that does not require forces to act faster than light.

Rather than explaining gravity as a force that acts[1] instantly as Newton did, he hypothesized that gravity is simply a consequence of space–time curvature. In order to understand space–time curvature, picture a two-dimensional surface. A ball, representing the sun, lies on the centre of the surface. The surface is curved so when another ball, representing a planet, is rolled in a straight line on the surface, the curve causes the ball to travel around, or orbit, the ball representing the sun.

In his theory, Einstein hypothesized that objects travel in geodesics through curved space–time. A geodesic is a straight line, or the shortest distance between two points. All objects, including planets, travel in geodesics in four-dimensional space–time. In order to understand why the theory is accurate yet the paths of planets do not look straight, we will use an analogy of an airplane travelling along a geodesic. The airplane takes the shortest path over the three-dimensional earth. When the path is charted on a two-dimensional map, it appears curved rather than straight. The airplanes' path on the map does not look straight because it is rendered in one less dimension than the space in which the airplane actually traveled. In the same way, the paths of planets seem curved to us because they travel along geodesics in four-dimensional space–time and we observe them in only three dimensions.

General relativity is the geometrical theory of gravitation, proposed by Albert Einstein[1] in 1915. It unifies special theory of relativity with Newton's law of universal gravitation. On the generalization of the special theory of relativity for the gravitational forces, the theoretical predictions were not able to explain the observed phenomena. According to special theory, the bending of light rays passing near the sun, should be 0.88 seconds of arc while in actual observation it is 1.75 seconds of arc. In addition, special theory fails when the velocity and gravitational field both are present as in the case of precession of perihelion of planet Mercury. According to special theory, the advance of the perihelion of Mercury is 7.2 seconds of arc per century, whereas the actual advance observed is 43 seconds of arc per century. Thus, in order to explain these observations, certain modifications had to be introduced in the framework of special relativity. Special relativity deals only with inertial systems and accordingly, all the physical laws are invariant under coordinate transformations, like Lorentz transformations. If we extend the above statement that all the physical laws in nature, are invariant under any coordinate transformation, *i.e.*, for non-inertial systems as well, there results general theory.

According to the general theory of relativity, the gravitational acceleration is caused by the curvature of space and time, which is produced by the mass-energy and momentum of the matter in space–time, as implied by Einstein's field equations. Undeniably, it is the most successful gravitational theory, fully supported by observations. The initial success was based on explaining the anomalous perihelion precession of Mercury as well as Sir Arthur Eddington's confirmation that massive objects bend light. Later on, modern tests, such as gravitational time dilation, the gravitational red-shift of light, signal delay, frame dragging, gravitational radiation and the existence of black holes, led to its glorious success.

However, in spite of its tremendous success, some reservations remain since it is incompatible with quantum mechanics and the singularities of black holes (at which the mathematics of general relativity breaks down). Because of these difficulties, numerous other theories had been proposed, the still-popular among them being the Brans–Dicke theory, which although does not solve the problems of quantum gravity and singularities, did have observational support in 1960. These observations had been refined and modern measurements show that Brans–Dicke type of deviation from general relativity, if it exists, must be very small.

3.2 PRINCIPLE OF EQUIVALENCE

The equivalence principle is a fundamental background concept of the general theory of relativity. It emerged in 16th and 17th centuries when Galileo demonstrated experimentally that the acceleration of a test mass due to gravitation is independent of the amount of mass being accelerated. This finding ultimately led to gravitational theory where the inertial and gravitational masses are identical.

Let us digress a little and examine briefly the definition of mass. Mass of a body is defined in two different ways, which are fundamentally different from each other. These are in the following:

1. The kinematical or inertial mass, m_i.
2. The gravitational mass, m_g.

The inertial mass is the coefficient, which measures the resistance of inertia of the body opposing the action of the force. Thus,

$$\frac{F}{\alpha} = m_i \tag{3.1}$$

where F is the force applied to the body and α, the resulting acceleration.

The gravitational mass is the coefficient, which determines the attracting force experienced by a body in the gravitational field. Newton's theory of gravitation instead of offering any explanation for the equality of inertial and gravitational masses, rather assumes it as a hypothesis. Thus,

$$m_i = m_g \tag{3.2}$$

Now
$$m_i \frac{d^2x}{dt^2} = m_g g$$

or
$$\frac{d^2x}{dt^2} = g \tag{3.3}$$

As is obvious, the acceleration acquired by a body, does not depend on its mass, provided only the inertial and gravitational masses are equal. Thus, we conclude that the rate of fall of all bodies is the same under gravitational force irrespective of their masses.

Einstein argued that the equality of two entirely different concepts of mass, cannot be a mere coincidence, rather it is a pointer to something more basic, which may unify the concepts of inertia and gravitation. He showed successfully that by an extension of the principle of relativity from uniform translatory motion to non-uniform relative motion, the equality of inertial and gravitational masses follows naturally. Further, the extension of the principle of relativity leads to general theory of relativity, since a mere change of the system of coordinates results into the creation of a gravitational field of force. The gravitation now becomes a property of space–time structure. Einstein called the extension of the principle of relativity to non-uniform motion as the *Principle of Equivalence* and stated it thus:

> *'We assume the complete physical equivalence of a gravitational field and a corresponding acceleration of the reference system.'*
>
> (Einstein, 1907)

That is, being at rest on the surface of earth is equivalent to being inside a space ship (far from any source of gravity) that is being accelerated by its engines. Einstein deduced that free fall is actually inertial motion. At the earth's surface, the force of gravity is counteracted by the mechanical (physical) resistance of earth's surface.

However, at present, three forms of the equivalence principle are in current use. These are shown in the following:

1. The weak equivalence principle,
2. The Einstein equivalence principle, and
3. The strong equivalence principle.

1. **The weak equivalence principle**, also known as the universality of free fall, is stated as follows:

   The trajectory of a falling body depends only on its initial position and velocity and not on its composition. Alternately stated, *it implies that all bodies at the same space–time point in a given gravitational field, will undergo the same acceleration.*

   This principle does not hold in the case of either large bodies or heavy bodies, since the former will experience the tidal forces and the presence of the latter, will alter the gravitational field around them. The weak equivalence principle is the closest to Einstein's original statement.

   Tests of this principle are based on the verification of the equivalence of gravitational mass and inertial mass. This is simply done by dropping two objects of different masses or composition in a vacuum and observing if they strike the ground at the same time. Such experiments have been performed by Simon Steven (1586) who dropped lead balls of different masses off the Delft church tower. No detectable difference in arrival times was found. Galileo Galilei (1610) rolled balls down inclined planes and again no detectable difference in times was found. Isaac Newton (1680) measured the period of pendulums of different mass but identical length and Friedrich Wilhelm Bessel (1832) repeated the similar experiment and no measurable difference was found.

   More refined tests employing a torsion balance to measure the tension on a wire, suspending a balance beam, between two nearly identical masses under the acceleration of gravity and the rotation of earth, were for the first time, performed by L. Eötvös (1906). His experimental accuracy was of the order of 10^{-9}. Roll, Krotkov and Dicke (1964) repeated the torsion balance experiment dropping aluminium and gold test masses, achieving an accuracy of 10^{-11}. However, the most accurate experiment to-date has been performed by Braginsky and Panov

(1971) employing a torsion balance, aluminium and platinum test masses and measuring their acceleration towards the sun and as a result could achieve an accuracy of one part in 10^{-12}. Later on, Eöt-Wash (1987) measured acceleration of different masses towards the earth, the sun and galactic centre and attained an accuracy of a few parts in a trillion.

Future satellite experiments—STEP (Satellite Test of Equivalence Principle) will test the weak equivalence principle in space to a much higher accuracy.

It may seem redundant to continue to test Einstein's theory of relativity as it is by large, the most elegant theory of gravity and is in agreement with almost all observations. However, no quantum theory of gravity is known and most of the suggestions violate one of the equivalence principles at some level. String theory, supergravity and even quintessence, seem to violate the weak equivalence principle because they contain many light scalar fields with long Compton wavelengths. These fields should generate fifth forces and variation of the fundamental constants. There are a number of mechanisms that have been suggested to reduce these violations of the equivalence principle to below observable levels.

2. **The Einstein Equivalence Principle:** It states that *the result of any local non-gravitational experiment in a laboratory is independent of the velocity of laboratory or its location in space–time.*

   Local has a special connotation: not only must the experiment not look outside the laboratory, but it must be small compared to variations in the gravitational field, tidal forces; so that the entire laboratory is moving inertially. This requires that suitable frames of reference all over the universe behave identically. It may be construed as an extension of the postulates of special relativity in that it requires that dimensionless physical constants as the fine-structure constant and electron-to-proton mass ratio, be constant.

   As tests of the Einstein equivalence principle, the electron–proton mass ratio as inferred from measurements of distant quasars (2002) show a variation at the 10^{-4} level. But the measurements are disputed by other researchers. The position independence of experiments is tested by the Pound–Rebka experiment on the gravitational red-shift of photon[2].

3. **The Strong Equivalence Principle:** It states that *the gravitational motion of a small test body depends on its initial position in space–time and velocity, and not on its acceleration.*

   Alternatively stated, *the outcome of any local experiment, whether gravitational or not, in a laboratory moving in an inertial frame of reference, is independent of the laboratory or its location in space–time.*

   The first statement is a version of the weak equivalence principle that applies to objects that exert gravitational force on themselves, such as stars, planets, black holes or Cavendish experiments. The second statement is the Einstein equivalence principle generalized to allow gravitational experiments and self-gravitating bodies. The freely falling object or laboratory, must be small so that tidal forces may be neglected. A freely falling body into a black hole will experience strong tidal forces, since it will experience a more powerful force on its parts closer to the black hole.

The strong principle implies that gravity is entirely geometrical in nature, that is, space-metric alone determines the effect of gravity and does not have any extra fields associated with it. Einstein's theory of gravity is the only theory, which satisfies the strong equivalence principle. Alternative theory as Brans–Dicke theory satisfies only the Einstein equivalence principle.

The strong equivalence principle can be tested by searching for a variation of Newton's gravitational constant G over the life of the universe, or equivalently variation in the masses of the fundamental particles. There are many independent observations limiting the possible variation of Newton's gravitational constant[3], but one of the best comes from lunar range finding, which suggests that the gravitational constant does not change by more than one part in 10^{11} per year.

3.2.1 Weak Equivalence Principle Tests and String Theory

The best modern limits, based on, for example, laser ranging of the moon to measure how fast it falls around earth, show the weak equivalence principle holds within a few parts in a trillion (10^{12}). This is fantastically accurate, yet the possibility remains that the equivalence principle could fail at some subtle level. 'It is a possibility we must investigate,' says physicist Clifford Will of Washington University in St. Louis, Missouri. 'Discovering even the slightest difference in how gravity acts on objects of different materials would have enormous implications.'

In fact, it could provide the first real evidence for string theory. String theory elegantly explains fundamental particles as different vibrations of infinitesimal strings and in doing so solves many lingering problems of modern physics. But string theory is highly controversial, in part because most of its predictions are virtually impossible to verify with experiments. If it is not testable, it is not science. The equivalence principle could offer one way to test string theory.

'Some variants of string theory predict the existence of a very weak force that would make gravity slightly different depending on an object's composition,' says Will. Finding a variation in gravity for different materials would not immediately prove that string theory is correct, but it would give the theory a dose of supporting evidence. This new facet of gravity, if it exists, would be so astonishingly weak that detecting it, is a tremendous challenge. Gravity itself is a relatively weak force—it is a trillion trillion trillion (10^{36}) times more feeble than electromagnetism. Theorists believe that new force would be at least ten million million (10^{13}) times weaker than gravity.

Just as magnetism acts on objects made of iron but not plastic, the new force would not affect all matter equally. The force's pull would vary depending on what the object is made of. For example, some versions of string theory suggest that this new force would interact with the electromagnetic energy contained in the material. Two atoms that have the same mass can contain different amounts of electromagnetic energy, if say, one has more protons, which have an electric charge, while the other has more neutrons, which have no charge. Traditional gravity would pull on both of these atoms equally, but if gravity includes this new force, the pull on these two atoms would differ ever so slightly.

No experiment to date has detected this tiny difference. But now three groups of scientists are proposing space-borne missions that would hunt for this effect with greater sensitivity than ever before. 'What you want to do is take two test masses made of different materials and watch for very small differences in how fast they fall,' Will says. 'On earth, an object can only fall for a short time before it hits the ground. But an object in orbit is literally falling around the earth, so it can fall continuously for a long time.' Tiny differences in the pull of gravity would accumulate over time, perhaps growing large enough to be detectable.

One test mission, called the STEP, is being developed by Stanford University and an international team of collaborators. STEP would be able to detect a deviation in the equivalence principle as small as one part in a million trillion (10^{18}). That is 100,000 times more sensitive than the current best measurement.

STEP's design uses four pairs of test masses instead of just one pair. The redundancy is to ensure that any difference seen in how the test masses fall, is truly caused by a violation of the equivalence principle, not by some other disturbance or imperfection in the hardware.

'When trying to measure such a minuscule effect, you have to eliminate as many external disturbances as possible,' Will explains. STEP's design places the test masses inside a large tank of liquid helium to insulate them from external temperature fluctuations, and surrounds the masses with a superconducting shell to shield them from magnetic and electrical interference. Microthrusters counteract the effects of atmospheric drag on the orbiting satellite, making the free fall of the test masses nearly perfect.

In this pristine environment, each pair of test masses should stay aligned with each other as they fall around the earth—that is, if the equivalence principle holds. But if this new component of gravity does exist, one test mass will fall at a slightly different rate than its partner, so the pair will drift slightly out of alignment over time.

Currently, STEP is still in the design phase. Another satellite-based experiment, the French-developed Micro-Satellite ă trainee Compensee pour l' observation du Principle d' Equivalence (MICROSCOPE), is scheduled to launch in 2016. MICROSCOPE will have two pairs of test masses instead of four, and will be able to detect a violation of the equivalence principle as small as one part in a million billion (10^{15}).

The third experiment is the Italian satellite Galileo Galilei (GG), which will operate in much the same way as STEP and MICROSCOPE, except that it uses only one pair of test masses. To improve its accuracy, the Galileo Galilei will spin about its central axis at a rate of 2 rotations per second. That way, any disturbances within the spacecraft will pull in all directions equally, thus cancelling themselves out. The experiment should be able to achieve a sensitivity of one part in a hundred million billion (10^{17}).

Whether any of these missions stands a chance of detecting a violation of the equivalence principle is hard to say. Finding no violation would still be helpful: it would rule out some variants of string theory, inching physicists towards the correct '*Theory of Everything*'. But finding a deviation, however, small, would be a great leap.

3.2.2 Anisotropy of Inertia. Mach's Principle

Regarding the origin of inertia, Newton and Mach had different viewpoints. According to Newton, the inertial forces like centrifugal force in a rotating frame of reference, arise from acceleration with respect to 'absolute space', whereas Mach argued that these were caused by acceleration with respect to the mass of celestial bodies. According to Mach, a large mass could produce a small change in the acceleration forces in its vicinity, whereas no such effects could result according to Newton.

Initially, Einstein supported Mach's idea, but according to the equivalence principle, the answer to the problem of inertia lies somewhere between both the viewpoints. The inertial frame, such as a freely falling coordinate system, is determined by the local gravitational field that results from the entire mass in the universe. However, once in an inertial frame, the laws of motion are not affected by the matter either through gravity or in any other manner. In our solar system, the mass of sun determines the motion of the freely falling earth, and once we fix the coordinate frame to the earth, as precisely demonstrated by Dicke, we cannot determine the gravitational field of the sun. This applies only locally. Gravitational fields could otherwise be detected through tidal effects over sufficiently large system, but here the equivalence principle ceases to apply.

The celestial bodies do play a role, since the gravitational field equations for $g_{\mu\nu}$ require the boundary conditions at infinity, and these are provided by the requirement that at distances far away from the sun, $g_{\mu\nu}$ merge with the cosmic gravitational field due to all the mass in the universe. The planetary orbits far from the sun, as observed, do not precess with respect to the typical stars.

It is the mean cosmic gravitational field, which is, in turn, determined by the mean density of the stars, which determines the inertial frames in the absence of matter, so their inertial frames are either at rest or in a state of uniform non-rotating motion with respect to the typical stars. When a large mass like the sun, is brought nearby, the inertial frames accelerate towards the mass, but the dynamical laws in the freely falling frames are still those of special relativity and are independent of the surrounding mass distribution. As such, the Mach's principle and Einstein's equivalence principle are in obvious opposition.

A possibility was raised by Cocconi and Salpeter[4,5], who argued that if inertial mass arises from an interaction with matter, then if the surrounding mass distribution is anisotropic, then inertial mass should mirror that anisotropy. At the time, the galactic centre was largest known source of mass anisotropy, so they suggested that a quadrupole anisotropy in the inertial frame should be detectable from studying variations in Zeeman transition lines, when the applied magnetic field is parallel to or perpendicular to the galactic matter.

Hughes, *et. al.*[6] and Driver, *et. al.*[7] performed magnetic resonance experiments on Li^7 nuclei in the ground state, since this offered a greater resolution. The ground state nuclear spin is 3/2 and in a strong magnetic field there are $2I+1=4$ magnetic sub-levels M_p which should be equally spaced, if the laws of nuclear physics are rotationally invariant. In this case, the three transitions among neighbouring states should have the same energy and the photon absorption coefficient should show a single sharp line at this energy. However, if the inertia is anisotropic, then the four magnetic substates will not be exactly equally spaced and instead of one there will be three closely spaced resonance lines. Hughes, *et. al.* did not find any such splitting greater than the line width of $5\cdot3\times10^{-21}$ MeV, occurred over a 12 hour period, during which the rotation of earth carried the magnetic field from $22°$ towards the galactic centre to $104°$ away from the galactic centre. Considering the nuclear structure of Li^7 to be a single $P_{3/2}$ proton in a central nuclear potential of other nucleons, then the anisotropy Δm in the proton mass must be such that

$$\Delta\left(\frac{p^2}{2m}\right) \cong \frac{\Delta m}{m}\left(\frac{p^2}{2m}\right) \leq 5\times3\times10^{-21}\ \text{MeV} \tag{3.4}$$

where $p^2/2m$ is the kinetic energy. As $p^2/2m$ is greater than $0.5\,\text{MeV}$, one can conclude that the anisotropy in the inertial mass is given by the inequality

$$\frac{\Delta m}{m} \leq 10^{-20} \tag{3.5}$$

The evidence from this experiment favors the equivalence principle rather than Mach's principle.

However, one explanation for this null experiment was offered by Dicke[8], noting that if there is anisotropic coupling to all forms of mass-energy, then the anisotropy should be unobservable at least locally. The argument is based on Cocconi and Salpeter's original proposal that inertial mass could be a second-order tensor. They used the velocity–momentum relationship $p_\mu = m_{\mu\nu}u^\nu$. Dicke expressed the mass as $m_{\mu\nu} = mf_{\mu\nu}$, taking it as an interaction with a universal tensor field.

He then derived equation of motion that employed $f_{\mu\nu}$ as metric tensor instead of the usual $g_{\mu\nu}$. Since $f_{\mu\nu}$ can always be transformed to a locally Minkowskian form, any anisotropy can be transformed away. Thus, if all mass energy couples to this field in the same way, the anisotropy is invisible. In this manner, Dicke asserts that the result actually upholds Mach's principle, since the anisotropy is universal. However, if the anisotropic field cannot be expressed as a second-order tensor or if the field couples differently to different forms of mass-energy, the argument does not hold.

However, recent experimental observations show an anisotropy in the Cosmic Microwave Background, indicating anisotropy of the matter at much greater distances. Since the inertial interaction acts as a power law of order unity, the effect of this matter would far outweigh the relatively local contribution from the galaxy[9]. Cross[10] extends the work of Cocconi and Salpeter to higher multipoles leading to unequal level splittings that should be measurable by magnetic resonance experiments on nuclei of appropriate spin.

3.2.3 Olbert's Paradox

Olbert's paradox (1826) may be stated as follows:

> *'A static, infinitely old universe with an infinite number of stars, distriubuted in an infinitely large space, would be bright rather than dark.'*

In order to show it, we consider ourselves to be at the centre of a spherical shell of radius r and thickness s. The volume of the shell is $4\pi r^2 s$ and it increases as the square of the distance r from us. Assuming that the distribution of the stars is uniform in the universe, the number of stars in the shell will be proportional to the volume of the shell. Thus, the number of stars, will increase as the square of the distance. As the light that we receive from the star is inversely proportional to the square of the distance from the star, the two effects will cancel out. Thus, each shell of a given thickness, will produce the same amount of light, irrespective of how far away it is. That is, the light of each shell adds to the total amount. Therefore, the more the shells, the more light and with infinitely many shells there would be a bright sky.

Alternately stated, if the universe is uniformly populated with stars, then every line of sight from the earth, sooner or later, will encounter a star and as a result the sky will appear to be uniformly bright. The line of sight will be very long, of course. Assuming that stars are only about a million miles in diameter and are separated from each other by say, 10 light-years, then the average line of sight would go on the order of 10^8 billion light-years before hitting a star.

Olbert himself proposed that interstellar dust will absorb light from distant stars. But, this is only a temporary solution, since sooner or later the dust will get heated up by the absorbed radiation and will radiate as much light as it absorbed.

The paradox is resolved by showing that the universe is not infinite. But, this lands in another problem that if it is finite, then what lies on the other side of the boundary.

The paradox was resolved with Hubble's finding that the universe is expanding. Hubble's law (1936 and 1952) that the average velocity of recession of all the galaxies from each other, increases linearly with distance in a manner that at a distance of around 10 billion light years, the galaxies will be receding with almost the speed of light and will not be visible by any means. This conclusion is arrived at from the Doppler red-shifts of the spectra of distant stars and galaxies.

Due to the expansion of the universe, the effective size of the universe is only 10 billion light years, much shorter than the average line of sight to a star. Only the stars in this volume, can be visible from earth. That is the reason why it gets dark at night.

3.3 PRINCIPLE OF COVARIANCE

If the principle of covariance is to be built in as the integral part of the theory of gravitation, then the concept of inertial frames of reference, has to be abandoned. In general relativity, we incorporate all kinds of motion and there is no reason to choose an inertial frame of reference over a non-inertial frame. In order to make use of the general idea of relativity, encompassing all kinds of motion, it is imperative that the laws of physics are expressed in a form, which is independent of the choice of the coordinate system, or in other words the laws of nature remain invariant with respect to any space–time coordinate system. This statement is called the *principle of covariance.* This is achieved through the use of tensor calculus, since the expression of a law by a tensor equation, preserves its form in all systems of coordinates.

In special theory of relativity, we described the basic laws of physics, in a four-dimensional space–time geometry described by the line element

$$ds^2 = c^2dt^2 - dx^2 - dy^2 - dz^2 \tag{3.6}$$

The Eq. (3.6) remains invariant for the Lorentz transformations from one Cartesian coordinate system (ct, x, y, z) to another coordinate system (ct', x', y', z'), which is moving in uniform relative motion with respect to the other. However, under transformation to either an accelerated frame of reference or curvilinear coordinate system, the Eq. (3.6) will not remain invariant. Thus, to be in conformity with the principle of covariance, the required expression for the line element for specifying the space–time geometry, has to be a covariant expression. Only then, the space–time continuum can be the basis for the development of the general theory of relativity.

The line element, Eq. (3.6) can be put in a tensor form

$$\begin{aligned} ds^2 &= g_{\mu\nu}dx^\mu dx^\nu \\ &= (dx^0)^2 - (dx^1)^2 - (dx^2)^2 - (dx^3)^2 \end{aligned} \tag{3.7}$$

where $x^0 = ct,\ x^1 = x,\ x^2 = y,\ x^3 = z$ and $g_{\mu\nu}$ is the metric tensor with constant components

$$g_{\mu\nu} = \begin{pmatrix} 1 & 0 & 0 & 0 \\ 0 & -1 & 0 & 0 \\ 0 & 0 & -1 & 0 \\ 0 & 0 & 0 & -1 \end{pmatrix} \tag{3.8}$$

Further, we consider a general coordinate transformation to a new set of axes, which are not necessarily Cartesian. Let the new axes be chosen as

$$x'^\mu = x'^\mu(x'^0, x'^1, x'^2, x'^3) \tag{3.9}$$

The differentials dx^μ in terms of dx'^μ are

$$dx^\mu = \frac{\partial x^\mu}{\partial x'^\alpha} dx'^\alpha \tag{3.10}$$

The line element in terms of the new coordinates dx'^{μ} becomes

$$
\begin{aligned}
ds^2 &= g_{\mu\nu}(x)dx^{\mu}dx^{\nu} \\
&= g_{\mu\nu}(x)\left(\frac{\partial x^{\mu}}{\partial x'^{\alpha}}dx'^{\alpha}\right)\left(\frac{\partial x^{\nu}}{\partial x'^{\beta}}dx'^{\beta}\right) \\
&= g_{\mu\nu}(x)\frac{\partial x^{\mu}}{\partial x'^{\alpha}}\frac{\partial x^{\nu}}{\partial x'^{\beta}}dx'^{\alpha}dx'^{\beta} \\
&= g'_{\alpha\beta}(x')dx'^{\alpha}dx'^{\beta} \qquad (3.11)
\end{aligned}
$$

where
$$g'_{\alpha\beta}(x') = g_{\mu\nu}(x)\frac{\partial x^{\mu}}{\partial x'^{\alpha}}\frac{\partial x^{\nu}}{\partial x'^{\beta}} \qquad (3.12)$$

Equation (3.12) is the law of transformation for a covariant tensor of rank two. Thus, the covariant expression for interval with respect to the general coordinate transformation is

$$ds^2 = g_{\mu\nu}(x)dx^{\mu}dx^{\nu} \qquad (3.13)$$

As is obvious, the components of the metric tensor $g_{\mu\nu}$ are no longer constants, but are functions of the coordinates. The line element of the special relativity is only a particular case of generally covariant line element given by Eq. (3.13), where the metric tensor $g_{\mu\nu}$ has constant values as given by Eq. (3.8).

3.4 NEWTONIAN EQUATION OF MOTION AS AN APPROXIMATION OF GEODESIC EQUATIONS

It is proposed to show that geodesic equations are reducible to Newtonian equations in case of a weak static field. The motion of a test particle is governed by geodesic equations

$$\frac{d^2x^{\mu}}{ds^2} + \Gamma^{\mu}_{\alpha\beta}\frac{dx^{\alpha}}{ds}\frac{dx^{\beta}}{ds} = 0 \qquad (3.14)$$

where $\Gamma^{\mu}_{\alpha\beta}$, the Christoffel symbols of the second kind, are

$$\Gamma^{\mu}_{\alpha\beta} = g^{\mu\nu}\Gamma_{\nu,\alpha\beta} \qquad (3.15)$$

where
$$\Gamma_{\nu,\alpha\beta} = \frac{1}{2}\left(\frac{\partial g_{\nu\alpha}}{\partial x^{\beta}} + \frac{\partial g_{\nu\beta}}{\partial x^{\alpha}} - \frac{\partial g_{\alpha\beta}}{\partial x^{\nu}}\right) \qquad (3.16)$$

In the case of special relativity, the line element is a flat space–time manifold and all the $g_{\mu\nu}$'s are constants and hence independent of coordinates. Thus, all the $\Gamma^{\mu}_{\alpha\beta}$ are zero and the equation of trajectory reduces to the equation of a straight line, *i.e.*,

$$\frac{d^2x^{\mu}}{ds^2} = 0 \qquad (3.17)$$

It is indeed remarkable to observe that the metric tensor $g_{\mu\nu}$ determines the geometry of the space–time manifold as well as the trajectory of the particle. It is the first derivatives of the $g_{\mu\nu}$'s with respect to the coodinates that determine the trajectory. As such the $g_{\mu\nu}$'s serve as gravitational potential in determining the equation of trajectory in analogy to Newton's equation of motion.

Next we seek the connection of the equation of trajectory, Eq. (3.14) with the Newton's equation of motion in a weak gravitational field. Since the field is static, *i.e.*, it does not change with time, so the velocity components can be taken as

$$\frac{dx^0}{ds} = 1, \quad \frac{dx^1}{ds} = \frac{dx^2}{ds} = \frac{dx^3}{ds} = 0 \tag{3.18}$$

where coordinates are Galilean coordinates

$$x^0 = ct, \quad x^1 = x, \quad x^2 = y, \quad x^3 = z \tag{3.19}$$

For a weak static field, we assume that $g_{\mu\nu}$'s are not constants but differ from their constant values by infinitesimal amounts *viz.*,

$$g_{\mu\nu} \to g'_{\mu\nu} = g_{\mu\nu} + \psi_{\mu\nu} \tag{3.20}$$

where $\psi_{\mu\nu}$'s are independent of time and are functions of space coordinates. As such $\partial g'_{\mu\nu} / \partial x^0 = 0$.

Now,

$$\begin{aligned}
\Gamma^{\mu}_{\alpha\beta} &= g'^{\mu\nu}\Gamma_{\nu,\alpha\beta} \\
&= g'^{\mu\nu}\frac{1}{2}\left(\frac{\partial g'_{\nu\alpha}}{\partial x^{\beta}} + \frac{\partial g'_{\nu\beta}}{\partial x^{\alpha}} - \frac{\partial g'_{\alpha\beta}}{\partial x^{\nu}}\right) \\
&= \frac{1}{2}(g^{\mu\nu} + \psi^{\mu\nu})\left(\frac{\partial \psi_{\nu\alpha}}{\partial x^{\beta}} + \frac{\partial \psi_{\nu\beta}}{\partial x^{\alpha}} - \frac{\partial \psi_{\alpha\beta}}{\partial x^{\nu}}\right) \\
&\cong \frac{1}{2}\left(\frac{\partial \psi_{\mu\alpha}}{\partial x^{\beta}} + \frac{\partial \psi_{\mu\beta}}{\partial x^{\alpha}} - \frac{\partial \psi_{\alpha\beta}}{\partial x^{\mu}}\right)
\end{aligned} \tag{3.21}$$

where the second order terms in ψ have been ignored. We have

$$\begin{aligned}
ds^2 &= (dx^0)^2 - (dx^1)^2 - (dx^2)^2 - (dx^3)^2 \\
&= c^2 dt^2 (1 - v^2/c^2)
\end{aligned}$$

and for the case $v \ll c$, it becomes

$$ds \sim cdt = dx^0 \tag{3.22}$$

Obviously, the terms dx^i / ds for $i = 1, 2, 3$ are negligible as compared to dx^0 / ds. Thus, Eq. (3.14) for the trajectory becomes for $i = 1, 2, 3$

$$\frac{d^2x^i}{ds^2} + \Gamma^i_{00} \frac{dx^0}{ds} \frac{dx^0}{ds} = 0$$

or

$$\frac{d^2x^i}{ds^2} = -\Gamma^i_{00} \tag{3.23}$$

Let us evaluate the term Γ^i_{00}. Then

$$\begin{aligned}
\Gamma^i_{00} &= g^{ij}\Gamma_{00,j} \\
&= g^{ii}\Gamma_{00,i} \\
&= g^{ii}\frac{1}{2}\left(2\frac{\partial g_{0i}}{\partial x^0} - \frac{\partial g_{00}}{\partial x^i}\right) \\
&= \frac{1}{2g_{ii}}\left[-\frac{\partial}{\partial x^i}(1+\psi_{00})\right] \\
&= \frac{1}{2}(-1+\psi_{ii})^{-1}\left(-\frac{\partial \psi_{00}}{\partial x^i}\right) \\
&= \frac{1}{2}(1-\psi_{ii})^{-1}\frac{\partial \psi_{00}}{\partial x^i} \\
&= \frac{1}{2}\frac{\partial \psi_{00}}{\partial x^i}
\end{aligned} \tag{3.24}$$

so Eq. (3.23) becomes

$$\frac{d^2x^i}{ds^2} = -\frac{1}{2}\frac{\partial \psi_{00}}{\partial x^i} \quad (i = 1, 2, 3) \tag{3.25}$$

Newtonian equations of motion are

$$\frac{d^2x^i}{dt^2} = -\frac{\partial \psi}{\partial x^i} \quad (i = 1, 2, 3) \tag{3.26}$$

where ψ is the potential function.

The Eqs. (3.25) and (3.26) become identical if

$$-\frac{c^2}{2}\frac{\partial \psi_{00}}{\partial x^i} = -\frac{\partial \psi}{\partial x^i}$$

Integrating, we obtain

$$\int \frac{\partial \psi_{00}}{\partial x^i}\, dx^i = \frac{2}{c^2}\int \frac{\partial \psi}{\partial x^i}\, dx^i$$

or

$$\int d\psi_{00} = \frac{2}{c^2}\int d\psi$$

or
$$\psi_{00} = \frac{2\psi}{c^2} + \text{const}$$

or
$$1 + \psi_{00} = \frac{2\psi}{c^2} + \text{const}$$

or
$$g_{00} = \frac{2\psi}{c^2} + k \tag{3.27}$$

Choosing ψ such that when $g_{00} = 1$, $\psi = 0$, so that $k = 1$.

Then
$$g_{00} = 1 + \frac{2\psi}{c^2} \tag{3.28}$$

Thus, we can identify $\frac{1}{2} g_{00}$ as the gravitational potential, the partial derivative of which w. r. t. x^i gives a corresponding component of force in the limit $v \ll c$, *i.e.*, the Newtonian approximation. In Newtonian theory, a single component of $g_{\mu\nu}$ plays the role of gravitational potential. However, in general relativity, the metric tensor $g_{\mu\nu}$ plays the role of gravitational potentials. Furthermore, in the relativistic theory of gravitation, the concepts of inertia and gravity get unified. In Eq. (3.14), the first term accounts for inertia and the second term stands for the force of gravity on the material particle. None of these terms taken individually is tensor in character, whereas the geodesic equation is a tensor equation, implying thereby that inertia and gravitation are connected irrevocably with each other.

Although Newton's theory has been shown as an approximation of Einstein's theory in the non-relativistic limit, most modern non-relativistic calculations are still carried out using Newton's theory, since Newton's theory is a much simpler theory than general relativity and the results it gives in the non-relativistic situations are appropriate.

3.5 HEURISTIC DERIVATION OF EINSTEIN'S FIELD EQUATIONS

According to Newton's theory of gravitation, the gravitational potential ϕ satisfies Poisson's equation

$$\nabla^2 \phi = 4\pi G \rho \tag{3.29}$$

where G is the gravitational constant and ρ is the density of matter. While discussing the trajectory of a particle in the non-relativistic approximation, we had seen that g_{00}, a component of the metric tensor, plays the role of the gravitational potential. In order to get the analogue of Eq. (3.29), in the relativistic theory, ϕ has to be replaced by the metric tensor $g_{\mu\nu}$. Furthermore, we see that the field equations in general relativity should be expressible in terms of the second derivatives of $g_{\mu\nu}$. We know that the covariant derivative of $g_{\mu\nu}$ are identically zero and the appropriate tensor, which contains second order derivatives, is the Ricci tensor $R_{\mu\nu}$, obtained from Riemann–Christoffel tensor $R_{\mu\nu\rho\lambda}$ by contracting twice, such as $R_{\mu\nu}$ and $R^{\mu}_{\mu} = R$. Therefore, the left-hand-side of the desired equation will be either $R_{\mu\nu}$ or its linear combination

$$R_{\mu\nu} + a g_{\mu\nu} \tag{3.30}$$

In addition, we have to require that the relativistic field equations must be invariant under the tensor law of transformation, implying thereby that both the sides must be expressed in terms of tensors. The term ρ must be replaced by a second rank tensor, such as $T_{\mu\nu}$, the energy–momentum tensor. This tensor is appropriate since its divergence vanishes automatically. All these requirements are realized in the equations

$$R_{\mu\nu} + ag_{\mu\nu}R + \Lambda g_{\mu\nu} = -kT_{\mu\nu} \tag{3.31}$$

where k is Einstein's constant of integration, a and Λ are scalar invariants. Since the covariant derivative of $g_{\mu\nu}$ is identically zero, we require that

$$(R_{\mu\nu} + ag_{\mu\nu}R)_{;\nu} = 0 \tag{3.32}$$

We already know that the covariant derivative of Einstein tensor $G_{\mu\nu} = R_{\mu\nu} - \frac{1}{2}g_{\mu\nu}R$ is zero, so the requirement of Eq. (3.32) is met if $a = -1/2$. Thus, the field equations of general relativity become

$$R_{\mu\nu} - \frac{1}{2}g_{\mu\nu}R + \Lambda g_{\mu\nu} = -kT_{\mu\nu} \tag{3.33}$$

Λ is cosmological constant introduced by Einstein. The two other forms of field equations are obtained by raising the index once and twice

$$R^{\nu}_{\mu} - \frac{1}{2}g^{\nu}_{\mu}R + \Lambda g^{\nu}_{\mu} = -kT^{\nu}_{\mu} \tag{3.34}$$

$$R^{\mu\nu} - \frac{1}{2}g^{\mu\nu}R + \Lambda g^{\mu\nu} = -kT^{\mu\nu} \tag{3.35}$$

Multiplying Eq. (3.33) by $g^{\mu\nu}$, we get

$$g^{\mu\nu}R_{\mu\nu} - \frac{1}{2}g_{\mu\nu}g^{\mu\nu}R + \Lambda g_{\mu\nu}g^{\mu\nu} = -kT_{\mu\nu}g^{\mu\nu}$$

or

$$R - \frac{1}{2}4R + 4\Lambda = -kT$$

or

$$R - 4\Lambda = kT \tag{3.36}$$

In empty space, $T = 0$ and the field equations become

$$R_{\mu\nu} = \Lambda g_{\mu\nu} \tag{3.37}$$

For empty space, Einstein chose the equations

$$R_{\mu\nu} = 0 \tag{3.38}$$

These are ten in number and all of these equations are not independent of each other. Since some of these satisfy the four Bianchi identities, these are reduced to 6 in number. As 10 independent components of $g_{\mu\nu}$ are to be determined from 6 equations uniquely, that is not possible.

3.6 EINSTEIN'S FIELD EQUATIONS BY VARIATIONAL TECHNIQUE

In classical mechanics and electromagnetism, we derive the field equations by variational technique. We follow the same approach in deriving Einstein's field equations, but before undertaking the exercise, one has to define a scalar invariant over the four-dimensional space between two given points.

It is known from the earlier article that gravitational field equations involve derivatives of the potentials $g_{\mu\nu}$ upto the second derivatives. Furthermore, we know that the scalar curvature $R = g^{\mu\nu} R_{\mu\nu}$ of the four-dimensional manifold, contains the first and second order partial derivatives of $g_{\mu\nu}$. As such, we assume the action integral as our starting point. Thus,

$$
\begin{aligned}
I &= \int_{\Omega} R\sqrt{(-g)}\, d^4x \\
&= \int g^{\mu\nu} R_{\mu\nu} \sqrt{(-g)}\, d^4x
\end{aligned} \tag{3.39}
$$

We assume that variation in $g_{\mu\nu}$ or its first derivatives vanish at the surface of the four-volume Ω. Applying the variational principle to Eq. (3.39), one gets

$$
\begin{aligned}
\delta I &= \delta \int g^{\mu\nu} R_{\mu\nu} \sqrt{(-g)}\, d^4x \\
&= \int (\delta R_{\mu\nu}) \sqrt{(-g)} g^{\mu\nu}\, d^4x + \int R_{\mu\nu} (\delta g^{\mu\nu} \sqrt{(-g)})\, d^4x = 0
\end{aligned} \tag{3.40}
$$

Now
$$R_{\mu\nu} = -\Gamma^{\alpha}_{\mu\nu,\alpha} + \Gamma^{\alpha}_{\mu\alpha,\nu} - \Gamma^{\alpha}_{\mu\nu}\Gamma^{\lambda}_{\alpha\lambda} + \Gamma^{\lambda}_{\mu\alpha}\Gamma^{\alpha}_{\nu\lambda} \tag{3.41}$$

and
$$\delta R_{\mu\nu} = -\delta\Gamma^{\alpha}_{\mu\nu,\alpha} + \delta\Gamma^{\alpha}_{\mu\alpha,\nu} - \Gamma^{\alpha}_{\mu\nu}(\delta\Gamma^{\lambda}_{\alpha\lambda}) - \Gamma^{\lambda}_{\alpha\lambda}(\delta\Gamma^{\alpha}_{\mu\nu}) + \Gamma^{\lambda}_{\mu\alpha}(\delta\Gamma^{\alpha}_{\nu\lambda}) + \Gamma^{\alpha}_{\nu\lambda}(\delta\Gamma^{\lambda}_{\mu\alpha}) \tag{3.42}$$

As stated earlier while treating Christoffel's 3-index symbols, these do not transform like tensors, whereas their variations like $\delta\Gamma^{\lambda}_{\alpha\lambda}$ do transform like tensors. In addition, it is possible to use a coordinate system, so that at any arbitrary point $\Gamma^{\lambda}_{\alpha\lambda} = 0$ and not their derivatives. In view of these conditions, Eq. (3.42) becomes

$$
\begin{aligned}
\delta R_{\mu\nu} &= -\delta\Gamma^{\alpha}_{\mu\nu,\alpha} + \delta\Gamma^{\alpha}_{\mu\alpha,\nu} \\
&= -\delta\left(\frac{\partial\Gamma^{\alpha}_{\mu\nu}}{\partial x^{\alpha}} - \frac{\partial\Gamma^{\alpha}_{\mu\alpha}}{\partial x^{\nu}}\right) \\
&= -\frac{\partial}{\partial x^{\alpha}}\delta\Gamma^{\alpha}_{\mu\nu} + \frac{\partial}{\partial x^{\nu}}\delta\Gamma^{\alpha}_{\mu\alpha} \\
&= -(\delta\Gamma^{\alpha}_{\mu\nu})_{,\alpha} + (\delta\Gamma^{\alpha}_{\mu\alpha})_{,\nu}
\end{aligned} \tag{3.43}
$$

In a coordinate system where $\Gamma^{\alpha}_{\mu\nu} = 0$, at a particular point, the ordinary and the covariant derivatives are identical. Thus, we get

$$\delta R_{\mu\nu} = -(\delta\Gamma^{\alpha}_{\mu\nu})_{;\alpha} + (\delta\Gamma^{\alpha}_{\mu\alpha})_{;\nu} \tag{3.44}$$

Therefore $g^{\mu\nu}\delta R_{\mu\nu}$ can be written as

$$
\begin{aligned}
g^{\mu\nu}\delta R_{\mu\nu} &= g^{\mu\nu}(-\delta\Gamma^{\alpha}_{\mu\nu})_{;\alpha} + g^{\mu\nu}(\delta\Gamma^{\alpha}_{\mu\alpha})_{;\nu} \\
&= -(g^{\mu\nu}\delta\Gamma^{\alpha}_{\mu\nu})_{;\alpha} + (g^{\mu\nu}\delta\Gamma^{\alpha}_{\mu\alpha})_{;\nu} \\
&= -(g^{\mu\nu}\delta\Gamma^{\alpha}_{\mu\nu})_{;\alpha} + (g^{\mu\alpha}\delta\Gamma^{\alpha}_{\mu\nu})_{;\alpha} \qquad (3.45)
\end{aligned}
$$

where we used the fact that the covariant derivative of $g^{\mu\nu}$ is identically zero. $g^{\mu\nu}\delta R_{\mu\nu}$ is a scalar, so the expression on the r.h.s. of Eq. (3.45) is also a scalar. Thus, according to quotient law

$$
\eta^{\alpha} \equiv -g^{\mu\nu}(\delta\Gamma^{\alpha}_{\mu\nu}) + g^{\mu\alpha}(\delta\Gamma^{\nu}_{\mu\nu}) \qquad (3.46)
$$

is a four-vector and

$$
\begin{aligned}
g^{\mu\nu}\delta R_{\mu\nu} &= \eta^{\alpha}_{\mu\nu,\alpha} \\
&= \frac{1}{\sqrt{(-g)}}\frac{\partial}{\partial x^{\alpha}}(\sqrt{(-g)}\eta^{\alpha}) \qquad (3.47)
\end{aligned}
$$

Finally, the first term in Eq. (3.40) becomes

$$
\int \sqrt{(-g)}g^{\mu\nu}\delta R_{\mu\nu}\, d^4x = \int_{\Omega} \frac{\partial}{\partial x^{\alpha}}(\sqrt{(-g)}\eta^{\alpha})\, d^4x \qquad (3.48)
$$

Applying Gauss's theorem, we convert the volume integral on the r.h.s. into a surface integral over the bounding surface of the four volume Ω, so

$$
\begin{aligned}
\int \sqrt{(-g)}g^{\mu\nu}\delta R_{\mu\nu}\, d^4x &= \int_S (\sqrt{(-g)}\eta^{\alpha})\, dS_{\alpha} \\
&= -\int_S \sqrt{(-g)}\left[g^{\mu\nu}(\delta\Gamma^{\alpha}_{\mu\nu}) - g^{\mu\alpha}(\delta\Gamma^{\nu}_{\mu\nu})\right] dS_{\alpha} \\
&= 0 \qquad (3.49)
\end{aligned}
$$

since the variations of 3-index symbols vanish at the boundary at infinity. In view if this simplification, Eq. (3.40) becomes

$$
\delta I = \int R_{\mu\nu}(\delta g^{\mu\nu}\sqrt{(-g)})\, d^4x = 0 \qquad (3.50)
$$

Now

$$
\begin{aligned}
\delta(g^{\mu\nu}\sqrt{(-g)}) &= (\delta g^{\mu\nu})\sqrt{(-g)} + g^{\mu\nu}\delta\sqrt{(-g)} \\
&= \delta g^{\mu\nu}\sqrt{(-g)} - \frac{1}{2}g^{\mu\nu}\frac{1}{\sqrt{(-g)}}\delta g \\
&= \delta g^{\mu\nu}\sqrt{(-g)} + \frac{1}{2}\frac{1}{g}g^{\mu\nu}\sqrt{(-g)}\,\delta g \qquad (3.51)
\end{aligned}
$$

Let us evaluate δg. Thus,

$$g = g_{\alpha\beta} A^{\alpha\beta} \tag{3.52}$$

where $A^{\alpha\beta}$ is the cofactor of $g_{\alpha\beta}$.

or

$$\frac{\partial g}{\partial g_{\alpha\beta}} = A^{\alpha\beta}$$

Also

$$\delta g = \frac{\partial g}{\partial g_{\alpha\beta}} \delta g_{\alpha\beta}$$

$$= A^{\alpha\beta} \delta g_{\alpha\beta} \tag{3.53}$$

From Eq. (3.52), we define

$$g^{\alpha\beta} = \frac{1}{g_{\alpha\beta}} = \frac{A^{\alpha\beta}}{g}$$

so from Eq. (3.53), one gets

$$\begin{aligned}\delta g &= g g^{\alpha\beta} \delta g_{\alpha\beta} \\ &= g g^{\mu\nu} \delta g_{\mu\nu}\end{aligned} \tag{3.54}$$

Now

$$\delta(g^{\mu\nu} g_{\mu\nu}) = g^{\mu\nu} (\delta g_{\mu\nu}) + (\delta g^{\mu\nu}) g_{\mu\nu} = 0$$

or

$$g^{\mu\nu} (\delta g_{\mu\nu}) = -(\delta g^{\mu\nu}) g_{\mu\nu} \tag{3.55}$$

Thus, Eq. (3.54) becomes

$$\delta g = -g(\delta g^{\mu\nu}) g_{\mu\nu} \tag{3.56}$$

Writing Eq. (3.50) in view of Eqs. (3.51) and (3.56), we get

$$\begin{aligned}\delta I &= \int_{\Omega} R_{\mu\nu} \delta(g^{\mu\nu} \sqrt{(-g)})\, d^4x \\ &= \int_{\Omega} R_{\mu\nu} \left[\delta g^{\mu\nu} \sqrt{(-g)} - \frac{1}{2} \sqrt{(-g)} g^{\mu\nu} g_{\alpha\beta} (\delta g^{\alpha\beta}) \right] d^4x \\ &= \int_{\Omega} \sqrt{(-g)}\, d^4x \left[R_{\mu\nu} - \frac{1}{2} g_{\mu\nu} R \right] \delta g^{\mu\nu} \\ &= 0\end{aligned} \tag{3.57}$$

As the variations $\delta g^{\mu\nu}$ are arbitrary, the only way the integrand will vanish is when

$$R_{\mu\nu} - \frac{1}{2} g_{\mu\nu} R = 0 \tag{3.58}$$

Thus, we obtain Einstein's field equations for empty space by variational technique.

3.7 FUNDAMENTAL HYPOTHESES AND POSTULATES OF GENERAL RELATIVITY

General relativity is based on the following set of fundamental principles, which guided its development. These are in the following:

1. The general principle of relativity—The laws of physics must be the same for all observers (accelerated or not).
2. Physical events are described in a four-dimensional space–time manifold, the metric of which is

$$ds^2 = g_{\mu\nu} dx^\mu dx^\nu \tag{3.59}$$

3. Principle of covariance—The laws of physics must take the same form in all coordinate systems.
4. Local Lorentz invariance—The laws of special relativity apply locally for all inertial observers.
5. The principle that inertial motion is geodesic motion. The world lines of particles unaffected by physical forces are time-like or null geodesics of space–time.
6. Space–time is curved. This permits gravitational effects, such as free fall to be described as a form of inertial motion.
7. Space–time curvature is created by stress-energy within the space–time. This is described in general relativity by Einstein's field equations, which in the presence of matter are

$$R_{\mu\nu} - \frac{1}{2} g_{\mu\nu} R + \Lambda g_{\mu\nu} = -kT_{\mu\nu} \tag{3.60}$$

where $T_{\mu\nu}$ is the energy–momentum tensor. In empty space, $T_{\mu\nu} = 0$ and the equations become

$$R_{\mu\nu} = 0 \tag{3.61}$$

The equivalence principle, which served as a starting point for the development of general relativity, ended up being a consequence of the general principle of relativity and the principle that inertial motion is geodesic motion.

Example 3.1

In a local region of space–time, an observer concludes that the Ricci curvature scalar is nearly constant $R \approx \frac{1}{a^2}$. Why the sign be positive. If the region of space–time is filled only with electromanetic energy, what is the value of R.

Solution

On contracting Einstein's field equations, one gets a relation between the trace of $T_{\mu\nu}$ and Ricci scalar is $R = -8\pi T_\mu^\mu$. In the local Lorentz frame in which T_μ^μ is diagonal,

$$T_\mu^\mu = -\rho + p_x + p_y + p_z$$

where p_x, p_y, p_z are the principal pressures. For all equations of state, $\rho \geq 3p$, so T_μ^μ should always be negative and R should be positive.

The electromagnetic stress–energy is traceless, thus, $R = 0$ if the only stress–energy is electromagnetic.

3.8 POISSON'S EQUATION AS APPROXIMATION OF EINSTEIN'S FIELD EQUTIONS. EVALUATION OF CONSTANT *k*

Einstein's field equations are

$$R_{\mu\nu} - \frac{1}{2} g_{\mu\nu} R + \Lambda g_{\mu\nu} = -kT_{\mu\nu} \tag{3.62}$$

where all the symbols have their usual meaning and k is an undetermined constant. It is proposed to fix its value by demanding that Eq. (3.62) yield the Newton's equation of gravitation in the non-relativistic approximation, *i.e.*, $v \ll c$. This approach seems quite plausible since Newton's equation describes gravitation satisfactorily in the non-relativistic approximation. Any improvement of the theory, must reduce to the Newton's theory in the first approximation. Newton's equation in the presence of matter is described by Poisson's equation

$$\nabla^2 \phi = 4\pi G \rho \tag{3.63}$$

where ϕ is the gravitational potential, G is the gravitational constant and ρ the matter density.

As Λ, the cosmological constant is a small quantity, we will ignore it for the present and the field equations become

$$R_{\mu\nu} - \frac{1}{2} g_{\mu\nu} R = -kT_{\mu\nu} \tag{3.64}$$

The energy–momentum tensor $T_{\mu\nu}$ of an incoherent matter, is given by

$$T_{\mu\nu} = \rho c^2 \frac{dx_\mu}{ds} \frac{dx_\nu}{ds} \tag{3.65}$$

Since the gravitational field is static, *i.e.*, does not depend on time, the velocity components become

$$\frac{dx_0}{ds} = 1; \quad \frac{dx_1}{ds} = \frac{dx_2}{ds} = \frac{dx_3}{ds} = 0 \tag{3.66}$$

The energy–momentum tensor $T_{\mu\nu}$, in the limit of Newtonian approximation, *viz.*, $v \ll c$ gives

$$T_{00} = \rho c^2 \frac{dx_0}{ds} \frac{dx_0}{ds} \cong \rho c^2 \tag{3.67}$$

the rest of the components being zero. Thus, Eq. (3.64) becomes

$$R_{00} - \frac{1}{2} g_{00} R = -kc^2 \rho \tag{3.68}$$

Further, we get

$$T = T_\mu^\mu = T_0^0 = c^2 \rho \tag{3.69}$$

and multiplying Eq. (3.64) by $g^{\mu\nu}$, we get

$$R-\frac{1}{2}R4=-kT$$

or
$$R=kT \tag{3.70}$$

In view of Eqs. (3.69) and (3.70), Eq. (3.68) becomes

$$R_{00}-\frac{1}{2}g_{00}kc^2\rho=-kc^2\rho \tag{3.71}$$

or approximately
$$R_{00}-\frac{1}{2}kc^2\rho=-kc^2\rho$$

or
$$R_{00}=-\frac{1}{2}kc^2\rho$$

Alternatively, we evaluate R_{00} from the tensor $R_{\mu\nu}$

$$R_{\mu\nu}=\frac{\partial\Gamma^{\alpha}_{\mu\alpha}}{\partial x^{\nu}}-\frac{\partial\Gamma^{\alpha}_{\mu\nu}}{\partial x^{\alpha}}+\Gamma^{\alpha}_{\mu\lambda}\Gamma^{\lambda}_{\nu\alpha}-\Gamma^{\alpha}_{\mu\nu}\Gamma^{\lambda}_{\alpha\lambda} \tag{3.72}$$

In the non-relativistic limit $v \ll c$, the $g_{\mu\nu}$'s differ infinitesimally from their flat space values and are independent of time. Thus, neglecting the product of the 3-index symbols, we obtain

$$R_{\mu\nu}=\frac{\partial\Gamma^{\alpha}_{\mu\alpha}}{\partial x^{\nu}}-\frac{\partial\Gamma^{\alpha}_{\mu\nu}}{\partial x^{\alpha}}$$

or
$$R_{00}=\frac{\partial\Gamma^{\alpha}_{0\alpha}}{\partial x^{0}}-\frac{\partial\Gamma^{\alpha}_{00}}{\partial x^{\alpha}}$$

Therefore,
$$R_{00}=-\frac{\partial\Gamma^{\alpha}_{00}}{\partial x^{\alpha}} \tag{3.73}$$

since $g_{\mu\nu}$ is not a function of t, hence $\frac{\partial\Gamma^{\alpha}_{0\alpha}}{\partial x^{0}}=0$. Combining Eq. (3.73) with Eq. (3.71), we obtain

$$\frac{\partial\Gamma^{\alpha}_{00}}{\partial x^{\alpha}}=\frac{1}{2}kc^2\rho \tag{3.74}$$

We had shown that for a weak gravitational field, Eq. (3.24)

$$\Gamma^{\alpha}_{00}=\frac{1}{2}\frac{\partial g_{00}}{\partial x^{\alpha}} \tag{3.75}$$

Further, since the gravitational field is static, so

$$\frac{\partial g_{00}}{\partial x^{0}}=0 \tag{3.76}$$

Rewriting Eq. (3.74) in the light of Eqs. (3.75) and (3.76), we obtain

$$\nabla^2\left(\frac{1}{2}c^2 g_{00}\right)=\frac{1}{2}kc^4\rho \tag{3.77}$$

But g_{00} is given by Eq. (3.27)

$$g_{00} = 1 + \frac{2\phi}{c^2} \tag{3.78}$$

Combining Eq. (3.78) with Eq. (3.77), we get

$$\nabla^2\phi = \frac{1}{2}kc^4\rho \tag{3.79}$$

Comparing it with Poisson's equation $\nabla^2\phi = 4\pi G\rho$, one obtains

$$k = \frac{8\pi G}{c^4} \tag{3.80}$$

Finally, the field equations become

$$R_{\mu\nu} - \frac{1}{2}g_{\mu\nu}R = -\frac{8\pi G}{c^4}T_{\mu\nu} \tag{3.81}$$

In relativistic units, the unit of length is 1 cm, but that of mass and time are such that the velocity of light and the gravitational constant are both unity, *i.e.*, $c = G = 1$. Thus, in relativistic units, k has the value

$$k = 8\pi \tag{3.82}$$

Example 3.2

A slightly generalized form of the Einstein's field equations is

$$R_{\mu\nu} - ag_{\mu\nu}R = -8\pi T_{\mu\nu}$$

where a is some dimensionless constant. Show that if a is different from 1/2, the field equations disagree with experiment even in the Newtonian limit.

Solution

According to Bianchi identities, the Einstein tensor $R_{\mu\nu} - \frac{1}{2}g_{\mu\nu}R$ is divergenceless, so the divergence of the given generalized field equation is

$$\left(\frac{1}{2} - a\right)R_{,\mu} = -8\pi T^{\nu}_{\mu,\nu}$$

On the other hand, if we contract our field equation first and then differentiate, we get

$$(1-4a)R_{,\mu} = -8\pi T_{,\mu}$$

so that the equation of motion must be

$$T^{\nu}_{\mu,\nu} = \kappa T_{,\mu} \qquad \kappa = \frac{1/2 - a}{1-4a}$$

In the Newtonian limt and for a fluid with density ρ and negligible pressure, the $\mu = 0$ component of the equation is

$$\frac{\partial\rho}{\partial t} + \nabla\cdot(\rho\mathbf{v}) = \kappa\frac{\partial\rho}{\partial t}$$

where **v** is the velocity of fluid flow. If κ were not zero, *i.e.*, if $a \neq \frac{1}{2}$, the above equation will differ from the Newtonian continuity equation and violate the law of conservation of mass.

References

1. Einstein, A. (1916). *Annalen der Phys.*, 49; 769. For English translation, see *The Principle of Relativity*, Dover Publication (1923).
2. Pound, R.V. and Rebka, G.A. (1959). *Phys. Rev. Lett.*, **3**, 4399; *ibid* **4**, 337.
3. Uzan, J.P. (2003). 'The Fundamental Constants and Their Variation: Observational Status and Theoretical Motivations', *Reviews of Modern Physics*, 75(5), 403.
4. Cocconi, G. and Salpeter, E. (1958). A Search for Anisotropic Inertia, *Nouvo Cimento* 10; 646.
5. Cocconi, G. and Salpeter, E. (1960). *Phys. Rev. Letters*, 4; 176.
6. Hughes, V.W., Robinson, H.G., and Beltran-Lopez (1960). *Phys. Rev. Letters*, 4; 342.
7. Drever, R.W.P., (1961). *Phil. Mag.* 6; 683.
8. Dicke, R.H. (1961). Experimental Tests of Mach's Principle, *Phys. Rev. Letters*, 7; 359.
9. Prestage, J.D., *et al.* (1985). Limits for Spatial Anisotropy by the Use of Nuclear-Spin-Polarized $^9Be^+$ Ion, *Phys. Rev. Letters*, 54; 2387.
10. Cross, Daniel J. Advisor Dr Robert Gilmore, (2004). *Unpublished*, 9 June.

Problems

3.1 Show that in the presence of electromagnetic field, Einstein's field equations become

$$R_{\mu\nu} = -8\pi T_{\mu\nu}$$

where $T_{\mu\nu}$ is the electromagnetic energy–momentum tensor defined as

$$T_{\mu\nu} = -F_{\mu\rho}F_{\nu\rho} + \frac{1}{4}\delta_{\mu\nu}F_{\alpha\beta}F^{\alpha\beta}$$

$F_{\mu\nu}$ is the electromagnetic field tensor.

3.2 Show that the two invariants of the energy momentum tensor $T_{\mu\nu}$ are
   (i) $\rho - 3p$
   (ii) $\rho^2 + 3p^2$

3.3 Show that the divergence of energy–momentum tensor $T_{\mu\nu}$ vanishes and $G = -8\pi T$ where $G_{\mu\nu}$ is the Einstein tensor, $G_{\mu\nu} = R_{\mu\nu} - \frac{1}{2}\delta_{\mu\nu}R$.

3.4 Derive the following conversion factors from the SI values of G and c.

$$G/c^2 = 7.425 \times 10^{-28}\,\mathrm{m\,kg^{-1}} = 1$$
$$c^5/G = 3.629 \times 10^{52}\,\mathrm{J\,s^{-1}} = 1$$

3.5 Express the following quantities in generalized units:
   (i) A density (typical of neutron stars) $\rho = 10^{17}\,\mathrm{kg\,m^{-3}}$.
   (ii) A pressure (also typical of neutron stars) $p = 10^{33}\,\mathrm{kg\,s^{-2}\,m^{-1}}$.
   (iii) The acceleration of gravity on Earth's surface $g = 9.80\,\mathrm{m\,s^{-2}}$.
   (iv) The luminosity of a supernova $L = 10^{41}\,\mathrm{J\,s^{-1}}$.

4

Einstein's Law of Gravitation in Empty Space—Schwarzschild Solution

4.1 INTRODUCTION

In this chapter, we will be seeking the solution of Einstein's law of gravitation for empty space surrounding an isolated and stable gravitating mass, by obtaining a line element of such a space. Einstein had solved this problem by the method of successive approximations but the rigorous solution of this problem was provided by Karl Schwarzschild[1] in 1916. The importance of this solution emanates from the fact that it is specially suited for studying the gravitational field of sun and in addition, in discussing the three crucial tests predicted by the general theory of relativity *vis-a-vis* the predictions of Newton's theory of gravitation. To make a mention of these tests, these are as follows:

- Advance of the perihelion of Mercury.
- Bending of a light ray passing near the Sun.
- Gravitational red shift of spectral lines.

4.2 A STATIC SPHERICALLY SYMMETRIC SPACE–TIME

Firstly, we consider the limiting form of the metric that represents the field outside of a spherically symmetric body. At a far away distance or as r becomes large, it will reduce to the form of the Minkowski metric. For spherical symmetry, the Minkowski metric in spherical coordinates, is

$$ds^2 = dt^2 - dr^2 - r^2 d\theta^2 - r^2 \sin^2\theta d\phi^2 \tag{4.1}$$

First we consider the requirement of time independence. A time-independent metric on time reversal, *i.e.*, on replacing $dt \to -dt$, preserves the form. Furthermore, the metric will not have any mixed terms, such as $dtdr$, $dt\,d\theta$, $dtd\phi$. In view of these restrictions, the general form can be written in the form

$$ds^2 = g_{00}dt^2 + g_{ij}dx^i dx^j \tag{4.2}$$

In addition, the components of the metric are time independent, *i.e.*,

$$\frac{\partial g_{\mu\nu}}{\partial t} = 0 \tag{4.3}$$

Such a metric is called a static metric.

A spherically symmetric metric has no preferred angular direction in space, which implies that on replacing $d\theta \to -d\theta$ and $d\phi = -d\phi$ does not alter the form of the metric. Like the time- independent case, it will not have the mixed terms, such as $drd\theta, drd\phi, d\theta d\phi,$ since such terms will be affected by the changes $d\theta \to -d\theta$ and $d\phi = -d\phi$. So the metric must have an entirely diagonal form. Also we have eliminated any explicit time dependence from the terms of the metric.

As we are imposing radial symmetry, each term in the metric can be multiplied by a function depending only on r. Therefore, the metric may be written as

$$ds^2 = A(r)\,dt^2 - B(r)\,dr^2 - C(r)r^2 d\theta^2 - D(r)r^2 \sin^2\theta d\phi^2 \tag{4.4}$$

Spherical symmetry conjoins the requirement that the angular terms assume the normal form $d\Omega^2$, so we take $C = D$ and Eq. (4.4) becomes

$$ds^2 = A(r)\,dt^2 - B(r)\,dr^2 - C(r)\,(r^2 d\theta^2 + r^2 \sin^2\theta d\phi^2) \tag{4.5}$$

To simplify it further, we change the radial coordinate to eliminate C. Calling the new radial coordinate ρ, which is defined as $\rho = \sqrt{C(r)}\,r$, the angular part of the metric in Eq. 4.5 becomes

$$\begin{aligned} C(r)(r^2 d\theta^2 + r^2\sin^2\theta d\phi^2) &= Cr^2 d\theta^2 + Cr^2 \sin^2\theta d\phi^2 \\ &= \rho^2 d\theta^2 + \rho^2 \sin^2\theta d\phi^2 \end{aligned} \tag{4.6}$$

Now

$$\begin{aligned} d\rho &= \frac{1}{2\sqrt{C}} dC(r) r + \sqrt{C} dr \\ &= \left(\frac{1}{2\sqrt{C}} \frac{dC}{dr} r + \sqrt{C}\right) dr \\ &= \sqrt{C}\left(\frac{r}{2C}\frac{dC}{dr} + 1\right) dr \end{aligned} \tag{4.7}$$

Squaring and solving for dr^2, we get

$$dr^2 = \frac{1}{C}\left(1 + \frac{r}{2C}\frac{dC}{dr}\right)^{-2} d\rho^2 \tag{4.8}$$

Recalling the coefficient function

$$B' = \frac{1}{C}\left(1 + \frac{r}{2C}\frac{dC}{dr}\right)^{-2} B$$

we obtain

$$Bdr^2 = B'd\rho^2 \tag{4.9}$$

In view of Eq. (4.9), rewriting Eq. (4.5), one gets

$$ds^2 = A'(\rho)dt^2 - B'd\rho^2 - \rho^2(d\theta^2 + \sin^2\theta d\phi^2) \tag{4.10}$$

where A' is a function of ρ.

Till now, the only condition we imposed was, that the coefficient functions go to unity as r gets large. So in essence, we have labels only. Dropping primes on the coefficient functions and changing $\rho \to r$, we rewrite the line element as

$$ds^2 = Adt^2 - Bdr^2 - r^2(d\theta^2 + \sin^2\theta d\phi^2) \tag{4.11}$$

Lastly, we impose one more condition. In order to conform to the metric, Eq. (4.1), we have to preserve the signature of space. This is done by writing the coefficient functions as exponentials, since exponentials are always positive. Thus, we set $A = e^{\nu(r)}$, $B = e^{\lambda(r)}$ and the Schwarzschild metric becomes

$$ds^2 = e^{\nu(r)}dt^2 - e^{\lambda(r)}dr^2 - r^2d\theta^2 - r^2\sin^2\theta d\phi^2 \tag{4.12}$$

4.3 SCHWARZSCHILD LINE-ELEMENT

It is intended to find a line-element for an interval in the vicinity of a gravitating point particle, for instance, the sun. In the empty space, the space–time would be flat and in spherical polar coordinates and time, the line element is given by

$$ds^2 = dt^2 - dr^2 - r^2d\theta^2 - r^2\sin^2\theta d\phi^2 \tag{4.13}$$

In this equation, the velocity of light c is taken to be unity in order to use relativistic units. The isolated particle is assumed to be static and spherically symmetrical and hence, the gravitational field will depend on r alone and not on θ, ϕ. The line-element would be spatially spherically symmetric about the point and static.

Thus, the line-element in the most general form, Eq. (4.12), is

$$ds^2 = -e^{\lambda}dr^2 - r^2(d\theta^2 + \sin^2\theta d\phi^2) + e^{\nu}dt^2 \tag{4.14}$$

where $\lambda = \lambda(r)$ and $\nu = \nu(r)$.

However, the gravitational field due to the particle decreases as we move away from it and the line-element, Eq. (4.14) will become the Galilean line-element at $r = \infty$. Thus, $\lambda = 0 = \nu$ at $r = \infty$.

Rewriting the line-element, Eq. (4.14) as

$$ds^2 = g_{\mu\nu}dx^{\mu}dx^{\nu} \tag{4.15}$$

with $x^0 = t,\ x^1 = r,\ x^2 = \theta,\ x^3 = \phi$. The metric tensor $g_{\mu\nu}$ is

$$g_{\mu\nu} = \begin{pmatrix} e^{\nu} & 0 & 0 & 0 \\ 0 & -e^{\lambda} & 0 & 0 \\ 0 & 0 & -r^2 & 0 \\ 0 & 0 & 0 & -r^2 \sin^2\theta \end{pmatrix} \tag{4.16}$$

The tensor $g^{\mu\nu}$ is

$$g^{\mu\nu} = \begin{pmatrix} e^{-\nu} & 0 & 0 & 0 \\ 0 & -e^{-\lambda} & 0 & 0 \\ 0 & 0 & -r^{-2} & 0 \\ 0 & 0 & 0 & -1/(r^2 \sin^2\theta) \end{pmatrix} \tag{4.17}$$

Let us evaluate the Christoffel's 3-index symbols $\Gamma^{\alpha}_{\mu\nu}, \Gamma_{\alpha,\mu\nu}$ for the static and spherically symmetric line-element, Eq. (4.15). Recalling that

$$\Gamma_{\alpha,\mu\nu} = 1/2(\partial_{\nu} g_{\alpha\mu} + \partial_{\mu} g_{\alpha\nu} - \partial_{\alpha} g_{\mu\nu}) \tag{4.18}$$

$$\Gamma^{\alpha}_{\mu\nu} = g^{\alpha\lambda}\Gamma_{\lambda,\mu\nu} = \Gamma^{\alpha}_{\mu\nu} \tag{4.19}$$

In view of the symmetry relation, Eq. (4.19), only 40 are independent out of the 64 three-index symbols. In addition, for the line-element, Eq. (4.15), 31Christoffel's 3-index symbols are zero and the following nine are non-zero. Further, it is clear that since $g_{\mu\nu}$ is diagonal, the 3-index symbols with all the three indices different, are zero.

$$\begin{aligned} &\Gamma^0_{10} = \frac{1}{2}\nu' && \Gamma^2_{12} = \frac{1}{r} \\ &\Gamma^1_{00} = \frac{1}{2}\nu' e^{\nu-\lambda} && \Gamma^2_{33} = -r\sin\theta\cos\theta \\ &\Gamma^1_{11} = \frac{1}{2}\lambda' && \Gamma^3_{13} = \frac{1}{r} \\ &\Gamma^1_{22} = -re^{-\lambda} && \Gamma^3_{32} = \cot\theta \\ &\Gamma^1_{23} = -r\sin^2\theta e^{-\lambda} && \end{aligned} \tag{4.20}$$

where

$$\nu' = d\nu/dr \quad \text{and} \quad \lambda' = d\lambda/dr. \tag{4.21}$$

Einstein's law of gravitation for empty space is

$$R_{\mu\nu} = 0 \tag{4.22}$$

where $R_{\mu\nu}$ is the contracted curvature tensor.

Making use of the 3-index symbols $T^{\alpha}_{\mu\nu}$, Eq. (4.20), we calculate the different components of $R_{\mu\nu}$, given by

$$R_{\mu\nu} = -\partial_{\alpha}\Gamma^{\alpha}_{\mu\nu} + \partial_{\nu}\Gamma^{\alpha}_{\mu\alpha} + \Gamma^{\rho}_{\mu\alpha}\Gamma^{\alpha}_{\nu\rho} - \Gamma^{\alpha}_{\mu\nu}\Gamma^{\rho}_{\alpha\rho}$$

Thus,

$$\begin{aligned}
R_{00} &= -\partial_{\alpha}\Gamma^{\alpha}_{00} + \partial_{0}\Gamma^{\alpha}_{0\alpha} + \Gamma^{\rho}_{0\alpha}\Gamma^{\alpha}_{0\rho} - \Gamma^{\alpha}_{00}\Gamma^{\rho}_{\alpha\rho} \\
&= -\partial_{\alpha}\Gamma^{\alpha}_{00} + \Gamma^{\rho}_{0\alpha}\Gamma^{\alpha}_{0\rho} - \Gamma^{\alpha}_{00}\Gamma^{\rho}_{\alpha\rho} \\
&= -\partial_{1}\Gamma^{1}_{00} + \Gamma^{1}_{00}\Gamma^{0}_{01} + \Gamma^{0}_{01}\Gamma^{1}_{00} - \Gamma^{1}_{00}\left(\Gamma^{0}_{10} + \Gamma^{1}_{11} + \Gamma^{2}_{12} + \Gamma^{3}_{13}\right) \\
&= -\partial_{1}\left(\frac{1}{2}\nu' e^{\nu-\lambda}\right) + \frac{1}{4}\nu'^{2} e^{\nu-\lambda} + \frac{1}{4}\nu'^{2} e^{\nu-\lambda} \\
&\quad -\frac{1}{2}\nu' e^{\nu-\lambda}\left(\frac{1}{2}\nu' + \frac{1}{2}\lambda' + \frac{1}{r} + \frac{1}{r}\right) \\
&= -\frac{1}{2}[\nu'' + \nu'(\nu' - \lambda')]e^{\nu-\lambda} + \frac{1}{2}\nu'^{2}e^{\nu-\lambda} - \frac{1}{4}\nu' e^{\nu-\lambda}(\nu' + \lambda') - \frac{\nu'}{r}e^{\nu-\lambda}
\end{aligned}$$

or

$$R_{00} = -\frac{1}{2}e^{\nu-\lambda}\left(\nu'' + \frac{1}{2}\nu'^{2} - \frac{1}{2}\nu'\lambda' + \frac{2}{r}\nu'\right) \tag{4.23}$$

Analogously, we calculate R_{11}.

$$\begin{aligned}
R_{11} &= -\partial_{\alpha}\Gamma^{\alpha}_{11} + \partial_{1}\Gamma^{\alpha}_{1\alpha} + \Gamma^{\rho}_{1\alpha}\Gamma^{\alpha}_{1\rho} - \Gamma^{\alpha}_{11}\Gamma^{\rho}_{\alpha\rho} \\
&= -\partial_{1}\Gamma^{1}_{11} + \partial_{1}\Gamma^{0}_{10} + \partial_{1}\Gamma^{1}_{11} + \partial_{1}\Gamma^{2}_{12} + \partial_{1}\Gamma^{3}_{13} + \Gamma^{0}_{1\alpha}\Gamma^{\alpha}_{10} \\
&\quad + \Gamma^{1}_{1\alpha}\Gamma^{\alpha}_{11} + \Gamma^{2}_{1\alpha}\Gamma^{\alpha}_{12} + \Gamma^{3}_{1\alpha}\Gamma^{\alpha}_{13} - \Gamma^{1}_{11}(\Gamma^{0}_{10} + \Gamma^{1}_{11} + \Gamma^{2}_{12} + \Gamma^{3}_{13}) \\
&= \partial_{1}\left(\Gamma^{0}_{10} + \Gamma^{2}_{12} + \Gamma^{3}_{13}\right) + \Gamma^{0}_{10}\Gamma^{0}_{10} + \Gamma^{1}_{11}\Gamma^{1}_{11} + \Gamma^{2}_{12}\Gamma^{2}_{12} + \Gamma^{3}_{13}\Gamma^{3}_{13} \\
&\quad - \Gamma^{1}_{11}\left(\Gamma^{0}_{10} + \Gamma^{1}_{11} + \Gamma^{2}_{12} + \Gamma^{3}_{13}\right) \\
&= \partial_{1}(1/2\nu' + 2/r) + \nu'^{2}/4 + \lambda'^{2}/4 + 2/r^{2} - \lambda'/2(\nu'/2 + \lambda'/2 + 2/r)
\end{aligned}$$

or

$$R_{11} = \frac{1}{2}\nu'' + \frac{1}{4}\nu'^{2} - \frac{1}{4}\nu'\lambda' - \frac{\lambda'}{r} \tag{4.24}$$

Similarly, we calculate R_{22}. Thus,

$$\begin{aligned}
R_{22} &= -\partial_{\alpha}\Gamma^{\alpha}_{22} + \partial_{2}\Gamma^{\alpha}_{2\alpha} + \Gamma^{\rho}_{2\alpha}\Gamma^{\alpha}_{2\rho} - \Gamma^{\alpha}_{22}\Gamma^{\rho}_{\alpha\rho} \\
&= \partial_{1}\Gamma^{1}_{22} + \partial_{2}\Gamma^{3}_{23} + \Gamma^{\rho}_{21}\Gamma^{1}_{2\rho} + \Gamma^{\rho}_{22}\Gamma^{2}_{2\rho} + \Gamma^{\rho}_{23}\Gamma^{3}_{2\rho} \\
&\quad - \Gamma^{1}_{22}\left(\Gamma^{0}_{10} + \Gamma^{1}_{11} + \Gamma^{2}_{12} + \Gamma^{3}_{13}\right) \\
&= -\partial_{1}(-re^{-\lambda}) + \partial_{2}(\cot\theta) + \Gamma^{2}_{21}\Gamma^{1}_{22} + \Gamma^{1}_{22}\Gamma^{2}_{21} + \Gamma^{3}_{23}\Gamma^{3}_{23} \\
&\quad + re^{-\lambda}(1/2\nu' + 1/2\lambda' + 2/r) \\
&= e^{-\lambda} - r\lambda' e^{-\lambda} - \operatorname{cosec}^{2}\theta - 2e^{-\lambda} + \cot^{2}\theta \\
&\quad + re^{-\lambda}(1/2\nu' + 1/2\lambda' + 2/r)
\end{aligned}$$

or
$$R_{22} = e^{-\lambda}(1+1/2r\nu'-1/2r\lambda')-1 \tag{4.25}$$

Lastly we calculate R_{33}. Thus,

$$\begin{aligned}
R_{33} &= -\partial_\alpha\Gamma^\alpha_{33} + \partial_3\Gamma^\alpha_{3\alpha} + \Gamma^\rho_{3\alpha}\Gamma^\alpha_{3\rho} - \Gamma^\alpha_{33}\Gamma^\rho_{\alpha\rho} \\
&= -\partial_1\Gamma^1_{33} - \partial_2\Gamma^2_{33} + \Gamma^\rho_{31}\Gamma^1_{3\rho} + \Gamma^\rho_{32}\Gamma^2_{3\rho} + \Gamma^\rho_{33}\Gamma^3_{3\rho} \\
&\quad -\Gamma^1_{33}\left(\Gamma^0_{10}+\Gamma^1_{11}+\Gamma^2_{12}+\Gamma^3_{13}\right)-\Gamma^2_{33}\left(\Gamma^3_{23}\right) \\
&= -\partial_1(-r\sin^2\theta e^{-\lambda}) - \partial_2(-\sin\theta\cos\theta) + \Gamma^3_{31}\Gamma^1_{33} + \Gamma^3_{32}\Gamma^2_{33} \\
&\quad +\Gamma^1_{33}\Gamma^3_{31} + \Gamma^2_{33}\Gamma^3_{32} - \Gamma^1_{33}(\nu'/2+\lambda'/2+2/r) - \Gamma^2_{33}(\cot\theta) \\
&= \sin^2\theta e^{-\lambda} - r\lambda'\sin^2\theta e^{-\lambda} + \cos 2\theta + 2(-\sin^2\theta e^{-\lambda} - \cot\theta\sin\theta\cos\theta) \\
&\quad + r\sin^2\theta e^{-\lambda}(\nu'/2+\lambda'/2+2/r) + \sin\theta\cos\theta\cot\theta \\
&= \sin^2\theta e^{-\lambda}(1-r\lambda') + \cos 2\theta + 1/2r\sin^2\theta e^{-\lambda}(\nu'+\lambda') \\
&\quad + 2\sin^2\theta e^{-\lambda} - 2\sin^2\theta e^{-\lambda} - 2\cos^2\theta + \cos^2\theta \\
&= \sin^2\theta e^{-\lambda}\left[1+1/2r\nu'-1/2r\lambda'\right] + \cos 2\theta - \cos^2\theta \\
&= \sin^2\theta e^{-\lambda}[1+1/2r\nu'-1/2r\lambda'] - \sin^2\theta
\end{aligned}$$

Therefore,
$$R_{33} = \sin^2\theta R_{22} \tag{4.26}$$

All the off-diagonal components of $R_{\mu\nu}$ are identically zero. The non-zero components of $R_{\mu\nu}$ are summarized as follows:

$$\begin{aligned}
R_{00} &= -1/2e^{\nu-\lambda}(\nu''+1/2\nu'^2-1/2\nu'\lambda'+2/r\nu') \\
R_{11} &= 1/2\nu''+1/4\nu'^2-1/4\nu'\lambda'-\lambda'/r \\
R_{22} &= e^{-\lambda}(1+1/2r\nu'-1/2r\lambda')-1 \\
R_{33} &= \sin^2\theta R_{22}
\end{aligned} \tag{4.27}$$

Now the field Eqs. (4.22), become in the light of Eqs. (4.27)

$$e^{-\lambda}(\nu''+1/2\nu'^2-1/2\nu'\lambda'+2\nu'/r)=0 \tag{4.28}$$

$$\nu''/2+\nu'^2/4-\nu'\lambda'/4-\lambda'/r=0 \tag{4.29}$$

$$e^{-\lambda}(1+1/2r\nu'-1/2r\lambda')-1=0 \tag{4.30}$$

$$[e^{-\lambda}(1+1/2r(\nu'-\lambda'))-1]\sin^2\theta=0 \tag{4.31}$$

Obviously, Eqs. (4.30) and (4.31) are identical. The three independent Eqs. (4.28), (4.29), and (4.30) enable us to determine the unknown functions ν and λ. Dividing Eq. (4.28) by $e^{-\lambda}$, we obtain

$$\nu''+1/2\nu'^2-1/2\nu'\lambda'+2\nu'/r=0 \tag{4.32}$$

Rewriting Eq. (4.29), we get

$$\nu'' + \nu'^2/2 - \nu'\lambda'/2 - 2\lambda'/r = 0 \tag{4.33}$$

Subtracting Eq. (4.33) from Eq. (4.32), we get

$$2/r(\nu' + \lambda') = 0 \tag{4.34}$$

Integrating Eq. (4.34), we obtain

$$\nu + \lambda = C \tag{4.35}$$

where C is a constant of integration. The gravitational field (disturbance from flat space–time) due to the particle, diminishes as we recede to the infinite distance. It implies that the line-element, Eq. (4.14), must assume the form of Galilean line-element at $r = \infty$. Thus, we must take $\nu = 0 = \lambda$ at $r = \infty$. Thus, $C = 0$ in Eq. (4.35).

Therefore,

$$\nu + \lambda = 0 \text{ or } \lambda = -\nu \tag{4.36}$$

In view of Eq. (4.36), the Eq. (4.30) becomes

$$e^{-\lambda}(1 - r\lambda') - 1 = 0$$

or

$$\frac{d}{dr}(re^{-\lambda}) = 1$$

or

$$re^{-\lambda} = r + C_1$$

or

$$e^{-\lambda} = 1 + C_1/r \tag{4.37}$$

Putting $C_1 = -2m$, we get

$$e^{\nu} = e^{-\lambda} = 1 - \frac{2m}{r} \tag{4.38}$$

The constant of integration is put equal to $-2m$, in order to facilitate the physical interpretation of m as the mass of gravitating particle. The line-element. Eq. (4.14), due to a static, isolated gravitating mass point is, in relativistic units $(c = G = 1)$

$$ds^2 = \left(1 - \frac{2m}{r}\right)dt^2 - \left(1 - \frac{2m}{r}\right)^{-1} dr^2 - r^2(d\theta^2 + \sin^2\theta d\phi^2) \tag{4.39}$$

In c.g.s. units, if the mass is M, the line-element is

$$ds^2 = \left(1 - \frac{2MG}{rc^2}\right)dt^2 - \left(1 - \frac{2MG}{rc^2}\right)^{-1} dr^2 - r^2(d\theta^2 + \sin^2\theta d\phi^2) \tag{4.40}$$

As it was obtained first by Schwarzschild in the closed form, it is called Schwarzschild's exterior solution. As is obvious, in the limit $r \to \infty$, it goes over to the metric of space–time of special relativity. This solution is of fundamental importance. Had we included the cosmological constant Λ, then for empty space, the solution corresponding to the field equations

$$R_{\mu\nu} = \Lambda g_{\mu\nu} \tag{4.41}$$

can be obtained by taking $T_{\mu\nu} = 0$ and the resulting metric is

$$ds^2 = \left(1 - \frac{2m}{r} - \frac{\Lambda r^2}{3}\right)dt^2 - \left(1 - \frac{2m}{r} - \frac{\Lambda r^2}{3}\right)^{-1} dr^2 - r^2(d\theta^2 + \sin^2\theta d\phi^2) \tag{4.42}$$

Comparing the two metrics, Eqs. (4.39) and (4.42), we see that the effect of term Λ on the field surrounding the attracting particle, increases with r, *i.e.*, the size of the region considered. But as we will see in the next section, the motion of planets is described with great accuracy by Eq. (4.40), we infer that the cosmological constant Λ, is so small that it does not produce tangible effects within regions of the size comparable to the solar system.

4.3.1 Schwarzschild Singularity

Schwarzschild solution, Eq. (4.39) has the following singularities:

- $r = 0$. But this type of singularity is usual and also occurs in the Newtonian theory of gravitation.
- $r = 2m$. The value of r is called the Schwarzschild's radius. At $r = 2m$, g_{00} becomes zero and g_{11} becomes infinite. For points $0 \le r \le 2m$, $ds^2 < 0$, *i.e.*, the interval is purely space-like.

Let us estimate the singular region for the sun. If it has the radius r, then

$$\begin{aligned} r &= 2m \qquad \text{(in relativistic units)} \\ &= \frac{2mG}{c^2} \qquad \text{(in c.g.s. units)} \\ &= 2.956\,\text{km} \end{aligned} \tag{4.43}$$

The physical radius of sun is 6.960×10^5 km, thus, the singular region is extreme deeply embedded in it. For a region inside, the Schwarzschild solution of Einstein's field equations, does not hold as $T_{\mu\nu}$ will not be zero. This is the situation long before the singular region starts. The Schwarzschild's radius is always much smaller than the physical radius of the body, say earth, electron.

However, it may be noted that Einstein and Rosen have shown that the singularity in the Schwarzschild solution, Eq. (4.39), can be removed by the transformation

$$r - 2m = u^2 \tag{4.44}$$

The line-element, Eq. (4.39), then becomes

$$ds^2 = -4(u^2 + 2m)du^2 - (u^2 + 2m)^2(d\theta^2 + \sin^2\theta d\phi^2) + \frac{u^2}{u^2 + 2m}dt^2 \tag{4.45}$$

The singularity is no more there in the line-element, but the singularity $r = 2m$ has been shifted from the line-element, Eq. (4.39) to the transformation, Eq. (4.44). The Jacobian of this transformation is zero at $u = 0$. But in general relativity, only non-singular transformations are allowed so that the Jacobian has to be non-zero and finite within its domain. Furthermore, the line-element, Eq. (4.45), is free of singularity only if m is positive, implying thereby that only positive values of mass are permissible. However, this is only a tentative argument.

4.3.2 Isotropic Coordinates

The form of Schwarzschild line-element depends on the coordinate system used. It is highly useful to employ other coordinate systems other than the spherical polar. An instance is the isotropic coordinate systems, which is obtained through the following transformation

$$r = \left(1 + \frac{m}{2r_1}\right)^2 r_1 \tag{4.46}$$

Then

$$dr = \left(1 - \frac{m^2}{4r_1^2}\right)dr_1$$

Also

$$1 - \frac{2m}{r} = 1 - \frac{2m}{\left(1 + \frac{m}{2r_1}\right)^2 r_1} = \frac{\left[\left(1 + \frac{m}{2r_1}\right)^2 r_1 - 2m\right]}{\left(1 + \frac{m}{2r_1}\right)^2 r_1}$$

$$= \frac{\left(1 - \frac{m}{2r_1}\right)^2 r_1}{\left(1 + \frac{m}{2r_1}\right)^2 r_1} = \frac{\left(1 - \frac{m}{2r_1}\right)^2}{\left(1 + \frac{m}{2r_1}\right)^2}$$

Further

$$\left(1 - \frac{2m}{r}\right)^{-1} dr^2 = \frac{\left(1 + \frac{m}{2r_1}\right)^2}{\left(1 - \frac{m}{2r_1}\right)^2}\left(1 + \frac{m}{2r_1}\right)^2\left(1 - \frac{m}{2r_1}\right)^2 dr_1^2 = \left(1 + \frac{m}{2r_1}\right)^4 dr_1^2$$

With these values, Schwarzschild line-element becomes

$$ds^2 = -\left(1+\frac{m}{2r_1}\right)^4 dr_1^2 - \left(1+\frac{m}{2r_1}\right)^4 r_1^2(d\theta^2+\sin^2\theta d\phi^2) + \frac{\left(1-\frac{m}{2r_1}\right)^2}{\left(1+\frac{m}{2r_1}\right)^2} dt^2$$

$$= -\left(1+\frac{m}{2r_1}\right)^4 [dr_1^2 + r_1^2(d\theta^2 + \sin^2\theta d\phi^2)] + \frac{\left(1-\frac{m}{2r_1}\right)^2}{\left(1+\frac{m}{2r_1}\right)^2} dt^2 \quad (4.47)$$

This is called isotropic line-element and the coordinates r_1, θ, ϕ are isotropic polar coordinates. We can express the line-element, Eq. (4.47) in Cartesian form through the transformations

$$x = r_1 \sin\theta \cos\phi$$
$$y = r_1 \sin\theta \sin\phi$$
$$z = r_1 \cos\theta$$

The line-element assumes the form

$$ds^2 = -\left(1+\frac{m}{2r_1}\right)^4 [dx^2 + dy^2 + dz^2] + \frac{\left(1-\frac{m}{2r_1}\right)^2}{\left(1+\frac{m}{2r_1}\right)^2} dt^2 \quad (4.48)$$

We get the path of a light ray by putting $ds = 0$. Thus,

$$\left(\frac{dx}{dt}\right)^2 + \left(\frac{dy}{dt}\right)^2 + \left(\frac{dz}{dt}\right)^2 = \frac{\left(1-\frac{m}{2r_1}\right)^2}{\left(1+\frac{m}{2r_1}\right)^6} \quad (4.49)$$

The velocity of light pulse at a distance r_1 from the origin is

$$\left(1-\frac{m}{2r_1}\right) \Big/ \left(1+\frac{m}{2r_1}\right)^3 \quad (4.50)$$

The velocity of light is same in all directions.

4.3.3 Birkhoff's Theorem

While deriving the Schwarzschild metric, the only input of information about the distribution of matter giving rise to the gravitational field, is in the following:

- spherically symmetric
- that it has zero density at the radius

Birkhoff showed that any spherically symmetric vacuum solution of Einstein field equations must also be static and agree with Schwarzschild solution. Therefore, the Schwarzschild metric results from a spherically symmetric mass, irrespective of whether the mass is static, collapsing, expanding or pulsating. This refers to the gravitational field in the space outside the mass, since we had taken $T_{\mu\nu} = 0$. As a consequence, the two of the most important conclusions of Newtonian gravity, therefore, apply to general relativity. These are as follows:

- The gravitational field due to a spherical body acts from a central point mass.
- The gravitational field inside a spherical shell vanishes.

Thus, Birkhoff's theorem may be stated as follows:

Every solution of the field equations corresponding to the field (static as well as non-static), which is spherically symmetric, satisfies the same boundary conditions as we have in Schwarzschild's exterior solution and comes out to be a transform of Schwarzschild's exterior solution, *viz.*

$$ds^2 = \left(1 - \frac{2m}{r}\right)dt^2 - \left(1 - \frac{2m}{r}\right)^{-1} dr^2 - r^2(d\theta^2 + \sin^2\theta d\phi^2) \tag{4.51}$$

We proceed to prove the theorem and take the line-element as

$$ds^2 = e^{\nu} dt^2 - e^{\lambda} dr^2 - r^2(d\theta^2 + \sin^2\theta d\phi^2) \tag{4.52}$$

where $\lambda = \lambda(r,t)$ and $\nu = \nu(r,t)$.

From the field equations without the cosmological constant

$$R_{\mu\nu} - \frac{1}{2}Rg_{\mu\nu} = -8\pi T_{\mu\nu}$$

we get

$$8\pi T_{\mu}^{\nu} = -\left(R_{\mu}^{\nu} - \frac{1}{2}Rg_{\mu}^{\nu}\right) \tag{4.53}$$

Putting the values of the tensor R_μ^ν and the scalar R, we get the field equations in the presence of matter

$$
\begin{aligned}
8\pi T_1^1 &= -e^{-\lambda}\left(\frac{\nu'}{r}+\frac{1}{r^2}\right)+\frac{1}{r^2} \\
8\pi T_2^2 = 8\pi T_3^3 &= -e^{-\lambda}\left(\frac{\nu''}{2}-\frac{\nu'\lambda'}{4}+\frac{\nu'^2}{4}+\frac{\nu'-\lambda'}{4}\right) \\
8\pi T_0^0 &= e^{-\lambda}\left(\frac{\lambda'}{r}-\frac{1}{r^2}\right)+\frac{1}{r^2} \\
8\pi T_\mu^\nu &= 0 \text{ for } \mu \neq \nu
\end{aligned}
\tag{4.54}
$$

Since for empty space $T_\mu^\nu = 0$, the Eqs. (4.54) become

$$e^{-\lambda}\left(\frac{\nu'}{r}+\frac{1}{r^2}\right)-\frac{1}{r^2}=0 \tag{4.55}$$

$$e^{-\lambda}\left(\frac{\nu''}{2}-\frac{\nu'\lambda'}{4}+\frac{\nu'^2}{4}+\frac{\nu'-\lambda'}{4}\right)=0 \tag{4.56}$$

$$e^{-\lambda}\left(\frac{\lambda'}{r}-\frac{1}{r^2}\right)+\frac{1}{r^2}=0 \tag{4.57}$$

The above equations help us to determine the unknown functions ν and λ. Adding Eqs. (4.55) and (4.57), one gets

$$e^{-\lambda}\left(\frac{\nu'+\lambda'}{r}\right)=0 \tag{4.58}$$

or

$$\nu'+\lambda'=0 \tag{4.59}$$

On integration, Eq. (4.59), gives

$$\nu+\lambda = f(t) \tag{4.60}$$

Assuming $f(t)=0$, does not cause any loss of generality, thus,

$$\nu+\lambda=0 \tag{4.61}$$

Writing Eq. (4.55) in view of Eq. (4.61), we obtain

$$e^{-\lambda}\left(-\frac{\lambda'}{r}+\frac{1}{r^2}\right)-\frac{1}{r^2}=0$$

or

$$e^{-\lambda}(r\lambda'-1)+1=0$$

or
$$-\frac{\partial}{\partial r}(re^{-\lambda})+1=0 \tag{4.62}$$

Integrating w.r.t. r, we get

$$-re^{-\lambda}+r=K$$

or
$$e^{-\lambda}=1-\frac{K}{r} \tag{4.63}$$

Taking $K=2m$, we get

$$e^{-\lambda}=1-\frac{2m}{r} \tag{4.64}$$

Rewriting Eq. (4.61), one has

$$\nu+\lambda=f(t) \tag{4.65}$$

where $f(t)$ is constant of integration.

From Eq. (4.65),

$$e^{\nu}=e^{f(t)}e^{-\lambda}=\left(1-\frac{2m}{r}\right)e^{f(t)}$$

or
$$e^{\nu}dt^2=\left(1-\frac{2m}{r}\right)e^{f(t)}dt^2 \tag{4.66}$$

Taking $t'=\int e^{f/2}dt$, we obtain

$$e^{\nu}dt^2=\left(1-\frac{2m}{r}\right)dt'^2 \tag{4.67}$$

Writing Eq. (4.51) in the light of Eqs. (4.64) and (4.67)

$$ds^2=\left(1-\frac{2m}{r}\right)dt'^2-\left(1-\frac{2m}{r}\right)^{-1}dr^2-r^2(d\theta^2+\sin^2\theta d\phi^2) \tag{4.68}$$

This is the Birkhoff's solution. As is obvious, Eq. (4.68) is a transform of Schwarzschild's exterior solution.

4.4 KILLING VECTOR

We know from particle physics that symmetries lead to conservation laws. To find symmetries, one makes use of a special type of vector called a *Killing vector.* If X is a Killing vector and a set of points is displaced by $X^{\alpha}dx_{\alpha}$ and all distance relationships do not change, then X is a Killing vector. Such distance preserving mapping is called an isometry. In short, moving along the direction of Killing vector, preserves the form of the metric.

A free particle moving in direction will not be subject to any force if the metric does not change. This leads to momentum conservation. Thus, if X is a Killing vector, then

$$X \cdot u = \text{const} \tag{4.69}$$

$$X \cdot p = \text{const} \tag{4.70}$$

along a geodesic, where u is the particle four-velocity and p, the particle four-momentum.

Killing vector X satisfies Killing equation, which is expressed in terms of covariant derivatives as

$$X_{\mu;\nu} + X_{\nu;\mu} = 0 \tag{4.71}$$

This equation is also valid for contravariant components as

$$X^{\mu}{}_{;\nu} + X^{\nu}{}_{;\mu} = 0 \tag{4.72}$$

Let us show that Killing's equation can be expressed in terms of the Lie derivative of the metric tensor.

The Lie derivative of the metric is

$$L_x g_{\mu\nu} = X^{\rho}\partial_{\rho} g_{\mu\nu} + g_{\rho\nu}\partial_{\mu} X^{\rho} + g_{\mu\rho}\partial_{\nu} X^{\rho} \tag{4.73}$$

The covariant derivative of X^{μ} is

$$X^{\mu}{}_{;\rho} = \partial_{\rho} X^{\mu} + \Gamma^{\mu}_{\nu\rho} x^{\nu} \tag{4.74}$$

We know that the covariant derivative of metric tensor $g_{\mu\nu;\rho} = 0$

Now
$$g_{\mu\nu;\rho} = \partial_{\rho} g_{\mu\nu} - \Gamma^{\sigma}_{\mu\rho} g_{\sigma\nu} - \Gamma^{\sigma}_{\nu\rho} g_{\mu\sigma} = 0$$

$$\Rightarrow \qquad g_{\mu\nu;\rho} = \Gamma^{\sigma}_{\mu\rho} g_{\sigma\nu} + \Gamma^{\sigma}_{\nu\rho} g_{\mu\sigma} \tag{4.75}$$

Using Eq. (4.75), we write the Lie derivative of the metric tensor

$$\begin{aligned} L_x g_{\mu\nu} &= X^{\rho}\partial_{\rho} g_{\mu\nu} + g_{\rho\nu}\partial_{\mu} X^{\rho} + g_{\mu\rho}\partial_{\nu} X^{\rho} \\ &= X^{\rho}\left(\Gamma^{\sigma}_{\mu\rho}\, g_{\sigma\nu} + \Gamma^{\sigma}_{\nu\rho} g_{\mu\sigma}\right) + g_{\rho\nu}\partial_{\mu} X^{\rho} + g_{\mu\rho}\partial_{\nu} X^{\rho} \\ &= g_{\sigma\nu}\Gamma^{\sigma}_{\mu\rho} X^{\rho} + g_{\mu\sigma}\Gamma^{\sigma}_{\nu\rho} X^{\rho} + g_{\rho\nu}\partial_{\mu} X^{\rho} + g_{\mu\rho}\partial_{\nu} X^{\rho} \end{aligned} \tag{4.76}$$

We will do some manipulation with a view to cast it in the form we require to write down the covariant derivative of X

$$X^{\rho}{}_{;\nu} = \partial_{\nu} X^{\rho} + \Gamma^{\rho}_{\nu\sigma} X^{\sigma} \tag{4.77}$$

The last term in Eq. (4.76) is $g_{\mu\rho}\partial_{\nu} X^{\rho}$. We require another term to write the covariant derivative. We know that a repeated index is a dummy index and we can rename it the way we like. Let us consider the second term in Eq. (4.76), *i.e.*, $g_{\mu\sigma}\Gamma^{\sigma}_{\nu\rho}X^{\rho}$. The indices ρ and σ are repeated, so these are dummy indices. We are free to switch them $\rho \leftrightarrow \sigma$ and rewrite this term as $g_{\mu\rho}\Gamma^{\rho}_{\nu\sigma}X^{\sigma}$. Thus, rewriting Eq. (4.76) and rearranging the terms to be in order for a covariant derivative, we obtain

$$\begin{aligned} L_x g_{\mu\nu} &= g_{\mu\rho}\partial_\nu X^\rho + g_{\sigma\nu}\Gamma^{\sigma}_{\mu\rho}X^\rho + g_{\mu\sigma}\Gamma^{\sigma}_{\nu\rho}X^\rho + g_{\rho\nu}\partial_\mu X^\rho \\ &= g_{\mu\rho}\partial_\nu X^\rho + g_{\mu\sigma}\Gamma^{\sigma}_{\nu\rho}X^\rho + g_{\sigma\nu}\Gamma^{\sigma}_{\mu\rho}X^\rho + g_{\rho\nu}\partial_\mu X^\rho \end{aligned} \tag{4.78}$$

Changing the indices $\rho \leftrightarrow \sigma$ in the second term of the above equation, we get

$$\begin{aligned} L_x g_{\mu\nu} &= g_{\mu\rho}\partial_\nu X^\rho + g_{\mu\rho}\Gamma^{\rho}_{\nu\sigma}X^\sigma + g_{\sigma\nu}\Gamma^{\sigma}_{\mu\rho}X^\rho + g_{\rho\nu}\partial_\mu X^\rho \\ &= g_{\mu\rho}\left(\partial_\nu X^\rho + \Gamma^{\rho}_{\nu\sigma}X^\sigma\right) + g_{\sigma\nu}\Gamma^{\sigma}_{\mu\rho}X^\rho + g_{\rho\nu}\partial_\mu X^\rho \\ &= g_{\mu\rho}X^\rho{}_{;\nu} + g_{\sigma\nu}\Gamma^{\sigma}_{\mu\rho}X^\rho + g_{\rho\nu}\partial_\mu X^\rho \end{aligned} \tag{4.79}$$

Lastly switching the indices $\rho \leftrightarrow \sigma$ in the second term in Eq. (4.79), we obtain

$$\begin{aligned} L_x g_{\mu\nu} &= g_{\mu\rho}X^\rho{}_{;\nu} + g_{\rho\nu}\partial_\mu X^\rho + g_{\rho\nu}\Gamma^{\rho}_{\mu\sigma}X^\sigma \\ &= g_{\mu\rho}X^\rho{}_{;\nu} + g_{\rho\nu}\left(\partial_\mu X^\rho + \Gamma^{\rho}_{\mu\sigma}X^\sigma\right) \\ &= g_{\mu\rho}X^\rho{}_{;\nu} + g_{\rho\nu}X^\rho{}_{;\mu} \end{aligned} \tag{4.80}$$

As the covariant derivative of the metric tensor vanishes, we move the metric inside the derivative and lower the indices. Thus,

$$\begin{aligned} L_x g_{\mu\nu} &= g_{\mu\rho}X^\rho{}_{;\nu} + (g_{\rho\nu}X^\rho)_{;\mu} \\ &= g_{\mu\rho}X^\rho{}_{;\nu} + X_{\nu;\mu} \end{aligned} \tag{4.81}$$

Since it is given that

$$L_x g_{\mu\nu} = 0$$

so

$$X_{\mu;\nu} + X_{\nu;\mu} = 0 \tag{4.82}$$

4.4.1 Conserved Current

It is to show how one can obtain an expression for conserved current with the use of Killing vector. Let X be a Killing vector and T, the stress-energy tensor.

We define the current as

$$J^\mu = T^{\mu\nu}X_\nu \tag{4.83}$$

In order to prove that J^μ is a conserved current, we proceed as follows:

The covariant derivative of J^{μ} is

$$\begin{aligned} J^{\mu}{}_{;\mu} &= (T^{\mu\nu}X_{\nu})_{;\mu} \\ &= (T^{\mu\nu}_{;\mu})X_{\nu} + T^{\mu\nu}(X_{\nu;\mu}) \end{aligned} \tag{4.84}$$

As the stress-energy tensor is conserved, $T^{\mu\nu}{}_{;\mu} = 0$, therefore,

$$J^{\mu}{}_{;\mu} = T^{\mu\nu}(X_{\nu;\mu}) \tag{4.85}$$

As the stress-energy tensor is symmetric, so we write Eq. (4.85) as

$$\begin{aligned} J^{\mu}{}_{;\mu} &= \frac{1}{2}(T^{\mu\nu}(X_{\nu;\mu}) + T^{\nu\mu}(X_{\mu;\nu})) \\ &= \frac{1}{2}T^{\mu\nu}(X_{\nu;\mu} + X_{\mu;\nu}) = 0 \end{aligned} \tag{4.86}$$

Thus, J^{μ} is a conserved quantity.

Example 4.1

(*a*) Show that linear combination of Killing vectors with constant coefficients, is a Killing vector.
(*b*) If ξ is a Killing vector, prove that $\xi_{\mu;\alpha\beta} = R_{\gamma\beta\alpha\mu}\xi^{\gamma}$

Solution

(*a*) If *u* and *v* are Killing vectors and *a* and *b* are constants, then

$$(au_{\alpha} + bv_{\alpha})_{;\beta} = au_{\alpha;\beta} + bv_{\alpha;\beta}$$

Therefore, if *u* and *v* satisfy Killing equation, so does $au + bv$.

(*b*) For any vector ξ

$$\xi_{\sigma;\rho\mu} - \xi_{\sigma;\mu\rho} = R^{\lambda}_{\sigma\mu\rho}\xi_{\lambda} \tag{i}$$

by the definition of $R_{\alpha\beta\gamma\delta}$. Add to Eq. (i), its two permutations and use the cyclic identity for the Reimann tensor

$$R^{\lambda}_{\sigma\rho\mu} + R^{\lambda}_{\mu\sigma\rho} + R^{\lambda}_{\rho\mu\sigma} = 0$$

to obtain the identity

$$0 = \xi_{\sigma;\rho\mu} - \xi_{\sigma;\mu\rho} + \xi_{\mu;\sigma\rho} - \xi_{\mu;\rho\sigma} + \xi_{\rho;\mu\sigma} - \xi_{\rho;\sigma\mu} \tag{ii}$$

For a Killing vector Eq. (ii) becomes

$$0 = \xi_{\sigma;\rho\mu} - \xi_{\sigma;\mu\rho} - \xi_{\mu;\rho\sigma} \tag{iii}$$

and substituting Eq. (iii) into Eq. (ii), we obtain

$$\xi_{\mu;\rho\sigma} = R^{\lambda}_{\sigma\rho\mu}\xi_{\lambda}$$

4.5 PARTICLE TRAJECTORIES IN SCHWARZSCHILD SPACE–TIME

Using the differential equations for the geodesics for finding the trajectories of particles and light, is relatively a cumbersome affair. Another method for such an exercise is based on the use of Killing vectors. We recall that each symmetry in the metric corresponds to a Killing vector. The dot product of the Killing vectors with the four velocity of the particle, yields the conserved quantities related to the motion of the particle.

The four velocity of a particle in the Schwarzschild space–time in terms of its components, is

$$u = \left(\frac{dt}{d\tau}, \frac{dr}{d\tau}, \frac{d\theta}{d\tau}, \frac{d\phi}{d\tau}\right) \tag{4.87}$$

where τ is the proper time. The four velocity satisfies the relation

$$u \cdot u = g_{\mu\nu}u^{\mu}u^{\nu} = 1 \tag{4.88}$$

Using the coordinate basis, the components of the metric tensor $g_{\mu\nu}$ from Eq. (4.39), are

$$g_{00} = \left(1 - \frac{2m}{r}\right), \quad g_{11} = -\frac{1}{\left(1 - \frac{2m}{r}\right)}, \quad g_{22} = -r^2, \quad g_{33} = -r^2\sin^2\theta \tag{4.89}$$

Expressing Eq. (4.88) in view of Eq. (4.89), we get

$$\begin{aligned} u \cdot u &= g_{\mu\nu}u^{\mu}u^{\nu} \\ &= \left(1 - \frac{2m}{r}\right)\left(\frac{dt}{d\tau}\right)^2 - \left(1 - \frac{2m}{r}\right)^{-1}\left(\frac{dr}{d\tau}\right)^2 - r^2\left(\frac{d\theta}{d\tau}\right)^2 - r^2\sin^2\theta\left(\frac{d\phi}{d\tau}\right)^2 \\ &= 1 \end{aligned} \tag{4.90}$$

On the basis of Eq. (4.90), we will find the orbits of the particle. We know that in classical mechanics, the orbit of a body in a central force field lies in a plane. It can be shown that a similar conclusion holds in the case of relativity as well. So, we can simplify the treatment through the choice of axes such that $\theta = \pi/2$ and $\dot{\theta} = 0$.

Further we construct two Killing vectors. The Schwarzschild metric is time independent and spherically symmetric, in addition, the metric is independent of the angular variable ϕ. In view of these criteria, we define the following Killing vectors. The Killing vector corresponding to the independence of the metric of t is

$$\xi = (1,0,0,0) \tag{4.91}$$

The independence of the metric from time is associated with the conservation of energy.

Another Killing vector corresponding to the independence of the metric from ϕ is

$$\eta = (0,0,0,1) \tag{4.92}$$

This vector is associated with the conservation of angular momentum.

Next we find the conserved quantities related to these Killing vectors by taking the dot product of each vector with the four-velocity.

The conserved energy per unit mass is

$$\begin{aligned} \varepsilon &= \xi \cdot u \\ &= g_{\mu\nu}\xi^{\mu}u^{\nu} \\ &= \left(1-\frac{2m}{r}\right)\frac{dt}{d\tau} \end{aligned} \tag{4.93}$$

Likewise, the conserved angular momentum per unit mass is

$$\begin{aligned} l &= \eta \cdot u \\ &= g_{\mu\nu}\eta^{\mu}u^{\nu} \\ &= -r^2 \sin^2\theta \frac{d\phi}{d\tau} \\ &= -r^2 \frac{d\phi}{d\tau} \quad \text{for } \theta = \pi/2 \end{aligned} \tag{4.94}$$

Rewriting Eq. (4.90) in the light of Eqs. (4.93) and (4.94), we obtain

$$\left(1-\frac{2m}{r}\right)^{-1}\varepsilon^2 - \left(1-\frac{2m}{r}\right)^{-1}\left(\frac{dr}{d\tau}\right)^2 - \frac{l^2}{r^2} = 1$$

or

$$\frac{\varepsilon^2}{2} - \frac{1}{2}\left(\frac{dr}{d\tau}\right)^2 - \frac{1}{2}\frac{l^2}{r^2}\left(1-\frac{2m}{r}\right) = \frac{1}{2} - \frac{m}{r} \tag{4.95}$$

Expressing it in terms of energy per unit mass, we get

$$\frac{\varepsilon^2 - 1}{2} = \frac{1}{2}\left(\frac{dr}{d\tau}\right)^2 + \frac{l^2}{2r^2}\left(1-\frac{2m}{r}\right) - \frac{m}{r} \tag{4.96}$$

Defining the effective potential as

$$V_{\text{eff}} = \frac{l^2}{2r^2}\left(1-\frac{2m}{r}\right)-\frac{m}{r} \tag{4.97}$$

and setting $E = \frac{\varepsilon^2-1}{2}$, we obtain

$$E = \frac{1}{2}\dot{r}^2 + V_{\text{eff}} \tag{4.98}$$

This describes a particle with energy E and unit mass moving in a force field given by the potential. V_{eff}. Rewriting the effective potential as

$$V_{\text{eff}} = -\frac{m}{r}+\frac{l^2}{2r^2}-\frac{l^2 m}{r^3} \tag{4.99}$$

The first two terms are the same as in Newtonian case. The third term is an additional term that arises as a modification of the potential in general relativity.

Further we find the minimum and maximum values of r. The first derivative of V_{eff} is

$$\frac{dV_{\text{eff}}}{dr} = \frac{m}{r^2}-\frac{l^2}{r^3}+\frac{3l^2 m}{r^4}$$

and equating it to zero, we get

$$mr^2 - l^2 r + 3l^2 m = 0 \tag{4.100}$$

Solving it for r, we get

$$\begin{aligned} r_{1,2} &= \frac{l^2}{2m}\left[1 \pm \sqrt{1-12m^2/l^2}\right] \\ &\cong \frac{l^2}{2m}[1 \pm (1-6m^2/l^2)] \end{aligned} \tag{4.101}$$

The two values for the circular orbits are

$$r_1 \approx \frac{l^2}{m} \tag{4.102}$$

which is a stable orbit, and

$$r_2 \approx \frac{l^2}{2m}\left(\frac{6m^2}{l^2}\right) = 3m \tag{4.103}$$

Putting $m = \frac{GM}{c^2}$, the orbit

$$r_2 = \frac{3GM}{c^2}$$

For the Sun, we get

$$r_2^{sun} = 3\frac{(6.67\times10^{-11}\ \text{m}^2\text{s}^2/\text{kg})(1.989\times10^{30}\ \text{kg})}{(3\times10^{10}\ \text{cm/s})^2}$$
$$= 4.422\ \text{km} \tag{4.104}$$

This orbit lies well inside the sun, whose equatorial radius is 695,000 km. Thus, the orbit corresponding to r_2 is unstable.

Let us examine Eq. (4.99) for V_{eff} for learning about the orbits for different values of r. Taking the case corresponding to the Schwarzschild radius, $r = 2m$ one gets

$$V_{\text{eff}}(r = 2m) = -\frac{m}{2m} + \frac{l^2}{2(2m)^2} - \frac{l^2 m}{(2m)^3}$$
$$= -1/2 \tag{4.105}$$

For large r,

$$V_{\text{eff}} = -m/r \tag{4.106}$$

which is the Newtonian potential. From Eq. (4.101), we see that if $l^2 < 12m^2$, the radius is a complex number, implying thereby that no stable orbit can exist. The orbiting body will hit the surface of the star. However, if the orbiting body happens to be approaching a black hole, the body will be devoured by the black hole. The effective potentials are plotted in Figs (4.1) and (4.2) for the case of general relativity and Newtonian case respectively. A comparison of these diagrams shows that as r gets large, the Newtonian and relativistic cases, converge. However, at smaller r, the differences are marked. Thus, the relativistic effects in the solar system will be more pronounced near the Sun. This was the precise reason for Einstein to consider the precession of the orbit of Mercury.

4.5.1 An Alternative Treatment of Planetary Orbits—Relativistic Analogue of Kepler's Laws

It is intended to obtain the differential equations governing the motion of a planet around the sun. The field surrounding the sun is taken as the field of an isolated particle at rest at the origin, where the space–time is given by the Schwarzschild line-element for empty space, *viz.*

$$ds^2 = (1-2m/r)dt^2 - (1-2m/r)^{-1}dr^2 - r^2 d\theta^2 - r^2\sin^2\theta d\phi^2 \tag{4.107}$$

In comparison to sun, the planets may be regarded as small free particles and their trajectories are given by geodesic equation

$$\frac{d^2x^\alpha}{ds^2} + \Gamma^\alpha_{\mu\nu}\frac{dx^\mu}{ds}\frac{dx^\nu}{ds} = 0 \tag{4.108}$$

Making use of the non-zero, 3-index symbols, Eq. (4.20), we set ourselves the task of determining the equations of motion for $\alpha = 0,1,2,3$.

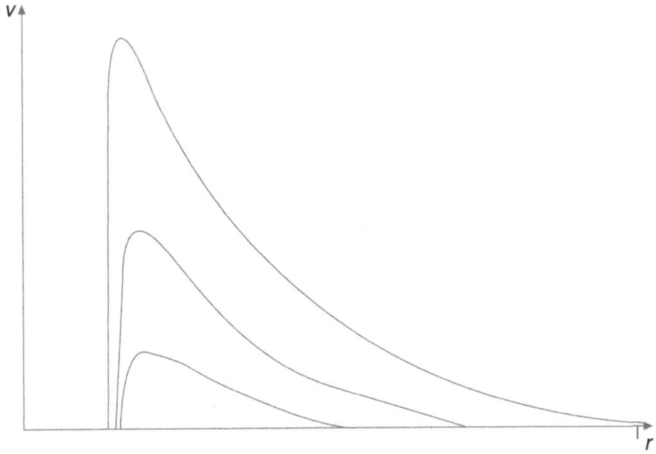

Figure 4.1 *Plots of V_{eff} for the relativistic case for three different values of l. The $1/r^3$ term is responsible for the marked differences at small values of r.*

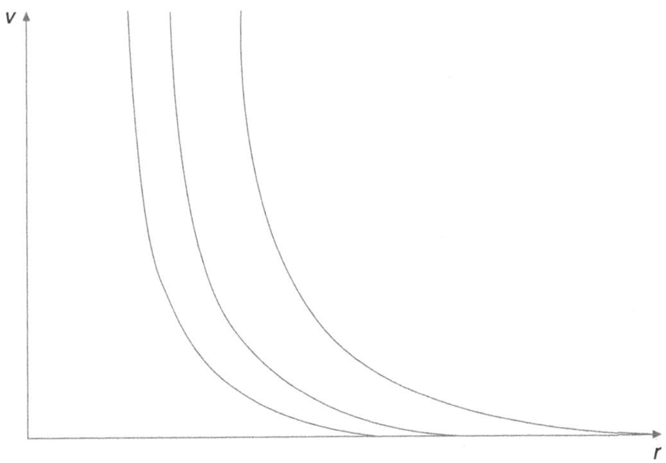

Figure 4.2 *Plots of V_{eff} for the Newtonian case for the same values of l. At large r, the behaviour is quite similar to the relativistic case. At smaller r, the behaviour is markedly different.*

For $\alpha = 0$,

$$\frac{d^2x^0}{ds^2} + \Gamma^0_{\mu\nu}\frac{dx^\mu}{ds}\frac{dx^\nu}{ds} = 0$$

or

$$\frac{d^2t}{ds^2} + \Gamma^0_{10}\frac{dr}{ds}\frac{dt}{ds} + \Gamma^0_{01}\frac{dt}{ds}\frac{dr}{ds} = 0$$

or

$$\frac{d^2t}{ds^2} + \nu'\frac{dr}{ds}\frac{dt}{ds} = 0 \tag{4.109}$$

For $\alpha = 1$, $$\frac{d^2x^1}{ds^2} + \Gamma^1_{\mu\nu}\frac{dx^\mu}{ds}\frac{dx^\nu}{ds} = 0$$

or $$\frac{d^2r}{ds^2} + \Gamma^1_{00}\left(\frac{dt}{ds}\right)^2 + \Gamma^1_{11}\left(\frac{dx^1}{ds}\right)^2 + \Gamma^1_{22}\left(\frac{dx^2}{ds}\right)^2 + \Gamma^1_{33}\left(\frac{dx^2}{ds}\right)^2 = 0$$

or $$\frac{d^2r}{ds^2} + \frac{\nu'}{2}e^{\nu-\lambda}\left(\frac{dt}{ds}\right)^2 + \frac{\lambda'}{2}\left(\frac{dr}{ds}\right)^2 - re^{-\lambda}\left(\frac{d\theta}{ds}\right)^2 - r\sin^2\theta e^{-\lambda}\left(\frac{d\phi}{ds}\right)^2 = 0 \tag{4.110}$$

Further, for $\alpha = 2$, we get

$$\frac{dx^2}{ds^2} + \Gamma^2_{\mu\nu}\frac{dx^\mu}{ds}\frac{dx^\nu}{ds} = 0$$

or $$\frac{d^2\theta}{ds^2} + \Gamma^2_{12}\frac{dx^1}{ds}\frac{dx^2}{ds} + \Gamma^2_{21}\frac{dx^2}{ds}\frac{dx^1}{ds} + \Gamma^2_{33}\frac{dx^2}{ds}\frac{dx^3}{ds} = 0$$

or $$\frac{d^2\theta}{ds^2} + \frac{2}{r}\frac{dr}{ds}\frac{d\theta}{ds} - r\sin\theta\cos\theta\left(\frac{d\phi}{ds}\right)^2 = 0 \tag{4.111}$$

Lastly for $\alpha = 3$, we have $$\frac{d^2x^3}{ds^2} + \Gamma^3_{\mu\nu}\frac{dx^\mu}{ds}\frac{dx^\nu}{ds} = 0$$

or $$\frac{d^2\phi}{ds^2} + \Gamma^3_{13}\frac{dx^1}{ds}\frac{dx^3}{ds} + \Gamma^3_{31}\frac{dx^3}{ds}\frac{dx^1}{ds} + \Gamma^3_{23}\frac{dx^2}{ds}\frac{dx^3}{ds} + \Gamma^3_{32}\frac{dx^3}{ds}\frac{dx^2}{ds} = 0$$

or $$\frac{d^2\phi}{ds^2} + \frac{2}{r}\frac{dr}{ds}\frac{d\phi}{ds} + 2\cot\theta\frac{d\theta}{ds}\frac{d\phi}{ds} = 0 \tag{4.112}$$

We choose the coordinates such that the planet moves initially in the plane $\theta = \pi/2$, so that $\sin\theta = 1$, $\cos\theta = 0$, and $d\theta/ds = 0$. Substituting these values in the Eqs. (4.109) and (4.112), we get

$$\frac{d^2t}{ds^2} + \nu'\frac{dr}{ds}\frac{dt}{ds} = 0 \tag{4.113}$$

$$\frac{d^2r}{ds^2} + \frac{\nu'}{2}e^{\nu-\lambda}\left(\frac{dt}{ds}\right)^2 + \frac{\lambda'}{2}\left(\frac{dr}{ds}\right)^2 - re^{-\lambda}\left(\frac{d\phi}{ds}\right)^2 = 0 \tag{4.114}$$

$$\frac{d^2\theta}{ds^2} = 0 \tag{4.115}$$

$$\frac{d^2\phi}{ds^2} + \frac{2}{r}\frac{dr}{ds}\frac{d\phi}{ds} = 0 \tag{4.116}$$

According to Eq. (4.115), the particle, which starts moving in the plane $\theta = \pi / 2$, continues to move in the same plane.

Rewriting Eq. (4.116), we get

$$r^2 \frac{d^2\phi}{ds^2} + 2r \frac{dr}{ds}\frac{d\phi}{ds} = 0$$

or

$$\frac{d}{ds}\left(r^2 \frac{d\phi}{ds} \right) = 0$$

Integrating this equation, we obtain

$$r^2 \frac{d\phi}{ds} = h \tag{4.117}$$

h is a constant of integration and is a measure of the angular momentum of the motion. Further, from Eq. (4.113), we have

$$e^{\nu} \frac{d^2t}{ds^2} + e^{\nu}\nu' \frac{dr}{ds}\frac{d\phi}{ds} = 0$$

or

$$\frac{d}{ds}\left(e^{\nu} \frac{dt}{ds} \right) = 0$$

Integrating this equation, we get

$$e^{\nu} \frac{dt}{ds} = k \tag{4.118}$$

where k is the constant of integration. Instead of working with Eq. (4.114), it is convenient to work with the line-element, Eq. (4.14), that gives another integral of motion. Thus, Eq. (4.14) for $\theta = \pi / 2$, becomes

$$1 = e^{\nu}\left(\frac{dt}{ds} \right)^2 - e^{-\nu}\left(\frac{dr}{ds} \right)^2 - r^2\left(\frac{d\phi}{ds} \right)^2$$

or

$$e^{\nu}\left(\frac{dt}{ds} \right) - e^{-\nu}\left(\frac{dr}{ds} \right)^2 - r^2\left(\frac{d\phi}{ds} \right)^2 - 1 = 0$$

Writing $\frac{dr}{ds} = \frac{dr}{d\phi}\frac{d\phi}{ds}$, and putting the values of $\frac{dt}{ds}$ and $\frac{d\phi}{ds}$ from Eqs. (4.117) and (4.118), we obtain

$$k^2 e^{-\nu} - \frac{h^2}{r^4}\left(\frac{dr}{d\phi} \right)^2 e^{-\nu} - \frac{h^2}{r^2} - 1 = 0$$

or
$$h^2\left(\frac{1}{r^2}\frac{dr}{d\phi}\right)^2+\left(1+\frac{h^2}{r^2}\right)\left(1-\frac{2m}{r}\right)-k^2=0 \tag{4.119}$$

where we multiplied throughout by e^{ν}.

Putting $r = 1/u$, one obtains

$$h^2\left(\frac{du}{d\phi}\right)^2+(1+h^2u^2)(1-2mu)-k^2=0$$

which on rearrangement is written as

$$\left(\frac{du}{d\phi}\right)^2+u^2=\frac{k^2-1}{h^2}+\frac{2m}{h^2}u+2mu^3 \tag{4.120}$$

Differentiating w.r.t. ϕ, we obtain

$$2\frac{du}{d\phi}\frac{d^2u}{d\phi^2}+2u\frac{du}{d\phi}=\frac{2m}{h^2}\frac{du}{d\phi}+6mu^2\frac{du}{d\phi}$$

Dividing by $2\frac{du}{d\phi}$, we obtain

$$\frac{d^2u}{d\phi^2}+u=\frac{m}{h^2}+3mu^2 \tag{4.121}$$

with $r^2\frac{d\phi}{ds}=h$

This is the required differential equation of the trajectory of a planet. R and ϕ are the spatial coordinates and ds is the element of the proper time as measured by a clock moving with the planet. Let us compare the Eq. (4.121) obtained in the relativistic theory of gravitation with Newtonian equations of the orbit

$$\frac{d^2u}{d\phi^2}+u=\frac{m}{h^2}$$
$$r^2\frac{d\phi}{dt}=h \tag{4.122}$$

The relativistic treatment introduces an additional term $3mu^2$, in the Newtonian equation of motion and the time coordinate dt is replaced by the proper time ds.

The ratio of the added term $3mu^2$ to $\frac{m}{h^2}$ is given by

$$\frac{3mu^2}{(m/h^2)}=3h^2u^2$$
$$=3\left(r\frac{d\phi}{ds}\right)^2$$
$$=3\text{ (transverse velocity)}^2 \tag{4.123}$$

$r\frac{d\phi}{ds}$ is the earth's transverse velocity and must be divided by c, for calculating the relativistic correction, Eq. (4.123). Thus,

$$3h^2u^2 = 3\left(\frac{r}{c}\frac{d\phi}{ds}\right)^2$$
$$= 3\times 10^{-6} \tag{4.124}$$

The relativistic correction is negligible. Nonetheless, there are three important consequences of Einstein's theory of gravitation, which can be tested experimentally and thereby provide means to compare it with Newton's theory. These are called classical tests and are as follows:

- Advance of perihelion of Mercury.
- Deflection of a ray of light passing near a heavy mass.
- Gravitational red-shift of spectral lines.

Let us study these tests one by one. In addition, there are some modern tests of the general relativity as well, which we will discuss alongwith these classical tests.

4.6 EXPERIMENTAL TESTS OF GENERAL RELATIVITY

4.6.1 Advance of the Perihelion of Mercury

The Equation (4.121) for the trajectory of a planet can be integrated in terms of elliptical functions, but we will use the direct method of successive approximations for solving it.

Rewriting Eq. (4.121), we get

$$\frac{d^2u}{d\phi^2} + u = \frac{m}{h^2} + 3mu^2 \tag{4.125}$$

We neglect the smallest term $3mu^2$ in the first approximation, so that

$$\frac{d^2u}{d\phi^2} + u = \frac{m}{h^2} \tag{4.126}$$

Its solution as obtained in Newtonian theory, is

$$u = \frac{m}{h^2}(1 + e\cos(\phi - \omega)) \tag{4.127}$$

where a and ω are constants of integration, giving the eccentricity and longitude of the perihelion. As the second approximation, we substitute Eq. (4.127) on the r.h.s. of Eq. (4.125) and obtain

$$\frac{d^2u}{d\phi^2} + u = \frac{m}{h^2} + 3m\frac{m^2}{h^4}[1 + e\cos(\phi - \omega)]^2$$
$$= \frac{m}{h^2} + \frac{3m^3}{h^4} + \frac{6m^3}{h^4}e\cos(\phi - \omega) + \frac{3}{2}\frac{m^3e^2}{h^4}[1 + \cos 2(\phi - \omega)] \tag{4.128}$$

Of the additional terms in Eq. (4.128), which can produce an effect within the range of observation, is $\frac{6m^2}{h^4} e\cos(\phi-\omega)$.

The P. I. of this term is

$$\frac{1}{1+D^2}\frac{6m^3}{h^4}e\cos(\phi-\omega)=\frac{6m^3e}{h^4}\frac{1}{1+D^2}\cos(\phi-\omega)$$
$$=\frac{6m^3e}{h^4}\frac{\phi}{2}\sin(\phi-\omega)=\frac{3m^3e}{h^4}\phi\sin(\phi-\omega) \tag{4.129}$$

Here, we made use of the formula $\frac{1}{1+D^2}\cos x=\frac{x}{2}\sin x$. Therefore, the complete solution of Eq. (4.128), as a result of second approximation, is

$$u=\frac{m}{h^2}\left[1+e\cos(\phi-\omega)\right]+\frac{3m^3e}{h^4}\phi\sin(\phi-\omega)$$
$$=\frac{m}{h^2}+\frac{me}{h^2}\left[\cos(\phi-\omega)+\frac{3m^2}{h^2}\phi\sin(\phi-\omega)\right] \tag{4.130}$$

Calling $\frac{3m^2\phi}{h^2}=\delta\omega$, since $\delta\omega$, is very small, $\sin\delta\omega=\delta\omega$, $\cos\delta\omega=1$, we obtain

$$u=\frac{m}{h^2}+\frac{me}{h^2}[\cos\delta\omega\cos(\phi-\omega)+\sin\delta\omega\sin(\phi-\omega)]$$
$$=\frac{m}{h^2}+\frac{me}{h^2}\cos(\phi-\omega-\delta\omega) \tag{4.131}$$

It shows that in one revolution of the planet, $\phi=2\pi$, the perihelion does not return to its initial position but advances by an angle

$$\delta\omega=\frac{3m^2}{h^2}2\pi=6\pi\frac{m^2}{h^2} \tag{4.132}$$

Thus, the consequence of the relativistic theory of gravitation, is that the orbits are not closed. However, when applied to the planets in the solar system, then this effect is negligibly small for all planets except Mercury. Making use of the standard relation, $m/h^2=1/l$, where l is the semi-latus rectum of the orbit, we have

$$h^2=ml=ma(1-e^2) \tag{4.133}$$

According to Kepler's third law for the time period T, we have

$$T=\frac{2\pi}{\sqrt{m}}a^{3/2} \tag{4.134}$$

where a is the semi-major axis of the orbit. Thus,

$$\delta\omega = \frac{12\pi^2 a^2}{T^2(1-e^2)}\phi \tag{4.135}$$

Taking velocity of light into account, we get

$$\delta\omega = \frac{12\pi^2 a^2}{c^2 T^2(1-e^2)}\phi \tag{4.136}$$

Taking $\phi = 2\pi$, we have

$$\delta\omega = \frac{24\pi^3 a^2}{c^2 T^2(1-e^2)} \tag{4.137}$$

The advance of perihelion as given by Eq. (4.137), is appreciable only for Mercury and very small for all other planets. For Mercury,

$$e = 0.2$$
$$a = 0.6\times 10^8 \text{ km}$$
$$c = 3\times 10^{10} \text{ cm/sec}$$
$$T = 88 \text{ days}$$

The number of revolutions per century is $= \dfrac{365\times 100}{88}$

Thus, $$\delta\omega = \frac{24\pi^3\times(0.6\times 10^{13})^2}{(3\times 10^{10})^2(88\times 24\times 3600)^2(1-.04)}\times\frac{365\times 100}{88}\times\frac{(180\times 3600)}{\pi}\text{ sec}$$
$$= 43 \text{ sec}$$

The advance for Mercury perihelion, Fig. (4.3), is known from time of the astronomer Leverrier (1890). It may be remarked that Newton's theory of gravitation also predicts the advance of the perihelion of any planet due to perturbations caused by other planets and several other causes.

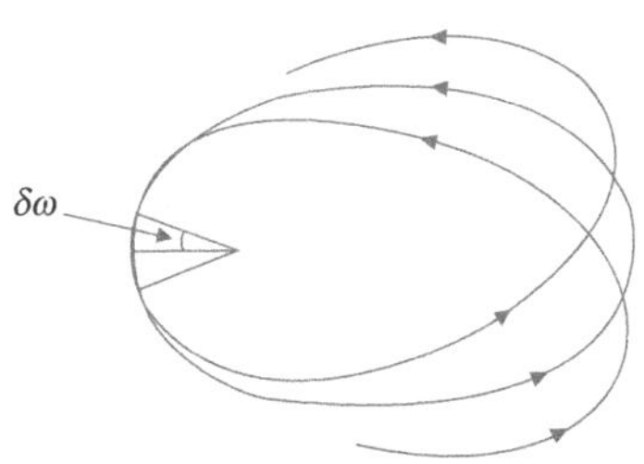

Figure 4.3 *The advance of the perihelion of Mercury.*

Table 4.1 *Advance of Perihelion of Planets*[2]

Planet	Eccentricity	$\delta\omega_{100}$(obs.)	$\delta\omega_{100}$(cal.)
Mercury	0.2050	43.11″±0.45	43.03″
Venus	0.0068	8.4″±4.8	8.64″
Earth	0.0167	5.0″±1.12	3.64″

However, after accounting for all these effects, there still remains an effect, which cannot be accounted for by Newton's theory. The residual effects for these planets are enlisted in Table 4.1, alongwith the calculated values of $\delta\omega_{100}$ on the basis of Eq. (4.137).

The agreement of the observed and calculated values of $\delta\omega_{100}$ for Mercury provides one of the experimental tests for general relativity.

4.6.2 Deflection of a Ray of Light Passing Near a Heavy Mass

Our concern is to investigate the trajectory of a light ray when it passes near a gravitating mass. We know that the orbit of a particle in the presence of a gravitating mass is given by Eq. (4.121)

$$\frac{d^2u}{d\phi^2}+u=\frac{m}{h^2}+3mu^2 \tag{4.138}$$

with

$$r^2\frac{d\phi}{ds}=h$$

The track of a light ray is given by the geodesic equations with the condition $ds=0$ and hence, $h\to\infty$. On substituting the value of h in Eq. (4.138), we get the differential equation for the trajectory of a light ray, *viz.*

$$\frac{d^2u}{d\phi^2}+u=3mu^2 \tag{4.139}$$

The homogeneous part of the equation has the solution

$$u=A\cos\phi+B\sin\phi \tag{4.140}$$

Boundary conditions are that at $\phi=0$, $u=1/R$ and applying these, we get

$$B=0,\quad A=1/R$$

so

$$u=\frac{1}{R}\cos\phi \tag{4.141}$$

Thus, on substituting Eq. (4.141) in Eq. (4.138), we get

$$\frac{d^2u}{d\phi^2}+u=\frac{3m}{R^2}\cos^2\phi \tag{4.142}$$

The P. I. of $\frac{3m}{R^2}\cos^2\phi$ is

$$\begin{aligned}\frac{1}{1+D^2}\frac{3m}{R^2}\cos^2\phi &= \frac{3m}{R^2}\frac{1}{1+D^2}\left(\frac{1+\cos 2\phi}{2}\right)\\ &= \frac{3m}{2R^2}\frac{1}{1+D^2}(e^0+\cos 2\phi)\\ &= \frac{3m}{2R^2}\left(\frac{e^0}{1+0}+\frac{\cos 2\phi}{1-2^2}\right)\\ &= \frac{3m}{2R^2}\left(1-\frac{\cos 2\phi}{3}\right)=\frac{m}{2R^2}(3-\cos 2\phi)\\ &= \frac{m}{2R^2}(3\cos^2\phi+3\sin^2\phi-\cos^2\phi+\sin^2\phi)\\ &= \frac{m}{R^2}(\cos^2\phi+2\sin^2\phi)=\frac{m}{r^2R^2}(r^2\cos^2\phi+2r^2\sin^2\phi)\end{aligned} \tag{4.143}$$

According to second approximation, the solution of Eq. (10.139), becomes

$$\frac{1}{r}=u=\frac{1}{R}\cos\phi+\frac{m(r^2\cos^2\phi+2r^2\sin^2\phi)}{r^2R^2}$$

Multiplying by rR, we get

$$R=r\cos\phi+\frac{m(r^2\cos^2\phi+2r^2\sin^2\phi)}{rR} \tag{4.144}$$

Introducing the Cartesian coordinates, which are assumed to be valid in the nearby Euclidean space surrounding the sun

$$x=r\cos\phi,\quad y=r\sin\phi$$

We have

$$R=x+\frac{m(x^2+2y^2)}{R\sqrt{x^2+y^2}}$$

or

$$x=R-\frac{m(x^2+2y^2)}{R\sqrt{x^2+y^2}} \tag{4.145}$$

The first approximation is

$$\frac{1}{r}=u=\frac{1}{R}\cos\phi$$

or
$$R = r\cos\phi$$

or
$$x = R \tag{4.146}$$

On comparing Eqs. (4.145) and (4.146), it is obvious that the second term $-\frac{m(x^2+2y^2)}{R\sqrt{x^2+y^2}}$ in Eq. (4.145) shows a deviation from the path $x = R$.

The equations of the asymptotes are obtained by taking $y \to \infty$, so these are

$$x = R - \frac{m}{R}(\pm 2y)$$

or
$$x = R + \frac{2my}{R} \quad \text{and} \quad x = R - \frac{2my}{R} \tag{4.147}$$

and
$$y = \frac{Rx}{2m} - \frac{R}{2m} \quad \text{and} \quad y = \frac{R^2}{2m} - \frac{Rx}{2m} \tag{4.148}$$

Let α be the angle between these asymptotes, so that

$$\tan\alpha = \frac{\frac{R}{2m} + \frac{R}{2m}}{1 - \frac{R}{2m} \times \frac{R}{2m}} = \frac{4mR}{4m^2 - R^2}$$

which gives $\sin\alpha = \frac{4mR}{4m^2+R^2}$.

Since $4m^2 \ll R^2$, we get

$$\sin\alpha = \frac{4m}{R}$$

or
$$\alpha = \frac{4m}{R} \quad \text{(in relativistic units)}$$

or
$$\alpha = \frac{4}{R}\frac{GM}{c^2} \quad \text{(in c.g.s. units)} \tag{4.149}$$

For the case of sun, $M = 1.99 \times 10^{33}$ gm, $R = 6.97 \times 10^{10}$ cm, so

$$\alpha = 1.75 \quad \text{seconds} \tag{4.150}$$

The Eq. (4.150) implies that a ray of light passing at grazing incidence, near a gravitating mass like sun, will be deflected by an angle 1.75 sec. Observations have been taken during the total eclipse of the sun and noting the apparent positions of the stars whose light has passed near the sun at grazing incidence. It has been concluded, on analyzing the data that the theoretical prediction agrees with the observation, thus satisfying the validity of general theory of relativity by Einstein.

Since light has gravitating mass, so a ray of light on passing near the sun will deviate from its original path. Actually, J. Soldner in 1801, calculated that a star viewed near the sun, would be displaced by 0.88 second on the basis of Newtonian theory. Let us treat the deflection of a light ray on the basis of Newtonian theory.

Let m be the mass of sun, Fig. (4.4), and a light ray emitted from a star be moving parallel to the y-axis and passing the mass m at a distance $x = R$.

The acceleration in the x-direction, is given by

$$\begin{aligned}\frac{d^2x}{dt^2} &= -\frac{m}{r^2}\cos\phi \\ &= -\frac{m}{r^2}\frac{x}{r} = -\frac{mx}{(x^2+y^2)^{3/2}}\end{aligned} \tag{4.151}$$

For the ray moving parallel to the y-axis, $\frac{dy}{dt} = 1$, in gravitational units.

Therefore,

$$\frac{d^2y}{dt^2} = 0 \tag{4.152}$$

$$\frac{dx}{dt} = \frac{dx}{dy}\frac{dy}{dt}$$

Now

$$\begin{aligned}\frac{d^2x}{dt^2} &= \frac{d}{dt}\left(\frac{dx}{dy}\frac{dy}{dt}\right) = \frac{d^2x}{dy^2}\left(\frac{dy}{dt}\right)^2 + \frac{dx}{dy}\frac{d^2y}{dt^2} \\ &= \frac{d^2x}{dy^2}1^2 + \frac{dx}{dy}0 \\ &= \frac{d^2x}{dy^2}\end{aligned}$$

Substituting it in Eq. (4.151), we get

$$\frac{d^2x}{dy^2} = -\frac{mx}{(x^2+y^2)^{3/2}} = -\frac{mR}{(R^2+y^2)^{3/2}}$$

or

$$\frac{d^2x}{dy^2} = -\frac{mR}{(R^2+y^2)^{3/2}} \tag{4.153}$$

Integrating w.r.t. y, we get

$$\frac{dx}{dy} = -\int \frac{mR\,dy}{(R^2 + y^2)^{3/2}}$$

Putting $y = R\tan\theta$, we get

$$\frac{dx}{dy} = -\int \frac{mR^2 \sec^2\theta\, d\theta}{R^3 \sec^3\theta}$$

$$= -\frac{m}{R}\int \cos\theta\, d\theta = -\frac{m}{R}\sin\theta + C$$

Integrating it again w.r.t. y, we obtain

$$x = -\frac{m}{R}\sqrt{(R^2 + y^2)} + Cy + C_1 \tag{4.154}$$

When $x = R,\ y = 0,\ \frac{dx}{dy} = 0$, so

$$C = 0 \quad \text{and} \quad C_1 = m + R$$

Therefore, Eq. (4.154) becomes

$$x = R + \left[m - \frac{m}{R}\sqrt{(R^2 + y^2)}\right] \tag{4.155}$$

The second term $m - \frac{m}{R}\sqrt{(R^2 + y^2)}$, indicates the deviation from the path $x = R$. Further, we take the asymptotes to Eq. (4.155) by taking y very large compared to x and get

$$x = R - \frac{m}{R}(\pm y)$$

i.e.,

$$x = R + \frac{my}{R} \quad \text{and} \quad x = R - \frac{my}{R}$$

$$y = \frac{Rx}{2m} - \frac{R^2}{2m} \quad \text{and} \quad y = -\frac{Rx}{2m} + \frac{R^2}{2m}$$

If β is the angle between these asymptotes, we get

$$\tan\beta = \frac{\frac{R}{m} - \left(-\frac{R}{m}\right)}{1 + \frac{R}{m}\left(-\frac{R}{m}\right)}$$

$$= \frac{2mR}{m^2 - R^2}$$

Since β is very small and $m^2 \ll R^2$, we get

$$\beta \cong \frac{2m}{R} \tag{4.156}$$

But

$$\alpha \cong \frac{4m}{R} = 2\beta$$

so

$$\alpha = 2\beta \tag{4.157}$$

Thus, the deflection of a light ray predicted by relativistic theory is twice that predicted by Newtonian theory.

Einstein's theory predicts that the light from a distant star will be slightly deflected in the gravitational field of a massive body. In 1919, Arthur Eddington[3] observed just such a deflection of starlight as it grazed the sun, Fig. (4.4). The measurement, in agreement with the relativity prediction, helped to make Einstein a world celebrity. The observations were made simultaneously in the city of Sobral, Ceara, Brazil, and in the west coast of Africa. The result was considered a spectacular news and made the front page of most major newspapers. When asked by his assistant what his reaction, would have been if general relativity had not been confirmed by Eddington and Dyson in 1919, Einstein famously made the quip, 'Then I would feel sorry for the dear lord. The theory is correct anyway.'

The early accuracy, however, was poor. Dyson *et al.* quoted an optimistically low uncertainty in their measurements, which is argued by some to have been plagued by systematic error and probably confirmation bias, although modern reanalysis of the data set, suggest that Eddington's analysis was accurate.

In 1801, Johann George Von Soldner had pointed out that Newtonian gravity predicts that starlight will bend around a massive object, but, as shown above, the predicted effect was only half of the value predicted by general relativity as calculated by Einstein in his 1911 paper. Nobel Laureate Phillip Lenard revived Soldner's results in an attempt to discredit Einstein. Eddington had been aware of the alternative prediction in 1919. However, considerable uncertainty remained in these measurements for almost fifty years, until observations started being made at radio-frequencies. By late, 1960s, it was definitely shown that the amount of deflection was the full value predicted by general relativity and not half that number.

In observational astronomy an **Einstein ring** is the deformation of the light from a source (such as a galaxy or star) into a ring through gravitational lensing of the source's light by an object with an extremely large mass (such as another galaxy, or a black hole). This occurs when the source, lens and observer are all aligned. The first complete Einstein ring, designated B1938+666, was discovered by collaboration between astronomers at the University of Manchester and NASA's Hubble Space Telescope in 1998. The Einstein ring is an example of the deflection of light from distant galaxies by more nearby objects.

Performing a new version of the test, astronomers from Harvard, M.I.T. and the Haystack observatory, have used antennas in Massachusetts and California, to measure the deflection of radio waves coming from extragalactic object 3C279, as they passed near the sun. The use of coordinated but widely spaced antennas (very long baseline interferometry) produces a much more accurate measurement than is possible with a single radio telescope. The ratio of, measured to predicted deflection[4], was 0.9998 ± 0.008.

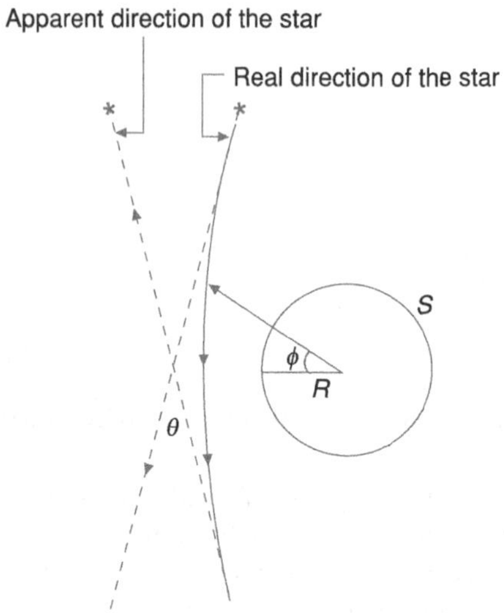

Figure 4.4 *Deviation of a ray of light passing near the sun.*

4.6.3 Gravitational Shift of Spectral Lines

The third significant prediction of general relativity entails that the wavelength of light depends on the gravitational potential of its source. We consider the emission of light by an atom, which is stationary on the surface of sun. The Schwarzschild line-element due to a mass m, is given by

$$\begin{aligned} ds^2 &= \left(1-\frac{2m}{r}\right)dt^2 - \left(1-\frac{2m}{r}\right)^{-1} dr^2 - r^2 d\theta^2 - r^2 \sin^2\theta d\phi^2 \\ &= g_{\mu\nu} dx^\mu dx^\nu \end{aligned} \tag{4.158}$$

The trajectory of a light ray is given by $ds = 0$, and the velocity of light, is

$$\frac{dr}{dt} = \left(1-\frac{2m}{r}\right) = g_{00} \tag{4.159}$$

The velocity is independent of time and this time independence of the velocity of light, implies that two successive light pulses, which are emitted at an interval of coordinate time Δt will on their reception by a stationary observer situated at the surface, will be separated by the same time interval.

The proper time interval $d\tau_s$ and the coordinate time interval dt_s are given by

$$d\tau_s = \sqrt{g_{00}}\, dt_s \tag{4.160}$$

The subscript refers to the values at the surface of sun. Analogously, at the surface of earth, the proper time interval $d\tau_e$ and coordinate time interval dt_e are given by

$$d\tau_e = \sqrt{g'_{00}}\, dt_e \tag{4.161}$$

g'_{00} is the value at the surface of earth.

Let us imagine that a stationary atom at sun's surface emits n waves of frequency ν_0 in proper time $\Delta\tau_s$. Then

$$n = \nu_0 \Delta\tau_s \tag{4.162}$$

A stationary observer at the earth's surface receives these waves. He receives n waves but the frequency and time interval are different. According to him,

$$n = \nu_e \Delta\tau_e \tag{4.163}$$

Here, ν_e is the frequency of waves received at earth. Comparing Eqs. (4.162) and (4.163), we get

$$\nu_e = \nu_0 \frac{\Delta\tau_s}{\Delta\tau_e} \tag{4.164}$$

However, from Eqs. (4.160) and (4.161), we have

$$\frac{\Delta\tau_s}{\Delta\tau_e} = \sqrt{\frac{g_{00}}{g'_{00}}} \frac{\Delta t_s}{\Delta t_e} = \sqrt{\frac{g_{00}}{g'_{00}}} \tag{4.165}$$

since for light ray, the coordinate times Δt_s and Δt_e are equal. Therefore, we get

$$\nu_e = \nu_0 \sqrt{\frac{g_{00}}{g'_{00}}} \tag{4.166}$$

Now
$$\begin{aligned} g_{00} &= 1 - \frac{2m}{r} \\ g_{00} &= 1 - \frac{2m}{r'} \end{aligned} \tag{4.167}$$

m is the mass of sun; r, the radius of sun while r', the distance of earth from the sun. Thus,

$$\begin{aligned} \nu_e &= \nu_0 \sqrt{\left(\frac{1-2m/r}{1-2m/r'}\right)} \\ &= \nu_0 (1 - m/r)(1 + m/r') \\ &= \nu_0 (1 - m/r + m/r') \end{aligned} \tag{4.168}$$

Since $r' \gg r$, one gets

$$\nu_e = \nu_0 (1 - m/r)$$

or

$$\frac{\nu_e - \nu_0}{\nu_0} = -\frac{m}{r}$$

or

$$\frac{\delta\nu}{\nu_0} = -\frac{m}{r} \tag{4.169}$$

Thus, the frequency of light on receipt at earth is decreased from the frequency on leaving the sun. Expressed in terms of wavelength, Eq. (4.169) becomes

$$\frac{d\lambda}{\lambda} = \frac{m}{r} \tag{4.170}$$

and expressed in c.g.s. units, it becomes

$$\frac{d\lambda}{\lambda} = \frac{mG}{c^2 r} \tag{4.171}$$

Normally, it is conventional to express it in velocity units by multiplying $d\lambda / \lambda$ by *c*, *i.e.*, for sun, we get

$$c\frac{d\lambda}{\lambda} = \frac{mG}{cR} \tag{4.172}$$

Putting $m = 1 \cdot 99 \times 10^{33}$ gm, $R = 6.97 \times 10^5$ km, we obtain

$$c d\lambda / \lambda = 0.64 \, \text{km sec}^{-1} \tag{4.173}$$

Einstein predicted the gravitational red shift of light from the equivalence principle in 1907, but it is very difficult to measure astrophysically. The predictions of general relativity have been verified by experiments of Adams *et al.* as well as Blamont and Roddier. The latest experiment on the gravitational red shift of solar spectral lines was done by Snider[5], on solar potassium absorption line at 7699Å making use of the atomic beam resonance scattering technique. There was good agreement with the predictions. However, it was conclusively tested when the Pound–Rebka experiment in 1959 measured the relative red shift of a source situated at the bottom of Harvard University's Jefferson tower using Mössbauer effect. This was one of the first precision experiments testing general relativity.

4.6.4 Terrestrial Verification of the Red Shift of Spectral Lines

The discovery of the phenomenon of γ-ray resonance, popularly known as Mössbauer effect, by Radolph L.Mössbauer[6] in 1958, rendered it possible to perform the experiment to verify the red shift of spectral lines on our earth itself; which had been confined to experiments on light passing near the sun or some other star.

We can easily appreciate as why a terrestrial experiment was not feasible from the following illustration. Let us imagine two clocks separated by a vertical distance Δh in the earth's gravitational field. The resulting frequency difference will be

$$\frac{\Delta\nu}{\nu}=\frac{\Delta\phi}{c^2}=\frac{1}{c^2}\left(-\frac{GM}{r}+\frac{GM}{(r+\Delta h)}\right)$$
$$=-\frac{GM}{c^2}\frac{\Delta h}{r^2}=-g\frac{\Delta h}{c^2} \tag{4.174}$$

Say for height $\Delta h=10$ m, it turns out to be

$$\frac{\Delta\nu}{\nu}=\frac{980\times10\times100}{9\times10^{20}}\cong10^{-15} \tag{4.175}$$

However, such a small shift was not possible to be detected even with the atomic clocks till 1958. Mössbauer discovered that under certain conditions, nuclei like Fe^{57} emit γ-rays with sharply defined frequency without exciting the lattice (zero phonon emission). The absorption of these recoillessly emitted γ-rays is also extremely sharp. Therefore, if two crystals of Fe^{57} are placed in the gravitational field of earth separated by height h, the radiation emitted by one will not be absorbed resonantly by the other, due to the slight shift in frequency of radiation. However, the shift can be compensated by Doppler tuning the absorber by moving it with a velocity v relative to the other and then the absorption will take place at the velocity

$$v=g\frac{\Delta h}{c} \tag{4.176}$$

An experiment was performed by Pound and Rebka[7] at Harvard University by placing the emitter and absorber at the bottom and top of a tower of height 74 ft. γ-rays emitted by the source at the bottom got red shifted in reaching the absorber at the top and the resonant absorption was achieved by Doppler tuning. The predicted shift is $\Delta\nu/\nu$ is 4.92×10^{-15}, whereas the measurements of velocity gave the result as $(5.13\pm0.51)\times10^{-15}$ *i.e.*, about (1.04 ± 0.1) times the calculated value. The experiment verified the existence of Einstein's gravitational red shift to a sufficient degree of accuracy.

4.6.5 Light Travel Time Delay Testing—Shapiro Time Delay

It was in 1964, that Irwin Shapiro proposed a new experiment, beyond the classical tests, to test general relativity. This test is performed in the solar system and is sometimes called the fourth 'classical' test of general relativity. According to him, there will be a relativistic time delay (*Shapiro delay*) in the round-trip travel time for radar signal reflecting off other planets[8]. A photon passing near the sun undergoes a curvature in its trajectory but the curvature is too small to have an observable delaying effect. However, the delay time increases progressively when the photon passes nearer to the sun, due to the sun's space–time curvature. The radar reflections from Mercury and Venus, taken just before and after it is eclipsed by the sun, confirm the predictions of general relativity at the 5% level[9]. The Cassini probe has carried out a similar experiment, which confirms the prediction to a precision of 1 part in 1000.

After having determined the Schwarzschild metric, it is straight forward to determine the paths, the particles and light rays are going to follow in the space–time. This will be accomplished by solving the geodesic equations for each coordinate. According to variational technique, we have

$$\delta\int ds = \delta\int [(1-2m/r)\dot{t}^2 - \frac{1}{(1-2m/r)}\dot{r}^2 - r\dot{\theta}^2 - r^2\sin^2\theta\dot{\phi}^2]ds = 0 \quad (4.177)$$

Defining the function F as

$$F \equiv (1-2m/r)\dot{t}^2 - \frac{1}{(1-2m/r)}\dot{r}^2 - r\dot{\theta}^2 - r^2\sin^2\theta\dot{\phi}^2 \quad (4.178)$$

The Euler–Lagrange equations are

$$\frac{d}{ds}\left(\frac{\partial F}{\partial \dot{x}^\alpha}\right) - \frac{\partial F}{\partial x^\alpha} = 0 \quad (4.179)$$

Taking the time coordinate, we get

$$\begin{aligned}\frac{\partial F}{\partial \dot{t}} &= 2(1-2m/r)\dot{t} \\ \Rightarrow \frac{d}{ds}\left(\frac{\partial F}{\partial \dot{t}}\right) &= \frac{d}{ds}[2(1-2m/r)\dot{t}] \\ &= \frac{d}{ds}[2(1-2m/r)]\dot{t} + 2(1-2m/r)\ddot{t} \\ &= \frac{4m}{r^2}\dot{r}\dot{t} + 2(1-2m/r)\ddot{t}\end{aligned} \quad (4.180)$$

Since there are no terms in F that contain t, so the geodesic equation for the time coordinate becomes

$$\ddot{t} + \frac{2m}{r(r-2m)}\dot{r}\dot{t} = 0 \quad (4.181)$$

Next considering the radial component, we get

$$\begin{aligned}\frac{\partial F}{\partial \dot{r}} &= -\frac{2}{(1-2m/r)}\dot{r} \\ \Rightarrow \frac{d}{ds}\left(\frac{\partial F}{\partial \dot{r}}\right) &= -\frac{2}{(1-2m/r)}\ddot{r} - \frac{4m}{r^2}\dot{r}^2(1-2m/r)^{-2}\end{aligned}$$

Also
$$\frac{\partial F}{\partial r} = \frac{2m}{r^2}\dot{t}^2 + (1-2m/r)^2\frac{2m}{r}\dot{r}^2 - 2r\dot{\theta}^2 - 2r\sin^2\theta\dot{\phi}^2$$

Thus, the geodesic equation for the radial coordinate is

$$\ddot{r}+\frac{m}{r^3}(r-2m)\dot{t}^2-\frac{m}{r(r-2m)}\dot{r}^2-(r-2m)(\dot{\theta}^2+\sin^2\theta\dot{\phi}^2)=0 \tag{4.182}$$

Further, we get

$$\begin{aligned}
&\frac{\partial F}{\partial \dot{\theta}}=-2r^2\dot{\theta}\\
\Rightarrow &\frac{d}{ds}\left(\frac{\partial F}{\partial \dot{\theta}}\right)=-4r\dot{r}\dot{\theta}-2r^2\ddot{\theta}\\
&\frac{\partial F}{\partial \theta}=-2r^2\sin\theta\cos\theta\dot{\phi}^2
\end{aligned}$$

Thus, the geodesic equation for θ is

$$\ddot{\theta}+\frac{2}{r}\dot{r}\dot{\theta}-\sin\theta\cos\theta\dot{\phi}^2=0 \tag{4.183}$$

Lastly, for the ϕ coordinate

$$\begin{aligned}
&\frac{\partial F}{\partial \dot{\phi}}=-2r^2\sin^2\theta\dot{\phi}\\
\Rightarrow &\frac{d}{ds}\left(\frac{\partial F}{\partial \dot{\phi}}\right)=-4r\dot{r}\sin^2\theta\dot{\phi}-4r^2\sin\theta\cos\theta\dot{\theta}\dot{\phi}-2r^2\sin^2\theta\ddot{\phi}\\
&\frac{\partial F}{\partial \phi}=0
\end{aligned}$$

Thus,

$$\ddot{\phi}+2\cot\theta\,\dot{\theta}\dot{\phi}+\frac{2}{r}\dot{r}\dot{\phi}=0 \tag{4.184}$$

From spherical symmetry a geodesic lies in a plane. Let it be the plane

$$\phi=\pi/2 \tag{4.185}$$

Integrating Eqs. (4.181) and (4.183), we get

$$\dot{t}(1-2m/r)=B \tag{4.186}$$

$$\dot{\theta}\,r^2=A \tag{4.187}$$

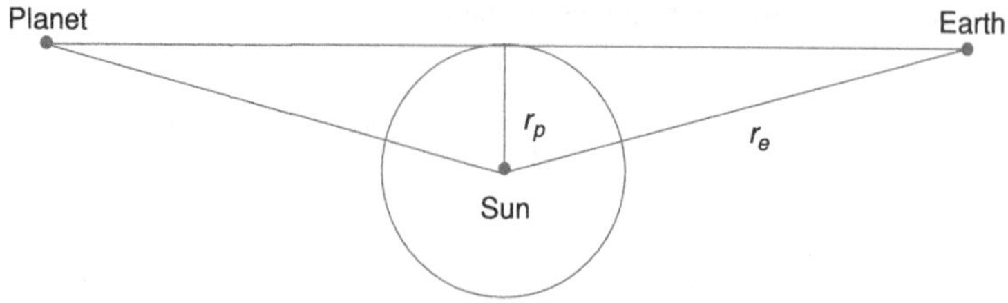

Figure 4.5 *Radar echo delay.*

where A and B are constants of integration. Physically, A and B are conserved quantities, since A is the angular momentum per unit mass and B is the energy per unit mass.

Substituting Eqs. (4.185)–(4.187) into Eq. (4.182) and integrating, we get

$$\frac{\dot{r}^2}{1-2m/r}+\frac{A^2}{r^2}-\frac{B^2}{1-2m/r}=-E=\begin{cases}0 & \text{for light}\\ -1 & \text{for inertial particles}\end{cases} \tag{4.188}$$

where E is constant of integration. The values of E are easily verifiable by substituting Eqs. (4.185)–(4.188) into the Schwarzschild metric, Eq. (4.42), giving $(ds/dp)^2 = E$, where p is a parameter along the geodesic. For inertial particles, we take $p = s$. For light $ds = 0$, so $E = 0$. For an inertial particle, $E = 1$.

A radar pulse can be sent from earth, reflected off the planet and detected on its return to earth. If this is carried out when the planet is about to pass behind the sun, then according to general relativity, the radar's return is delayed by the sun's gravitational field.

In order to calculate the time t for light pulse to go from the earth to sun, Fig. (4.5), r_p is the distance of closest approach to the sun and the velocity (dr/dt) has to vanish at $r = r_p$.

$$\frac{dr}{dp}=\frac{dr}{dt}\frac{dt}{dp}=\frac{dr}{dt}B(1-2m/r)^{-1} \tag{4.189}$$

where use has been made of Eq. (4.186).

Setting $r = r_p$ and $r|_{r=r_p}=0$ in Eq. (4.188), one gets

$$A^2B^{-2}=r_p^2(1-2m/r)^{-1} \tag{4.190}$$

Dividing Eq. (4.188) by B^2, substituting Eqs. (4.189) and (4.190), separating the variables and integrating, we get

$$t=\int_{r_p}^{r_e}\left(\frac{(1-2m/r)^{-2}}{1-\frac{1-2m/r}{1-2m/r_p}\left(\frac{r_p}{r}\right)^2}\right)^{1/2}dr \tag{4.191}$$

Approximating the integral by retaining only the first order terms in the small quantities m/r and m/r_p, we get

$$\frac{(1-2m/r)^{-2}}{1-\dfrac{1-2m/r}{1-2m/r_p}\left(\dfrac{r_p}{r}\right)^2} \approx \frac{(1+2m/r)^2}{1-(1-2m/r)(1+2m/r_p)(r_p/r)^2}$$

$$\approx \frac{1+4m/r}{1-(1-2m/r+2m/r_p)(r_p/r)^2}$$

$$\approx \frac{1+4m/r}{\left(1-\left(\dfrac{r_p}{r}\right)^2\right)\left(1-\dfrac{2mr_p}{r(r+r_p)}\right)}$$

$$\approx \left(1-\left(\frac{r_p}{r}\right)^2\right)^{-1}\left(1+\frac{4m}{r}\right)\left(1+\frac{2mr_p}{r(r+r_p)}\right)$$

$$\approx \left(1-\left(\frac{r_p}{r}\right)^2\right)^{-1}\left(1+\frac{4m}{r}+\frac{2mr_p}{r(r+r_p)}\right)$$

Thus, Eq. (4.191) becomes

$$t = \int_{r_e}^{r_p}\left(1-\left(\frac{r_p}{r}\right)^2\right)^{-1}\left(1+\frac{4m}{r}+\frac{2mr_p}{r(r+r_p)}\right)dr \tag{4.192}$$

On integration, it yields

$$t = \left(r_e^2 - r_p^2\right)^{1/2} + 2m\ln\frac{r_e+\left(r_e^2-r_p^2\right)^{1/2}}{r_p} + m\left(\frac{r_e-r_p}{r_e+r_p}\right)^{1/2} \tag{4.193}$$

Taking $r_p \ll r_e$, it becomes

$$t = \left(r_e^2 - r_p^2\right)^{1/2} + 2m\ln(2r_e/r_p) + m \tag{4.194}$$

The time $(r_e^2 - r_p^2)^{1/2}$, is by theorem of Pythagoras, the time required for the signal to travel in a straight line in a flat space–time from earth to sun. The other two terms represent the delay in time caused by the gravitational field of the sun.

One is to add analogous delay terms for the transit from sun to the planet and double to include the return trip. For Mercury, a delay of 2.4×10^{-4} sec was found.

The 'Shapiro delay' has been observed following highly accurate radio studies of a nearby pulsar. Willem van Straten of Swinburne University of Technology in Australia and colleagues, detected this effect while mapping the motion of the pulsar at the Parks Observatory[10]. Pulsars are rapidly spinning neutron stars that earn their name from the beams of radiation they emit, which appear as pulses to a stationary observer. Binary pulsar-systems in which a pulsar orbits another object, emit very regular pulses of radiation because their orbital and rotation periods are extremely regular. These properties make them excellent tools for probing the effects of general relativity.

At 450 light years from earth, PSRJ0437-4715 is the closest known binary pulsar, in which a pulsar orbits a white dwarf. Its proximity allows astronomers to measure the radio output from different angles, due to its motion relative to the earth. Using this geometrical method, van Straten and colleagues calculated the orbit of the pulsar in three dimensions. This also revealed the centre-of mass of the system, from which the team calculated the masses of the pulsar and the white dwarf. Next, van Straten and colleagues studied variations in the pattern of the radio pulses arriving on earth. If the pulses emitted throughout the orbital period of the pulsar travelled through equivalent regions of space on their journey to earth, they should all arrive at equal intervals.

But after eliminating geometrical effects, van Straten's team found that the pulses took longer to arrive when the plane of the binary system was in their line of sight. Conversely, when viewing the plane 'face on', the pulses were not delayed. This phenomenon arises because when the plane is 'edge on', the signal from the pulsar passes through a region of space that is distorted by the gravity of the white dwarf. This distortion means that the signal takes a longer route to earth, and therefore arrives later. This is Shapiro delay.

General relativity also states that binary systems should gradually slow down, and emit the excess rotational energy as gravitational energy. This predicted increase in orbital period has been observed in previous experiments, but attempts to detect gravitational waves have so far failed. The length of Shapiro delay was consistent with the amount of energy the binary system should be losing through gravitational waves. 'To our knowledge, this verification of the predicted space–time distortion is the first confirmation-outside the solar system-in which the orbital inclination was determined independently of general relativity', say the authors.

The study also means that pulsar PSRJ0437-4715 has the most accurately known location of any astronomical object. Now this system is known to exhibit the Shapiro delay, it is likely to be the subject of many more cosmological studies.

4.6.6 Orbital Effects

General theory of relativity has predictions for orbiting bodies, which are not implied by Newton's theory of gravity. The first difference is in its prediction that apsides of orbits will precess on their own. An early successful test of general relativity was its success in explaining the anomalous perihelion precession of Mercury as a classical test. Furthermore, perihelion precession has been observed in binary pulsar systems.

Another related effect is geodetic precession. This is a precession of the poles of a spinning object due to the effects of parallel transport in a curved space–time. This effect is not expected by Newton's theory of gravity. It has been verified by the Gravity B probe experiment to a precision of better than one percent[11].

Lastly, there is orbital decay due to the emission of gravitational radiation by a co-rotating system. This is continual decrease in their orbital periods in closely orbiting stars. This effect has been observed in binary pulsar systems.

4.6.7 Frame Dragging

Einstein's theory of general relativity has been highly successful at explaining how matter and light behave in strong gravitational fields, and has been successfully tested using a wide variety of astrophysical observations. General theory predicts that massive rotating objects should drag space–time around themselves as they rotate, in a phenomenon referred to as frame dragging. Frame dragging is like what happens if a bowling ball spins in a thick fluid, such as molasses. As the ball spins, it pulls the molasses around itself. Similarly, as the earth rotates, it pulls space–time in its vicinity around itself. This will shift the orbits of satellites near the earth.

The frame-dragging effect was first derived using general relativity by Austrian physicists, Joseph Lense and Hans Thirring in 1918. The predicted effect is exceedingly small-about one part in a few trillion. In order to detect it, it is necessary to look at a very massive object, or develop an instrument that is greatly sensitive.

Researchers believe that they have detected the effect by precisely measuring shifts in the orbits of two Earth-orbiting laser-ranging satellites, the Laser Geodynamics Satellite I (LAGEOS I), a NASA spacecraft, and LAGEOS II, a joint NASA/Italian Space Agency (ASI) spacecraft. It has recently been observed around distant celestial objects with intense gravitational fields, such as black holes and neutron stars. The new research around earth is the first detection and measurement of this phenomenon.

Gravity Probe B (GP-B) is a satellite-based relativity mission launched by NASA in 2004, with a view to measure the frame dragging effect to a precision of 1% as well as the geodesic effect. The data analysis for frame dragging is being done by the Physics Department, Stanford University and the outcome will be known in due course of time.

4.6.8 Gravitational Lensing

It is one of the most important tests. Gravitational lensing occurs when one distant object is in front of or close to being in front of another much more distant object. The bending of light by the nearer object can affect how the distant object appears. The first case of gravitational lensing was the discovery of two nearby images of the same pulsar. Now very many cases of distant galaxies and quasars being affected by gravitational lensing have been found. The sources that can be most precisely analyzed are distant radio sources such as quasars. A technique called Very Long Baseline Interferometry (VLBI), enables to attain high positional accuracy (from milli-arcsecond to micro-arcsecond) by combining radio telescopes across the earth. In this technique, radio observations couple the phase information of the radio signal observed in telescopes spread over large distances. The observations of deflections of radio waves by the sun, to a high degree of accuracy to the 0.04% level, have confirmed the predictions of general relativity. A number of corrections have to be applied for determining the precise location of the telescopes on earth. These are the earth's nutation, atmospheric refraction, tectonic displacement and tidal waves. Another disturbing effect is refraction of radio waves by the solar corona. However, this effect has a characteristic spectrum, whereas gravitational distortion is independent of wavelength. Therefore, taking measurements at several frequencies, can assist in subtracting this source of error.

4.6.9 Global Positioning System

Global Positioning System (GPS) is neither designed nor operated as a test of relativity but it has to account for two major relativistic influences upon its rate of time keeping: *a* special relativistic correction for its orbital speed and a general relativistic correction for its orbital altitude. It is a case of gravitational time dilation, under which clocks will run slower at lower gravitational potentials (deeper within a gravity well). It is another way of perceiving that decrease in frequency of the gravitationally red-shifted light. The Gravity Probe A (GP-A) satellite, launched in 1976, showed that gravity and velocity affect the ability to synchronize the rates of clocks orbiting a central mass. The Hafele–Keating experiment used atomic clocks in circumnavigating aircraft to test the general relativity and special relativity together[12].

Relativistic Effects on Clocks Aboard GPS Satellites

Firstly, with respect to an earth bound clock, the transmitting clock is subject to time dilation due to its speed and as a consequence the orbiting clock travelling at speed v, will run slow, compared to a stationary clock, by a factor of

$$\frac{1}{\gamma} = \sqrt{1 - v^2 / c^2} \approx 1 - v^2 / 2c^2 \tag{4.195}$$

provided $v \ll c$, as is the case of a satellite. Thus, for one second of the proper time as recorded by the stationary clock, the orbiting clock will lose $v^2/2c^2 = K/E_0$ seconds, where K and E_0 are the kinetic and rest energies of the clock, respectively.

Secondly, a clock at the higher gravitational potential of orbit runs faster than the clock at the earth surface. A photon travelling downward in earth's gravitational field, loses potential energy $h\nu\Delta\phi/c^2$ and gains an equal amount of kinetic energy, where h is Planck's constant; ν, the frequency of the photon and $\Delta\phi$ is the change in earth's gravitational potential. Thus, the downward falling photon is gravitationally blue-shifted by

$$\Delta\nu = \nu \frac{\Delta\phi}{c^2} \tag{4.196}$$

If the clock's ticking is synchronized to a light pulse, the orbiting clock at the earth's surface will tick faster, on account of this frequency shift towards the blue wavelength. Thus, for every one second of earth's time, the clock at high altitude gains $\Delta\phi / c^2 = \Delta U / E_0$ seconds. U is the gravitational potential energy of the clock.

Combining both the relativistic effects, we get for the time lost by the orbiting clock Δt

$$\Delta t = \left(\frac{K - U}{E_0} \right) \tau \tag{4.197}$$

τ is the time elapsed on the surface bound clock.

Furthermore, we estimate the size of these corrections to time for a GPS satellite, located at an altitude of r. According to Newton's second law, we get

$$a = \frac{F}{m} \Rightarrow \frac{v^2}{r} = \frac{Gm_E}{r^2}$$

or

$$v^2 = \frac{gr_E^2}{r} \tag{4.198}$$

where earth's gravitational potential is $g = Gm_E / r_E^2 = 9.8\ \text{m/s}^2$. Thus, the fractional time loss due to satellite's orbital speed is $-gr_E^2 / (2rc^2)$ per second. Furthermore, the general relativistic fractional time gain due to the satellite's altitude is

$$\begin{aligned}\frac{\Delta\phi}{c^2} &= \frac{1}{c^2}\left[-\frac{Gm_E}{r} + \frac{Gm_E}{r_E}\right] \\ &= \frac{gr_E}{c^2}\left[1 - \frac{r_E}{r}\right]\end{aligned} \tag{4.199}$$

It is obvious from comparing Eqs. (4.198) and (4.199), the gravitational effect is greater than the speed effect. The dominant G.S. correction is general, not special, relativistic.

If we consider satellites in progressively lower altitude orbits, their speeds will increase according to Eq. (4.198), whereas the gravitational potential difference $\Delta\phi$ in Eq. (4.199) will decrease. These opposing corrections will cancel at an altitude when these balance each other and the clock will run synchronously with the earth bound clock. This will occur when

$$\frac{v^2}{2c^2} = \frac{\Delta\phi}{c^2}$$

or

$$\frac{gr_E^2}{2r} = gr_E\left[1 - \frac{r_E}{r}\right]$$

or

$$r = 1.5r_E \tag{4.200}$$

Thus, the clock will run synchronously at an altitude, $r = 1.5r_E$ *i.e.*, at altitude of half the radius of earth.

References

1. Schwarzschild, K. (1916). *Sitz. Proc. Akad. Wiss.*, 426.
2. Duncombe, (1956). *Aust. Jour,* **6,** 1, 174.
3. Dyson, F.J., Eddington, A.S., Davidson, C. (1920). "A determination of the deflection of light by the sun's gravitational field, observations made at the total eclipse of May 29,1919", *Philos. Trans. Roy. Soc. London,* 220A, 291–333.

4. Lebach, D.E. *et al.,* (1975). *Phys. Rev. Letters,* 75, 1439.
5. Snider, J. L. (1972). *Phys Rev*, 28, 853.
6. Mössbauer, R.L., *Z. Physik,* 151, 124 (1958); *Nature Wissenschaften*, 45:538(1958); *Naturforsch*, 14a;211 (1959).
7. Pound, R.V. and Rebka G.A. (1960). "Gravitational Red Shift in Nuclear Resonance:, *Phys. Rev. Lett,* 3, 439(1959)," Apparent weight of photons", *ibid.*, 4, 337.
8. Shapiro, I. I. (1964). " Fourth test of General relativity", *Phys. Rev. Lett.* 13 (26), 789.
9. Shapiro, I. I. *et al.* (1971) "Fourth Test of General Relativity: New Radar Result", *Phys. Rev. Lett.* 26(18), 1132–35.
10. van Straten, W. *et al,* (2001). *Nature,* 412, 158.
11. http://Einstein Stanford.edu./content/press_releases/SU/pr-aps-041807.pdf.
12. Hafele, J. Keating, R. (1972). "Around the world atomic clocks: Predicted relativistic time gains", *Science,* 177, 166–168.

Problems

4.1 Calculate the angle through which a light beam grazing the sun will bend. Given that

$$\text{Mass of the sun} = 2\times10^{33}\,\text{gm}$$
$$\text{Radius of the sun} = 6.96\times10^{10}\,\text{cm}$$
$$\text{Gravitational constant} = 6.67\times10^{-8}\,\text{c.units}$$

4.2 Prove that $GM = c^2m$ where G is the constant of gravitation; M, the mass of the attracting particle and m, the constant occurring in Schwarzschild solution.

4.3 The mass of sun is 2×10^{33} gm. Show that in gravitational units it becomes 1.47 km.

4.4 Calculate the perihelion precession of planet Mercury in one century, given that for its orbit around the sun,

$$\text{semi-major axis} = 36.2\times10^{6}\,\text{miles}$$
$$\text{eccentricity} = 0.2056$$
$$\text{time period} = 88\text{ days}$$

4.5 A clock at the surface of a star runs slow by 20%. Calculate the radius of the star in km and light years. (Surface gravity = 1000 cm/sec^2)

4.6 What is the *coordinate velocity* of light in the Schwarzschild space at coordinate r,
  (i) in the radial direction?
  (ii) in the transverse direction? Are there any physical consequences of this?

4.7 Gravity evidently affects the rate of a pendulum clock. In a stronger field, does a given pendulum go faster or slower? Is this in any way connected with gravitational time dilation?

4.8 A light ray from a distant star touches the earth tangentially. By what angle will it be deflected?

5

Einstein's Field Equations for Non-Empty Space

5.1 INTRODUCTION

After having discussed the case of Einstein's equations in empty space, we discuss how these equations are formulated in the presence of matter. Further we assume that the system is spherically symmetric and isotropic about a point in space. While treating the case of static line-element with spherical symmetry, both outside as well as inside the sphere of matter, we will derive the Schwarzschild's exterior and interior solutions. Lastly, we discuss the conservation laws in general relativity.

5.2 STATIC LINE-ELEMENT WITH SPHERICAL SYMMETRY

Let us consider a static distribution of matter, which exhibits spherical symmetry. The line-element with spherical symmetry is expressed in spherical polar coordinates as

$$ds^2 = e^{\nu} dt^2 - e^{\lambda} dr^2 - r^2 d\theta^2 - r^2 \sin^2 \theta d\phi^2 \tag{5.1}$$

where $\nu = \nu(r)$ and $\lambda = \lambda(r)$.

The 3-index symbols for this line-element are already listed in Eq. (4.20). Einstein's field equations in the presence of matter are

$$-8\pi T_{\mu\nu} = R_{\mu\nu} - \frac{1}{2} g_{\mu\nu} R + \Lambda g_{\mu\nu} \tag{5.2}$$

Written in terms of mixed tensor, these are

$$-8\pi T_{\mu}^{\nu} = R_{\mu}^{\nu} - \frac{1}{2} g_{\mu}^{\nu} R + \Lambda g_{\mu}^{\nu} \tag{5.3}$$

Making use of non-zero components of $R_{\mu\nu}$, Eq. (4.27), and Eq. (4.17) for $g^{\mu\nu}$, we get

$$\begin{aligned}
R_0^0 &= g^{00} R_{00} = -\frac{1}{2} e^{-\lambda} \left(\nu'' + \frac{1}{2} \nu'^2 - \frac{1}{2} \nu'\lambda' + \frac{2}{r} \nu' \right) \\
R_1^1 &= g^{11} R_{11} = -e^{-\lambda} \left(\frac{1}{2} \nu'' + \frac{1}{4} \nu'^2 - \frac{1}{4} \nu'\lambda' - \frac{1}{r} \lambda' \right) \\
R_2^2 &= g^{22} R_{22} = -\frac{1}{r^2} e^{-\lambda} \left(1 + \frac{1}{2} r\nu' - \frac{1}{2} r\lambda' \right) + \frac{1}{r^2} \\
R_3^3 &= g^{33} R_{33} = -\frac{e^{-\lambda}}{r^2} \left(1 + \frac{1}{2} r\nu' - \frac{1}{2} r\lambda' \right) + \frac{1}{r^2} = R_2^2
\end{aligned} \tag{5.4}$$

Next we calculate scalar curvature R. Thus,

$$\begin{aligned} R &= R^{\mu}_{\mu} \\ &= R^0_0 + R^1_1 + R^2_2 + R^3_3 \\ &= -e^{-\lambda}\left(\nu'' + \frac{1}{2}\nu'^2 - \frac{1}{2}\nu'\lambda' + 2\frac{\nu' - \lambda'}{r} + \frac{2}{r^2}\right) + \frac{2}{r^2} \end{aligned} \tag{5.5}$$

Further, we calculate the Einstein tensor $R_{\mu\nu} - \frac{1}{2}g_{\mu\nu}R$. Writing it in the mixed tensor form, we get

$$\begin{aligned} R^0_0 - \frac{1}{2}g^0_0 R &= -e^{-\lambda}\left(\frac{\lambda'}{r} - \frac{1}{r^2}\right) - \frac{1}{r^2} \\ R^1_1 - \frac{1}{2}g^1_1 R &= e^{-\lambda}\left(\frac{\nu'}{r} + \frac{1}{r^2}\right) - \frac{1}{r^2} \\ R^2_2 - \frac{1}{2}g^2_2 R &= e^{-\lambda}\left(\frac{1}{2}\nu'' + \frac{1}{4}\nu'^2 - \frac{1}{4}\nu'\lambda' + \frac{1}{2}\frac{\nu' - \lambda'}{r}\right) \\ R^3_3 - \frac{1}{2}g^3_3 R &= R^2_2 - \frac{1}{2}g^2_2 R \end{aligned} \tag{5.6}$$

Writing the field equations in the presence of matter, Eq. (5.3), in the light of Eq. (5.6), we obtain

$$\begin{aligned} 8\pi T^0_0 &= e^{-\lambda}\left(\frac{\lambda'}{r} - \frac{1}{r^2}\right) + \frac{1}{r^2} - \Lambda \\ 8\pi T^1_1 &= -e^{-\lambda}\left(\frac{\lambda'}{r} + \frac{1}{r^2}\right) + \frac{1}{r^2} - \Lambda \\ 8\pi T^2_2 &= -e^{-\lambda}\left(\frac{1}{2}\nu'' + \frac{1}{4}\nu'^2 - \frac{1}{4}\nu'\lambda' + \frac{1}{2}\frac{\nu' - \lambda'}{r}\right) - \Lambda \\ 8\pi T^3_3 = 8\pi T^2_2 &= -e^{-\lambda}\left(\frac{1}{2}\nu'' + \frac{1}{4}\nu'^2 - \frac{1}{4}\nu'\lambda' + \frac{1}{2}\frac{\nu' - \lambda'}{r}\right) - \Lambda \end{aligned} \tag{5.7}$$

We make the simplifying assumption that matter consists of an incompressible perfect fluid and $T^{\mu\nu}$ is expressed as

$$T^{\mu\nu} = (\rho_0 + p_0)\frac{dx^{\mu}}{ds}\frac{dx^{\nu}}{ds} - g^{\mu\nu}p_0 \tag{5.8}$$

ρ_0 and p_0 are the proper density and pressure of the fluid. Expressed in mixed tensor form, Eq. (5.8) becomes

$$T^{\nu}_{\mu} = (\rho_0 + p_0)g_{\alpha\mu}\frac{dx^{\alpha}}{ds}\frac{dx^{\nu}}{ds} - g^{\nu}_{\mu}p_0 \tag{5.9}$$

For a static distribution of mass, all the velocity components of fluid are zero,

$$\frac{dr}{ds} = \frac{d\theta}{ds} = \frac{d\phi}{ds} = 0$$

and
$$\frac{dt}{ds} = e^{-\nu/2} \tag{5.10}$$

Let us calculate the different components of the energy-momentum tensor. Thus,

$$\begin{aligned}
T_0^0 &= (\rho_0 + p_0) g_{\alpha 0} \frac{dx^\alpha}{ds} \frac{dx^0}{ds} - p_0 \\
&= (\rho_0 + p_0) \frac{dx^\alpha}{ds} \frac{dx^0}{ds} g_{\alpha 0} - p_0 \\
&= (\rho_0 + p_0) e^{-\nu} e^{\nu} - p_0 \\
&= \rho_0
\end{aligned} \tag{5.11}$$

$$\begin{aligned}
T_1^1 &= (\rho_0 + p_0) g_{\alpha 1} \frac{dx^\alpha}{ds} \frac{dx^1}{ds} - p_0 \\
&= (\rho_0 + p_0) g_{11} \frac{dx^1}{ds} \frac{dx^1}{ds} - p_0 \\
&= -p_0
\end{aligned} \tag{5.12}$$

Similarly,
$$T_2^2 = -p_0$$
$$T_3^3 = -p_0$$

The energy-momentum tensor T_μ^ν becomes

$$T_\mu^\nu = \begin{pmatrix} \rho_0 & 0 & 0 & 0 \\ 0 & -p_0 & 0 & 0 \\ 0 & 0 & -p_0 & 0 \\ 0 & 0 & 0 & -p_0 \end{pmatrix} \tag{5.13}$$

Writing the Eqs. (5.7) in view of Eq. (5.13), we get

$$\begin{aligned}
8\pi\rho_0 &= e^{-\lambda}\left(\frac{\lambda'}{r} - \frac{1}{r^2}\right) + \frac{1}{r^2} - \Lambda \\
8\pi p_0 &= e^{-\lambda}\left(\frac{\nu'}{r} + \frac{1}{r^2}\right) - \frac{1}{r^2} + \Lambda \\
8\pi p_0 &= e^{-\lambda}\left(\frac{1}{2}\nu'' + \frac{1}{4}\nu'^2 - \frac{1}{4}\nu'\lambda' + \frac{1}{2}\frac{\nu' - \lambda'}{r}\right) + \Lambda \\
8\pi p_0 &= e^{-\lambda}\left(\frac{1}{2}\nu'' + \frac{1}{4}\nu'^2 - \frac{1}{4}\nu'\lambda' + \frac{1}{2}\frac{\nu' - \lambda'}{r}\right) + \Lambda
\end{aligned} \tag{5.14}$$

Adding the first and second equations of the set of Eq. (5.14), we get

$$8\pi(\rho_0 + p_0) = e^{-\lambda}\left(\frac{\lambda' + \nu'}{r}\right)$$

or
$$8\pi(\rho_0 + p_0)\frac{\nu'}{2} = e^{-\lambda}\left(\frac{\lambda'\nu' + \nu'^2}{2r}\right) \tag{5.15}$$

The second equation of the set of Eqs. (5.14) becomes, on differentiating both the sides w.r.t. r

$$8\pi \frac{dp_0}{dr} = e^{-\lambda}\left(\frac{\nu''}{r} - \frac{\nu'}{r^2} - \frac{2}{r^3} - \frac{\lambda'\nu'}{r} - \frac{\lambda'}{r^2}\right) + \frac{2}{r^3} \tag{5.16}$$

Adding Eqs. (5.15) and (5.16), we get

$$8\pi\left[\frac{dp_0}{dr} + (\rho_0 + p_0)\frac{\nu'}{2}\right] = e^{-\lambda}\left(\frac{\nu''}{r} - \frac{\nu'\lambda'}{2r} - \frac{\lambda' + \nu'}{r^2} + \frac{\nu'^2}{2r} - \frac{2}{r^3}\right) + \frac{2}{r^3}$$

or
$$8\pi\left[\frac{dp_0}{dr} + (\rho_0 + p_0)\frac{\nu'}{2}\right] = \frac{2}{r}\left(e^{-\lambda}\left(\frac{\nu''}{2} - \frac{\nu'\lambda'}{4} - \frac{\lambda' + \nu'}{2r} + \frac{\nu'^2}{4} - \frac{1}{r^2}\right) + \frac{1}{r^2}\right) \tag{5.17}$$

For a perfect fluid, the radial stress T_1^1 and the transverse stresses T_2^2, T_3^3 are equal. Equating the second and third equations of the set of Eqs. (5.14), we obtain

$$e^{-\lambda}\left(\frac{\nu''}{2} + \frac{\nu'^2}{4} - \frac{\lambda'\nu'}{4} + \frac{\nu' - \lambda}{2r}\right) + \Lambda = e^{-\lambda}\left(\frac{\nu'}{r} + \frac{1}{r^2}\right) - \frac{1}{r^2} + \Lambda$$

or
$$e^{-\lambda}\left(\frac{\nu''}{2} - \frac{\lambda'\nu'}{4} + \frac{\nu'^2}{4} - \frac{\lambda' + \nu'}{2r} - \frac{1}{r^2}\right) + \frac{1}{r^2} = 0 \tag{5.18}$$

In view of Eq. (5.18), Eq. (5.17) becomes

$$8\pi\left[\frac{dp_0}{dr} + (\rho_0 + p_0)\frac{\nu'}{2}\right] = 0$$

or
$$\frac{dp_0}{dr} + (\rho_0 + p_0)\frac{\nu'}{2} = 0 \tag{5.19}$$

Finally, for determining the static line-element, due to a spherically symmetric distribution of a perfect fluid, we have the following relations:

$$\begin{aligned}
8\pi p_0 &= e^{-\lambda}\left(\frac{\nu'}{r} + \frac{1}{r^2}\right) - \frac{1}{r^2} + \Lambda \\
8\pi\rho_0 &= e^{-\lambda}\left(\frac{\lambda'}{r} - \frac{1}{r^2}\right) + \frac{1}{r^2} - \Lambda \\
8\pi p_0 &= e^{-\lambda}\left(\frac{1}{2}\nu'' + \frac{1}{4}\nu'^2 - \frac{1}{4}\nu'\lambda' + \frac{1}{2}\frac{\nu' - \lambda'}{r}\right) + \Lambda \\
\frac{dp_0}{dr} &= -\frac{1}{2}(\rho_0 + p_0)\nu'
\end{aligned} \tag{5.20}$$

We seek the static line-element arising from the static and spherically symmetric distribution of the perfect fluid in the following two cases:

1. Outside the system in the empty space, called Schwarzschild's exterior solution.
2. Inside the system, called Schwarzschild's interior solution.

The line-element, Eq. (5.1), is the same in both the cases, but the functions $\nu(r)$ and $\lambda(r)$ will be different in the two cases.

5.3 SCHWARZSCHILD'S EXTERIOR SOLUTION

All the components of the energy-momentum tensor T^{ν}_{μ}, in the empty space outside the static system, will be zero and the Eq. (5.14) become

$$e^{-\lambda}\left(\frac{\nu'}{r}+\frac{1}{r^2}\right)-\frac{1}{r^2}+\Lambda=0 \tag{5.21}$$

$$e^{-\lambda}\left(\frac{\nu''}{2}+\frac{\nu'^2}{4}-\frac{\nu'\lambda'}{4}+\frac{\nu'-\lambda'}{2r}\right)+\Lambda=0 \tag{5.22}$$

$$e^{-\lambda}\left(\frac{\lambda'}{r}-\frac{1}{r^2}\right)+\frac{1}{r^2}-\Lambda=0 \tag{5.23}$$

From Eq. (5.23), we get

$$e^{-\lambda}(r\lambda'-1)+1-\Lambda r^2=0$$

or

$$-\frac{\partial}{\partial r}(re^{-\lambda})+1-\Lambda r^2=0$$

Integrating w.r.t. r, we get

$$e^{-\lambda}=1-\frac{\Lambda}{3}r^2-\frac{K}{r} \tag{5.24}$$

where K is a constant of integration. Taking $K=2m$, as it enables us to interpret the mass of the gravitating particle as m, we get

$$e^{-\lambda}=e^{\nu}=1-\frac{2m}{r}-\frac{\Lambda}{3}r^2 \tag{5.25}$$

Writing Eq. (5.1) with the help of Eq. (5.25), the line-element for exterior solution is

$$ds^2=-\left(1-\frac{2m}{r}-\frac{\Lambda}{3}r^2\right)^{-1}dr^2-r^2(d\theta^2+\sin^2\theta\, d\phi^2)+\left(1-\frac{2m}{r}-\frac{\Lambda}{3}r^2\right)dt^2 \tag{5.26}$$

5.4 SCHWARZSCHILD'S INTERIOR SOLUTION

The line-element inside the sphere of matter will inevitably depend on the properties of the fluid constituting the sphere. Schwarzschild made the assumption that the sphere is made of an incompressible fluid having a constant density ρ_0. We have the following relations, Eq. (5.20)

$$8\pi p_0 = e^{-\lambda}\left(\frac{\nu'}{r}+\frac{1}{r^2}\right)-\frac{1}{r^2}+\Lambda \tag{5.27}$$

$$8\pi\rho_0 = e^{-\lambda}\left(\frac{\lambda'}{r}-\frac{1}{r^2}\right)+\frac{1}{r^2}-\Lambda \tag{5.28}$$

$$\frac{dp_0}{dr} = -\frac{\rho_0+p_0}{2}\nu' \tag{5.29}$$

The Eqs. (5.27)–(5.29) have to satisfy the following boundary conditions:

1. At the boundary of the sphere, the pressure is zero.
2. The density ρ_0 is constant and uniform throughout the sphere.

Since the density is constant, we can easily integrate the Eq. (5.28). Thus,

$$-e^{-\lambda}(r\lambda'-1) = 1-(8\pi\rho_0+\Lambda)r^2$$

or

$$\frac{d}{dr}(re^{-\lambda}) = 1-(8\pi\rho_0+\Lambda)r^2 \tag{5.30}$$

$$e^{-\lambda} = 1-\frac{(8\pi\rho_0+\Lambda)}{3}r^2+\frac{\alpha}{r} \tag{5.31}$$

α is a constant of integration. As the last term in Eq. (5.31) will lead to a singularity at $r=0$, we put $\alpha = 0$, and thereby, get

$$e^{-\lambda} = 1-\frac{(8\pi\rho_0+\Lambda)}{3}r^2$$

or

$$=1-\frac{r^2}{R_0^2} \tag{5.32}$$

where $R_0^2 = \frac{3}{8\pi\rho_0+\Lambda}$.

The solution for ν is yet to be found. Taking Eq. (5.29), we have

$$\frac{dp_0}{\rho_0+p_0} = -\frac{d\nu}{2}$$

Integrating it, we get

$$\log(\rho_0+p_0) = -\frac{\nu}{2}+\text{const}$$

or

$$(\rho_0+p_0) = \text{const}\, e^{-\nu/2} \tag{5.33}$$

Adding Eqs. (5.27) and (5.28), we get

$$8\pi(\rho_0 + p_0) = e^{-\lambda}\left(\frac{\nu' + \lambda'}{r}\right)$$

But

$$e^{\nu/2} 8\pi(\rho_0 + p_0) = \text{const}$$

so

$$e^{\frac{\nu}{2}}\left(\frac{\nu' + \lambda'}{r}\right)e^{-\lambda} = \text{const} \tag{5.34}$$

But

$$e^{-\lambda}\left(\frac{\nu' + \lambda'}{r}\right) = e^{-\lambda}\frac{\lambda'}{r} + e^{-\lambda}\frac{\nu'}{r}$$

$$= \frac{2}{R_0^2} + \frac{\nu'}{r}\left(1 - \frac{r^2}{R_0^2}\right) \tag{5.35}$$

for $-e^{-\lambda}\lambda' = -\frac{2r}{R_0^2}$. Thus, Eq. (5.34) becomes

$$e^{\frac{\nu}{2}}\left[\frac{2}{R_0^2} + \frac{\nu'}{r}\left(1 - \frac{r^2}{R_0^2}\right)\right] = C \tag{5.36}$$

Putting $e^{\nu/2} = u$, so $\frac{du}{dr} = \frac{\nu'}{2}e^{\nu/2}$, thus

$$\frac{du}{dr} + \frac{r}{R_0^2 - r^2}u = \frac{rC}{R_0^2 - r^2} \tag{5.37}$$

The solution of

$$\frac{du}{dr} + uP(r) = Q(r)$$

is

$$ue^{\int P\,dr} = \int Qe^{\int P\,dr}\,dr + \text{const}$$

The solution of Eq. (5.37) turns out to be

$$\frac{u}{\sqrt{R_0^2 - r^2}} = \frac{C_2}{\sqrt{R_0^2 - r^2}} + C_3$$

or

$$u = C_2 + C_3\sqrt{R_0^2 - r^2}$$

or

$$e^{\frac{\nu}{2}} = \alpha - \beta\sqrt{1 - \frac{r^2}{R_0^2}} \tag{5.38}$$

where $\frac{C_3}{R_0} = -\beta$, $C_2 = \alpha$.

Finally,
$$e^{-\lambda}=1-\frac{r^2}{R_0^2} \tag{5.39}$$

$$e^{\frac{\nu}{2}}=\alpha-\beta\sqrt{1-\frac{r^2}{R_0^2}} \tag{5.40}$$

$$\frac{1}{R_0^2}=\frac{8\pi\rho_0+\Lambda}{3} \tag{5.41}$$

The Schwarzschild's interior solution is

$$ds^2=\left[\alpha-\beta\left(1-\frac{r^2}{R_0^2}\right)^{1/2}\right]^2 dt^2-\left(1-\frac{r^2}{R_0^2}\right)^{-1} dr^2-r^2d\theta^2-r^2\sin^2\theta d\phi^2 \tag{5.42}$$

Let us evaluate the pressure p_0 at a point r inside the sphere. From Eq. (5.27), we have

$$\begin{aligned}8\pi p_0&=e^{-\lambda}\frac{\nu'}{r}+\frac{e^{-\lambda}}{r^2}-\frac{1}{r^2}+\Lambda\\&=e^{-\lambda}\frac{\nu'}{r}+\frac{1}{r^2}-\frac{1}{R_0^2}-\frac{1}{r^2}+\Lambda\end{aligned}$$

or
$$8\pi p_0=e^{-\lambda}\frac{\nu'}{r}-\frac{1}{R_0^2}+\Lambda \tag{5.43}$$

and
$$e^{\frac{\nu}{2}}\frac{\nu'}{2}=\frac{r}{R_0^2}\frac{\beta}{\sqrt{1-r^2/R_0^2}}$$

or
$$\frac{\nu'}{r}=\frac{2\beta/R_0^2}{\left[\alpha-\beta\sqrt{1-r^2/R_0^2}\right]\left[\sqrt{1-r^2/R_0^2}\right]}$$

or
$$e^{-\lambda}\frac{\nu'}{r}=\frac{2\beta(1-r^2/R_0^2)}{R_0^2\sqrt{1-r^2/R_0^2}}\frac{1}{\left[\alpha-\beta\sqrt{1-r^2/R_0^2}\right]}$$

or
$$e^{-\lambda}\frac{\nu'}{r}-\frac{1}{R_0^2}=\frac{3\beta\sqrt{(1-r^2/R_0^2)}-\alpha}{R_0^2\left[\alpha-\beta\sqrt{1-r^2/R_0^2}\right]}$$

Using Eq. (5.43), we get
$$8\pi p_0=\frac{3\beta\sqrt{\left(1-r^2/R_0^2\right)}-\alpha}{R_0^2\left[\alpha-\beta\sqrt{1-r^2/R_0^2}\right]}+\Lambda \tag{5.44}$$

However, the cosmological factor Λ becomes important only for distances greater than r_1 where r_1 is the radius of the massive body. Thus for $r \ll r_1$, we take $\Lambda=0$. The pressure $p_0=0$ for $r=r_1$. Thus we get from Eq. (5.44):

$$\alpha = 3\beta\sqrt{\left(1-r_1^2/R_0^2\right)} \tag{5.45}$$

The constants α and β can be determined by requiring that the exterior, Eq. (5.26) and interior, Eq. (5.42) solutions become identical at $r=r_1$, the boundary of the sphere. We may neglect the term Λ, such that

$$1-\frac{r_1^2}{R_0^2}=1-\frac{2m}{r_1}=\left[\alpha-\beta\sqrt{1-\frac{r_1^2}{R_0^2}}\right]^2$$

Substituting from Eq. (5.45), we get

$$1-\frac{r_1^2}{R_0^2}=1-\frac{2m}{r_1}=4\beta^2\left(1-\frac{r_1^2}{R_0^2}\right)$$

or $$1-\frac{r_1^2}{R_0^2}=1-\frac{2m}{r_1};\quad 1-\frac{r_1^2}{R_0^2}=4\beta^2\left(1-\frac{r_1^2}{R_0^2}\right)$$

which yields $$m=\frac{r_1^3}{2R_0^2};\quad 4\beta^2=1$$

or $$\frac{4\pi}{3}r_1^3\rho_0=\frac{r_1^3}{2R_0^2};\quad 4\beta^2=1$$

or $$\frac{8\pi\rho_0}{3}=\frac{1}{R_0^2};\quad \beta=\frac{1}{2}$$

This result agrees with the Eq. (5.32)

$$\frac{1}{R_0^2}=\frac{8\pi p_0+\Lambda}{3}$$

where $\Lambda=0$ for $r=r_1$.

Therefore, we get $$\alpha=\frac{3}{2}\sqrt{1-\frac{r_1^2}{R_0^2}}$$

$$\beta=\frac{1}{2} \tag{5.46}$$

$$\frac{1}{R_0^2}=\frac{8\pi\rho_0}{3}$$

The interior solution will be real only if

$$\frac{2m}{r_1}<1$$

or

$$\frac{r_1^2}{R_0^2}<1$$

or

$$r_1^2 < \frac{3}{8\pi\rho_0} \tag{5.47}$$

The Eq. (5.47) provides an upper limit on the possible size of a sphere of a given density and on the mass of a sphere of a given radius. Although these limits are quite liberal but these have not led to any controversy with the astrophysical data.

5.5 CONSERVATION LAWS IN CURVED SPACE

In general relativity, the stress-energy or energy-momentum tensor $T^{\mu\nu}$ acts as a source of gravitational field. It is related to the curvature of space–time through the Einstein's field equations, without the cosmological constant

$$R_{\mu\nu}-\frac{1}{2}g_{\mu\nu}R=-8\pi T_{\mu\nu} \tag{5.48}$$

The covariant derivative of $T_{\mu\nu}$ is zero, that is

$$T_{\mu\nu;\nu}=0 \tag{5.49}$$

or when expressed in mixed form, we get

$$T^{\nu}_{\mu;\nu} \equiv \frac{\partial T^{\nu}_{\mu}}{\partial x^{\nu}}+\Gamma^{\nu}_{\rho\nu}T^{\rho}_{\mu}-\Gamma^{\rho}_{\mu\nu}T^{\nu}_{\rho}=0 \tag{5.50}$$

The ordinary four-divergence of T^{ν}_{μ} from Eq. (5.50), is

$$\frac{\partial T^{\nu}_{\mu}}{\partial x^{\nu}}=-\Gamma^{\nu}_{\rho\nu}T^{\rho}_{\mu}+\Gamma^{\rho}_{\mu\nu}T^{\nu}_{\rho} \tag{5.51}$$

As it depends on Christoffel's 3-index symbols that are non-zero for a non-Euclidean space, it is obvious that

$$\frac{\partial T^{\nu}_{\mu}}{\partial x^{\nu}}\neq 0 \tag{5.52}$$

Thus any of the above Eqs. (5.50), (5.51), or (5.52) implies that in general relativity, the energy-momentum tensor T^{ν}_{μ} is not conserved. This being the case, it is nonetheless, possible to express the l.h.s. of Eq. (5.50) in the form of four-divergence of certain quantities which could refer to quantities that are conserved. How this is achieved, is described here.

Consider the following form of $T^{\nu}_{\mu;\nu}$

$$\sqrt{-g}\left(T^{\nu}_{\mu;\nu}\right)=\sqrt{-g}\left(T^{\nu}_{\mu,\nu}+\Gamma^{\nu}_{\rho\nu}T^{\rho}_{\mu}-\Gamma^{\rho}_{\mu\nu}T^{\nu}_{\rho}\right) \tag{5.53}$$

where $T^{\nu}_{\mu,\nu}=\dfrac{\partial T^{\nu}_{\mu}}{\partial x^{\nu}}$.

Now
$$\Gamma^{\rho}_{\rho\nu} = \left\{ {\rho \atop \rho\nu} \right\} = \frac{1}{2} g^{\rho\sigma}\left(\frac{\partial g_{\rho\sigma}}{\partial x^{\nu}} + \frac{\partial g_{\nu\sigma}}{\partial x^{\rho}} - \frac{\partial g_{\rho\nu}}{\partial x^{\sigma}} \right)$$
$$= \frac{1}{2} g^{\rho\sigma} \frac{\partial g_{\rho\sigma}}{\partial x^{\nu}}$$
$$= \frac{1}{2}\frac{1}{g}\frac{\partial g}{\partial x^{\nu}}$$
$$= \frac{1}{\sqrt{-g}}\frac{\partial\sqrt{-g}}{\partial x^{\nu}} \tag{5.54}$$

Therefore,
$$\sqrt{-g}\,\Gamma^{\rho}_{\rho\nu} = \frac{\partial\sqrt{-g}}{\partial x^{\nu}} \tag{5.55}$$

Rewriting Eq. (5.53) in the light of Eq. (5.55), one gets

$$\sqrt{-g}\,(T^{\nu}_{\mu;\nu}) \equiv 0 = (T^{\nu}_{\mu}\sqrt{-g})_{,\nu} - \sqrt{-g}\,\Gamma^{\rho}_{\mu\nu}T^{\nu}_{\rho} \tag{5.56}$$

As is obvious, the first term on the r.h.s. of Eq. (5.56) has the form of the four-divergence and if we can express the other term $\sqrt{-g}\,\Gamma^{\rho}_{\mu\nu}T^{\nu}_{\rho}$ in the four-divergence form, the problem is solved. This is how it is accomplished.

We eliminate the tensor T^{ν}_{ρ} from the second term on the r.h.s. of Eq. (5.56) with the help of Einstein's field equations

$$-kT^{\nu}_{\mu} = R^{\nu}_{\mu} - \frac{1}{2}g^{\nu}_{\mu}R$$
$$\equiv G^{\nu}_{\mu}$$

or
$$T^{\nu}_{\mu} = -\frac{1}{k}G^{\nu}_{\mu} \tag{5.57}$$

where G^{ν}_{μ} is the Einstein tensor. Thus, the second term in Eq. (5.56), becomes

$$\sqrt{-g}\,\Gamma^{\rho}_{\mu\nu}T^{\nu}_{\rho} = -\frac{1}{k}\sqrt{-g}\,\Gamma^{\rho}_{\mu\nu}G^{\nu}_{\rho}$$
$$= -\frac{1}{k}\sqrt{-g}\,g^{\rho\sigma}\Gamma_{\sigma,\mu\nu}G^{\nu}_{\rho}$$
$$= -\frac{1}{k}\sqrt{-g}\,\Gamma_{\sigma,\mu\nu}G^{\nu\sigma}$$
$$= -\frac{1}{2k}\sqrt{-g}\,(\Gamma_{\sigma,\mu\nu} + \Gamma_{\nu,\mu\sigma})G^{\nu\sigma}$$
$$= -\frac{1}{2k}\sqrt{-g}\,\frac{\partial g_{\nu\sigma}}{\partial x^{\mu}}G^{\nu\sigma} \tag{5.58}$$

The Euler–Lagrange equations from the variational principle

$$\delta\int R\sqrt{-g}\,d^4x=0 \tag{5.59}$$

will not alter if we add a four-divergence of a four vector to the integrand.

Now
$$\begin{aligned} R\sqrt{-g} &= g^{\mu\nu}R_{\mu\nu}\sqrt{-g} \\ &= g^{\mu\nu}\sqrt{-g}\left(\Gamma^{\sigma}_{\mu\sigma,\nu}-\Gamma^{\sigma}_{\mu\nu,\sigma}-\Gamma^{\rho}_{\rho\sigma}\Gamma^{\sigma}_{\mu\nu}+\Gamma^{\rho}_{\mu\sigma}\Gamma^{\sigma}_{\nu\rho}\right) \end{aligned} \tag{5.60}$$

We rearrange the first two terms on the r.h.s. of Eq. (5.60) as

$$\begin{aligned} \left(\Gamma^{\sigma}_{\mu\sigma,\nu}-\Gamma^{\sigma}_{\mu\nu,\sigma}\right)g^{\mu\nu}\sqrt{-g} &= \left(\sqrt{-g}g^{\mu\nu}\Gamma^{\sigma}_{\mu\sigma}\right)_{,\nu}-\left(\sqrt{-g}g^{\mu\nu}\Gamma^{\sigma}_{\mu\nu}\right)_{,\sigma} \\ &\quad -\left(\sqrt{-g}g^{\mu\nu}\right)_{,\nu}\Gamma^{\sigma}_{\mu\sigma}+\left(\sqrt{-g}g^{\mu\nu}\right)_{,\sigma}\Gamma^{\sigma}_{\mu\nu} \end{aligned} \tag{5.61}$$

The first two terms on the r.h.s. are the four-divergences and as stated earlier, their addition to the integrand will not make any difference so far as the Euler–Lagrange equations are concerned. Therefore, Eq. (5.61), may, in effect, be written as

$$\begin{aligned} \left(\Gamma^{\sigma}_{\mu\sigma,\nu}-\Gamma^{\sigma}_{\mu\nu,\sigma}\right)g^{\mu\nu}\sqrt{-g} &= -\left(\sqrt{-g}g^{\mu\nu}\right)_{,\nu}\Gamma^{\sigma}_{\mu\sigma}+\left(\sqrt{-g}g^{\mu\nu}\right)_{,\sigma}\Gamma^{\sigma}_{\mu\nu} \\ &= -\left(\frac{\partial\sqrt{-g}}{\partial x^{\nu}}g^{\mu\nu}+\sqrt{-g}\frac{\partial g^{\mu\nu}}{\partial x^{\nu}}\right)\Gamma^{\sigma}_{\mu\sigma}+\left(\frac{\partial\sqrt{-g}}{\partial x^{\sigma}}g^{\mu\nu}+\sqrt{-g}\frac{\partial g^{\mu\nu}}{\partial x^{\sigma}}\right)\Gamma^{\sigma}_{\mu\nu} \\ &= \sqrt{-g}g^{\mu\nu}\left(-\Gamma^{\rho}_{\rho\nu}\Gamma^{\sigma}_{\mu\sigma}+\Gamma^{\rho}_{\rho\sigma}\Gamma^{\sigma}_{\mu\nu}\right)-\sqrt{-g}\left(\frac{\partial g^{\mu\nu}}{\partial x^{\nu}}\Gamma^{\sigma}_{\mu\sigma}-\frac{\partial g^{\mu\nu}}{\partial x^{\sigma}}\Gamma^{\sigma}_{\mu\nu}\right) \end{aligned} \tag{5.62}$$

In writing the last step, use was made of Eq. (5.55) and then the terms were rearranged.

Further we simplify the terms in the second parenthesis of Eq. (5.62), using the fact that the covariant derivative of $g^{\mu\nu}$ is zero, *i.e.*,

$$g^{\mu\nu}{}_{;\sigma}=\frac{\partial g^{\mu\nu}}{\partial x^{\sigma}}+\Gamma^{\mu}_{\sigma\rho}g^{\rho\nu}+\Gamma^{\nu}_{\sigma\rho}g^{\rho\mu}=0$$

or

$$\frac{\partial g^{\mu\nu}}{\partial x^{\sigma}}=-\Gamma^{\mu}_{\sigma\rho}g^{\rho\nu}-\Gamma^{\nu}_{\sigma\rho}g^{\rho\mu} \tag{5.63}$$

Writing Eq. (5.62) in view of Eq. (5.63). We get

$$\begin{aligned} \left(\Gamma^{\sigma}_{\mu\sigma,\nu}-\Gamma^{\sigma}_{\mu\nu,\sigma}\right)\sqrt{-g}g^{\mu\nu} &= \sqrt{-g}g^{\mu\nu}\left[-\Gamma^{\rho}_{\rho\nu}\Gamma^{\sigma}_{\mu\sigma}+\Gamma^{\rho}_{\rho\sigma}\Gamma^{\sigma}_{\mu\nu}\right]-\sqrt{-g}\left[\left(-\Gamma^{\mu}_{\nu\rho}\Gamma^{\sigma}_{\mu\sigma}g^{\rho\nu}-\Gamma^{\nu}_{\nu\rho}\Gamma^{\sigma}_{\mu\sigma}g^{\rho\mu}\right)\right. \\ &\quad \left.+\left(\Gamma^{\mu}_{\sigma\rho}\Gamma^{\sigma}_{\mu\nu}g^{\rho\nu}+\Gamma^{\nu}_{\sigma\rho}\Gamma^{\sigma}_{\mu\nu}g^{\rho\mu}\right)\right] \end{aligned} \tag{5.64}$$

Substituting Eq. (5.64) in the r.h.s. of Eq. (5.60), one obtains

$$R\sqrt{-g} = \sqrt{-g}g^{\mu\nu}\Big[\underset{I}{-\Gamma^{\rho}_{\rho\nu}\Gamma^{\sigma}_{\mu\sigma}} + \underset{II}{\Gamma^{\rho}_{\rho\sigma}\Gamma^{\sigma}_{\mu\nu}}\Big]$$
$$-\sqrt{-g}\Big[\underset{I}{\left(-\Gamma^{\mu}_{\nu\rho}\Gamma^{\sigma}_{\mu\sigma}g^{\rho\nu} - \Gamma^{\nu}_{\nu\rho}\Gamma^{\sigma}_{\mu\sigma}g^{\rho\mu}\right)}$$
$$+\underset{III}{\left(\Gamma^{\mu}_{\sigma\rho}\Gamma^{\sigma}_{\mu\nu}g^{\rho\nu} + \Gamma^{\nu}_{\sigma\rho}\Gamma^{\sigma}_{\mu\nu}g^{\rho\mu}\right)}\Big]$$
$$+\sqrt{-g}g^{\mu\nu}\Big[\underset{II}{-\Gamma^{\rho}_{\rho\sigma}\Gamma^{\sigma}_{\mu\nu}} + \underset{III}{\Gamma^{\rho}_{\mu\sigma}\Gamma^{\sigma}_{\nu\rho}}\Big] \tag{5.65}$$

The like terms labelled as I, II, III cancel and the remaining terms after re-arranging the dummy indices are

$$\begin{aligned} R\sqrt{-g} &= -\sqrt{-g}\left[-\Gamma^{\mu}_{\nu\rho}\Gamma^{\sigma}_{\mu\sigma}g^{\rho\nu} + \Gamma^{\mu}_{\sigma\rho}\Gamma^{\sigma}_{\mu\nu}g^{\rho\nu}\right] \\ &= -\sqrt{-g}g^{\rho\nu}\left[-\Gamma^{\mu}_{\nu\rho}\Gamma^{\sigma}_{\mu\sigma} + \Gamma^{\mu}_{\sigma\rho}\Gamma^{\sigma}_{\mu\nu}\right] \\ &\equiv L', \text{say} \end{aligned} \tag{5.66}$$

where L' is a new Lagrangian.

Writing the variational principle in terms of L', we get

$$\begin{aligned} \delta I &= \delta\int g^{\mu\nu}R_{\mu\nu}\sqrt{-g}d^4x \\ &= \delta\int \sqrt{-g}\,Rd^4x \\ &= \delta\int L'd^4x = 0 \end{aligned} \tag{5.67}$$

It may be remarked that L' is not invariant with respect to the coordinate transformations but the equations resulting from the integral $\int L'd^4x$ are covariant equations.

L' is a function of $g^{\mu\nu}$ and its derivatives $\partial g^{\mu\nu}/\partial x^{\sigma} = g^{\mu\nu}{}_{,\sigma}$. The Euler–Lagrange equations from Eq. (5.67) are

$$\frac{\partial L'}{\partial g^{\mu\nu}} - \frac{\partial}{\partial x^{\rho}}\left(\frac{\partial L'}{\partial\left(\frac{\partial g^{\mu\nu}}{\partial x^{\rho}}\right)}\right) = 0 \tag{5.68}$$

and these are equivalent to Einstein's field equations resulting from the original function given in Eqs. (5.59) and (5.60). Therefore, we get

$$\sqrt{-g}G_{\mu\nu} = \frac{\partial L'}{\partial g^{\mu\nu}} - \frac{\partial}{\partial x^{\rho}}\left(\frac{\partial L'}{\partial g^{\mu\nu}{}_{,\rho}}\right) \tag{5.69}$$

Multiplying by $g^{\mu\nu}{}_{,\alpha}$ and summing over $\mu\nu$, we get

$$\sqrt{-g}G_{\mu\nu}(g^{\mu\nu}{}_{,\alpha})=\frac{\partial L'}{\partial g^{\mu\nu}}(g^{\mu\nu}{}_{,\alpha})-\partial_{\rho}\left(\frac{\partial L'}{g^{\mu\nu}{}_{,\rho}}\right)(g^{\mu\nu}{}_{,\alpha})$$

$$=\frac{\partial L'}{\partial g^{\mu\nu}}(g^{\mu\nu}{}_{,\alpha})-\partial_{\rho}\left[\frac{\partial L'}{g^{\mu\nu}{}_{,\rho}}g^{\mu\nu}{}_{,\alpha}\right]+\frac{\partial L'}{g^{\mu\nu}{}_{,\rho}}(\partial_{\rho}\partial_{\alpha}g^{\mu\nu})$$

$$=\frac{\partial L'}{\partial g^{\mu\nu}}\partial_{\alpha}g^{\mu\nu}+\frac{\partial L'}{g^{\mu\nu}{}_{,\rho}}\partial_{\alpha}(\partial_{\rho}g^{\mu\nu})-\partial_{\rho}\left[\frac{\partial L'}{g^{\mu\nu}{}_{,\rho}}g^{\mu\nu}{}_{,\alpha}\right]$$

$$=\frac{\partial L'}{\partial x^{\alpha}}-\left(\frac{\partial L'}{g^{\mu\nu}{}_{,\rho}}g^{\mu\nu}{}_{,\alpha}\right)_{,\rho} \tag{5.70}$$

In the last step, we wrote $\frac{\partial L'}{\partial x^{\alpha}}$ for the first two terms in the line above.

Next writing

$$\frac{\partial L'}{\partial x^{\alpha}}=\delta^{\rho}_{\alpha}\frac{\partial L'}{\partial x^{\rho}}=g^{\rho}_{\alpha}\frac{\partial L'}{\partial x^{\rho}} \tag{5.71}$$

we rewrite Eq. (5.70) as

$$\sqrt{-g}G_{\mu\nu}(g^{\mu\nu}{}_{,\alpha})=\left[g^{\rho}_{\alpha}L'-\frac{\partial L'}{\partial g^{\mu\nu}{}_{,\rho}}g^{\mu\nu}{}_{\alpha}\right]_{,\rho} \tag{5.72}$$

Introducing a set of new quantities, t^{ρ}_{α}

$$\sqrt{-g}\,t^{\rho}_{\alpha}\equiv-g^{\rho}_{\alpha}L'+\frac{\partial L'}{\partial g^{\mu\nu}{}_{,\rho}}(g^{\mu\nu}{}_{,\alpha}) \tag{5.73}$$

we write Eq. (5.72) as

$$\sqrt{-g}G_{\mu\nu}(g^{\mu\nu}{}_{,\alpha})=-\left[\sqrt{-g}\,t^{\rho}_{\alpha,\rho}\right] \tag{5.74}$$

This result will finally accomplish the job we started with. However, this has to be cast in a slightly different form.

Noting that

$$g^{\mu\rho}g_{\mu\sigma}=\delta^{\rho}_{\sigma} \tag{5.75}$$

Therefore,

$$(g^{\mu\nu}g_{\mu\sigma})_{,\alpha}=0=g^{\mu\nu}{}_{,\alpha}g_{\mu\sigma}+g^{\mu\nu}g_{\mu\sigma,\alpha} \tag{5.76}$$

Writing $G_{\mu\nu}(g^{\mu\nu}{}_{,\alpha})$ differently, we have

$$\begin{aligned}G_{\mu\nu}(g^{\mu\nu}{}_{,\alpha}) &= g_{\mu\rho}g_{\nu\sigma}G^{\rho\sigma}(g^{\mu\nu}{}_{,\alpha})\\ &= (g^{\mu\nu}{}_{,\alpha})g_{\mu\rho}g_{\nu\sigma}G^{\rho\sigma}\\ &= -g^{\mu\nu}(g_{\mu\rho,\alpha})g_{\nu\sigma}G^{\rho\sigma}\\ &= -g^{\mu\nu}g_{\nu\sigma}G^{\rho\sigma}(g_{\mu\rho,\alpha})\\ &= -\delta^{\mu}_{\sigma}G^{\rho\sigma}(g_{\mu\rho,\alpha})\\ &= -G^{\mu\rho}(g_{\mu\rho,\alpha})\\ &= -G^{\mu\nu}(g_{\mu\nu,\alpha})\end{aligned} \tag{5.77}$$

Returning to Eq. (5.56)

$$\left(\sqrt{-g}T^{\nu}_{\mu}\right)_{,\nu} - \sqrt{-g}\Gamma^{\rho}_{\mu\nu}T^{\nu}_{\rho} = 0 \tag{5.78}$$

and writing it in view of Eq. (5.58), we get

$$\left(\sqrt{-g}T^{\nu}_{\mu}\right)_{,\nu} + \frac{1}{2k}\sqrt{-g}\,\frac{\partial g_{\nu\sigma}}{\partial x^{\mu}}G^{\nu\sigma} = 0 \tag{5.79}$$

Further substitution of Eq. (5.77) yields

$$\left(\sqrt{-g}T^{\nu}_{\mu}\right)_{,\nu} - \frac{1}{2k}\sqrt{-g}\,\frac{\partial g^{\nu\sigma}}{\partial x^{\mu}}G_{\nu\sigma} = 0 \tag{5.80}$$

Lastly, substituting Eq. (5.74), we get

$$\left(\sqrt{-g}T^{\nu}_{\mu}\right)_{,\nu} + \frac{1}{2k}\left(\sqrt{-g}t^{\nu}_{\mu}\right)_{,\nu} = 0 \tag{5.81}$$

Putting $k = 8\pi$, Eq. (5.81) becomes

$$\sqrt{-g}\left[T^{\nu}_{\mu} + \frac{1}{16\pi}t^{\nu}_{\mu}\right]_{,\nu} = 0 \tag{5.82}$$

Defining

$$P^{\nu}_{\mu} = \sqrt{-g}\left[T^{\nu}_{\mu} + \frac{1}{16\pi}t^{\nu}_{\mu}\right] \tag{5.83}$$

Equation (5.82) gives four differential conservation laws ($\mu = 0, 1, 2, 3$)

$$P^{\nu}_{\mu,\nu} = 0 \tag{5.84}$$

The four quantities $\int P^{0}_{\mu}d^{3}x$ are constants of motion.It may be remarked that the quantities P^{ν}_{μ} do not form the components of a tensor, but as these satisfy the four conservation laws, these are called the components of the stress-energy pseudo tensor of general relativity.

Further, the quantities t^{ν}_{μ}, Eq. (5.73) contain the first derivatives of the metric tensor, $g_{\mu\nu}$ and therefore are not tensors. For that very reason, P^{ν}_{μ} is not a tensor. However, it is possible to transform to a coordinate system where the first derivatives of $g_{\mu\nu}$ are zero locally, the components of t^{ν}_{μ} vanish and P^{ν}_{μ} reduces to $\sqrt{-g}T^{\nu}_{\mu}$. t^{ν}_{μ} is called the pseudo-energy tensor of the gravitational field. It is not unique since it depends on the choice of the coordinate system.

Problems

5.1 Find the metric of the Schwarzschild's interior solution in isotropic coordinates

$$ds^2 = e^{\nu(r(\rho))}c^2dt^2 - e^{2F(\rho)}[d\rho^2 + \rho^2(d\theta^2 + \sin^2\theta d\varphi^2)]$$

by performing a coordinate transformation with

$$r = \frac{2a\rho}{(a^2 + A\rho^2)}$$

and an appropriate a.

5.2 Show that the Schwarzschild's interior solution is coformally flat.

5.3 In Maxwell's equations, the scalar potential ϕ and the three-vector potential A_i are defined as components of a one form, $A_0 = \phi$, A_i(one form) = A_i(three-vector). A gauge transformation is the replacement $\phi \to \phi - \frac{\partial \chi}{\partial t}$, $\mathbf{A} \to \mathbf{A} + \nabla\chi$. This leaves the electric and magnetic fields unchanged. The Lorentz gauge is a gauge in which $\frac{1}{c}\frac{\partial \phi}{\partial t} + \nabla_i A^i = 0$. Write both the gauge transformation and the Lorentz gauge condition in four-tensor form. Draw the analogy with similar equations in linearized gravity.

6

Gravitational Waves

6.1 INTRODUCTION

General relativity and electrodynamics have profound similarities and yet fundamental differences[1,2]. In this connection, it may be interesting to mention some historical analogies between the two fields.

The tremendous success of Maxwell's equations did not rest only on their incorporating all the laws of electricity and magnetism in addition to Lorentz force equation, but also on their basis in 1873 James Clerk Maxwell was able to predict the existence of electromagnetic waves consisting of the periodic variation of electric and magnetic fields and carrying energy and propagating with speed c in vacuum. It was in 1887 that Heinrich Hertz both generated and detected the electromagnetic waves. Then in 1901, Guglieomo Marconi transmitted and received electromagnetic signals across the Atlantic Ocean.

In the twentieth century, a new era of great changes in our knowledge of the universe, other than the visible light, was opened by the detection and study of the electromagnetic waves. This led to the discovery of cosmic radio waves in 1930 and subsequently to the knowledge about colliding galaxies, quasars as large as our solar system but having luminosities order of magnitude larger than our galaxy, enormous jets from galactic nuclei and quasars of lengths of hundreds of thousands of light years, rotating pulsars and cosmic microwave background radiation, the relic of the big bang. The accretion disks about black holes and neutron stars were revealed by the studies of x-rays. In addition, the microwave, infrared, ultraviolet, x-rays and gamma-rays added considerably to our knowledge about the universe.

Identically in 1916, Albert Einstein predicted the existence of space–time curvature generated by mass-energy propagating with speed c on a flat and empty space–time, the gravitational waves[3].

It was in 1941 that Landau and Lifshitz[4] showed the emission of gravitational waves by a self-gravitating system of slowly moving bodies. Likewise to the case of electromagnetic waves, serious doubts were there about the reality of the existence of gravitational waves till 1957 when Hermann Bondi on the basis of a Gedanken experiment showed that gravitational waves do carry energy[5].

In an analogous manner as electromagnetic waves other than visible light that is, radio, millimetre, infrared, ultraviolet, x-rays and gamma-rays astronomy opened new vistas and brought radical changes in our knowledge of universe.

Gravitational waves presently is a hot topic that promises to play a significant role in astrophysics, cosmology and theoretical physics. Their direct observation may become a reality with the technological developments in enhancing the sensitivity of instruments for their detection. When detected directly, it will give birth to an entirely new field: gravitational wave astronomy. It will revolutionize our knowledge of universe by making-possible the observation of hitherto unseen phenomena, such as coalescence of compact objects (neutron stars and black holes), fall of stars into supermassive black holes, stellar core collapses, big bang relics and the unexpected.

It is to seek an approximate solution of Einstein's field equations in the presence of matter giving rise to energy–momentum tensor

$$R_{\mu\nu} - \frac{1}{2} g_{\mu\nu} R + \Lambda g_{\mu\nu} = -8\pi T_{\mu\nu} \tag{6.1}$$

The $g_{\mu\nu}$ tensor has ten components, which are connected with the different components of $T_{\mu\nu}$. The principle of superposition does not hold as the field equations, Eq. (6.1), are non-linear. The non-linearity arises due to the fact that the gravitational field itself serves as a part of its source, since the gravitational field possesses energy and this energy being equivalent to mass, in turn, produces gravitational field. The field produced by two bodies, is not the sum of the individual fields. One does not notice these effects in our solar system, because it is a region of weak gravitational field and that is the reason why the linear Newtonian theory serves as a good approximation. Mathematically speaking, these effects are there and this is another way that Einstein's theory is different from Newton's.

For a strong field, no general solution of the field equations can be found; but for sufficiently weak fields, as we will show, the field equations become linear and so admit a standard solution by Einstein's approximate method of successive approximations. We will be taking the case of weak gravitational field in detail next. Solving the linearized field equations, we will learn that gravitational effects propagate as waves travelling at the speed of light.

6.2 WEAK GRAVITATIONAL FIELD, LINEARIZED FIELD EQUATIONS

Rewriting the field equations Eq. (6.1), with the cosmological constant equal to zero, we obtain

$$-8\pi T_{\mu\nu} = R_{\mu\nu} - \frac{1}{2} g_{\mu\nu} R \tag{6.2}$$

The line-element is given by

$$ds^2 = g_{\mu\nu} dx^\mu dx^\nu \tag{6.3}$$

We make the assumption that the gravitational field is very weak and so the coordinates deviate very little from the Galilean values. Thus,

$$g_{\mu\nu} = \delta_{\mu\nu} + h_{\mu\nu} \tag{6.4}$$

where $\delta_{\mu\nu}$ are the constant Galilean values (± 1 and 0) and $h_{\mu\nu}$ are small perturbations. Furthermore, the terms $h_{\mu\nu}$ and their derivatives with respect to the coordinates are quantities of the first order and we neglect their squares and higher powers. Contracting Eq. (6.2), we get

$$R - \frac{1}{2}(4)R = -8\pi T \tag{6.5}$$

where $R = R^\mu_\mu$ and $T = T_{\mu\nu} g^{\mu\nu} = T^\mu_\mu$.

Therefore, $$R = 8\pi T \tag{6.6}$$

We get from Eq. (6.2), in view of Eq. (6.6)

$$R_{\mu\nu} = -8\pi\left(T_{\mu\nu} - \frac{1}{2}g_{\mu\nu}T\right) = -8\pi T^*_{\mu\nu} \tag{6.7}$$

where $T^*_{\mu\nu} = T_{\mu\nu} - \frac{1}{2}g_{\mu\nu}T$.

The Riemann-Christoffel tensor Eq. (2.182) is

$$R_{\lambda\mu\nu\sigma} = \frac{1}{2}\left(\frac{\partial^2 g_{\lambda\sigma}}{\partial x^\nu \partial x^\mu} + \frac{\partial^2 g_{\mu\nu}}{\partial x^\sigma \partial x^\lambda} - \frac{\partial^2 g_{\mu\sigma}}{\partial x^\nu \partial x^\lambda} - \frac{\partial^2 g_{\lambda\nu}}{\partial x^\sigma \partial x^\mu}\right) + g_{\rho\alpha}(\Gamma^\rho_{\mu\nu}\Gamma^\alpha_{\lambda\sigma} - \Gamma^\alpha_{\mu\sigma}\Gamma^\alpha_{\lambda\nu}) \tag{6.8}$$

The contracted Riemann-Christoffel tensor $R_{\mu\nu}$ is obtained as

$$R_{\mu\nu} = g^{\lambda\sigma}R_{\lambda\mu\nu\sigma} = \frac{1}{2}g^{\lambda\sigma}\left(\frac{\partial^2 g_{\lambda\sigma}}{\partial x^\nu \partial x^\mu} + \frac{\partial^2 g_{\mu\nu}}{\partial x^\sigma \partial x^\lambda} - \frac{\partial^2 g_{\mu\sigma}}{\partial x^\nu \partial x^\lambda} - \frac{\partial^2 g_{\lambda\nu}}{\partial x^\sigma \partial x^\mu}\right) + g^{\lambda\sigma}g_{\rho\alpha}(\Gamma^\rho_{\mu\nu}\Gamma^\alpha_{\lambda\sigma} - \Gamma^\rho_{\mu\sigma}\Gamma^\alpha_{\lambda\nu}) \tag{6.9}$$

Neglecting the second-order terms, which occur in terms like $\Gamma^\rho_{\mu\nu}\Gamma^\alpha_{\lambda\sigma}$ etc, in the second bracket, we get

$$R_{\mu\nu} = \frac{1}{2}g^{\lambda\sigma}\left(\frac{\partial^2 g_{\lambda\sigma}}{\partial x^\nu \partial x^\mu} + \frac{\partial^2 g_{\mu\nu}}{\partial x^\sigma \partial x^\lambda} - \frac{\partial^2 g_{\mu\sigma}}{\partial x^\nu \partial x^\lambda} - \frac{\partial^2 g_{\lambda\nu}}{\partial x^\sigma \partial x^\mu}\right) \tag{6.10}$$

Substituting for $g_{\mu\nu}$, Eq. (6.4), and retaining terms of first order, we get

$$R_{\mu\nu} = \frac{1}{2}\delta^{\lambda\sigma}\left(\frac{\partial^2 h_{\lambda\sigma}}{\partial x^\nu \partial x^\mu} + \frac{\partial^2 h_{\mu\nu}}{\partial x^\sigma \partial x^\lambda} - \frac{\partial^2 h_{\mu\sigma}}{\partial x^\nu \partial x^\lambda} - \frac{\partial^2 h_{\lambda\nu}}{\partial x^\sigma \partial x^\mu}\right) \tag{6.11}$$

Putting $h^\nu_\mu = \delta^{\nu\alpha}h_{\mu\alpha}, h = h^\mu_\mu = \delta^{\mu\nu}h_{\mu\nu}$, in Eq. (6.11), after slight rearranging, we get

$$\begin{aligned}
R_{\mu\nu} &= \frac{1}{2}\left(\frac{\partial^2 h_{\mu\nu}}{\partial x^\sigma \partial x^\sigma} + \frac{\partial^2 h}{\partial x^\nu \partial x^\mu} - \frac{\partial^2 h^\lambda_\mu}{\partial x^\nu \partial x^\lambda} - \frac{\partial^2 h^\sigma_\nu}{\partial x^\sigma \partial x^\mu}\right) \\
&= \frac{1}{2}\left(\frac{\partial^2 h_{\mu\nu}}{\partial x^\sigma \partial x^\sigma} + \frac{1}{2}\left(\frac{\partial^2}{\partial x^\nu \partial x^\mu} + \frac{\partial^2}{\partial x^\nu \partial x^\mu}\right)h - \frac{\partial^2 h^\lambda_\mu}{\partial x^\nu \partial x^\lambda} - \frac{\partial^2 h^\sigma_\nu}{\partial x^\sigma \partial x^\mu}\right) \\
&= \frac{1}{2}\left(\frac{\partial^2 h_{\mu\nu}}{\partial x^\sigma \partial x^\sigma} + \frac{1}{2}\frac{\partial^2}{\partial x^\nu \partial x^\lambda}(\delta^\lambda_\mu h) - \frac{\partial^2 h^\lambda_\mu}{\partial x^\nu \partial x^\lambda} + \frac{1}{2}\frac{\partial^2}{\partial x^\mu \partial x^\sigma}(\delta^\sigma_\nu h) - \frac{\partial^2 h^\sigma_\nu}{\partial x^\sigma \partial x^\mu}\right) \\
&= \frac{1}{2}\left(\frac{\partial^2 h_{\mu\nu}}{\partial x^\sigma \partial x^\sigma} - \frac{\partial^2}{\partial x^\nu \partial x^\lambda}\left(h^\lambda_\mu - \frac{1}{2}\delta^\lambda_\mu h\right) - \frac{\partial^2}{\partial x^\sigma \partial x^\mu}\left(h^\sigma_\nu - \frac{1}{2}\delta^\sigma_\nu h\right)\right)
\end{aligned} \tag{6.12}$$

Defining

$$\psi_{\mu}^{\nu} = h_{\mu}^{\nu} - \frac{1}{2}\delta_{\mu}^{\nu} h \tag{6.13}$$

and expressing Eq. (6.12) in terms of Ψ_{μ}^{ν}, we obtain

$$R_{\mu\nu} = \Box^2 \frac{1}{2} h_{\mu\nu} - \frac{1}{2}\frac{\partial}{\partial x^{\nu}}\left(\frac{\partial}{\partial x^{\lambda}}\Psi_{\mu}^{\lambda}\right) - \frac{1}{2}\frac{\partial}{\partial x^{\mu}}\left(\frac{\partial}{\partial x^{\sigma}}\Psi_{\nu}^{\sigma}\right) \tag{6.14}$$

where $\Box^2 \equiv \frac{\partial^2}{\partial t^2} - \nabla^2$

The Eq. (6.7) becomes in the light of Eq. (6.14)

$$\Box^2 h_{\mu\nu} - \frac{\partial}{\partial x^{\nu}}\left(\frac{\partial}{\partial x^{\lambda}}\Psi_{\mu}^{\lambda}\right) - \frac{\partial}{\partial x^{\mu}}\left(\frac{\partial}{\partial x^{\sigma}}\Psi_{\nu}^{\sigma}\right) = -16\pi T_{\mu\nu}^{*} \tag{6.15}$$

The Eq. (6.15) can be further simplified by the following argument. The transformation Eq. (6.4) is valid in any arbitrary coordinate system. Our initial choice was based on the requirement that $g_{\mu\nu}$ differ from the Galilean values $\delta_{\mu\nu}$ by infinitely small quantities. We effect a further infinitesimal change of the coordinates so that there are still four conditions on $h_{\mu\nu}$, provided these conditions and the requirement that $h_{\mu\nu}$ are infinitesimally small, do not contradict. We make the choice of the following four conditions on the coordinates

$$\frac{\partial}{\partial x^{\nu}}\Psi_{\mu}^{\nu} = 0 = \partial_{\nu}\left(h_{\mu}^{\nu} - \frac{1}{2}\delta_{\mu}^{\nu} h\right) \tag{6.16}$$

Therefore Eq. (6.15) becomes

$$\Box^2 h_{\mu\nu} = -16\pi T_{\mu\nu}^{*} \tag{6.17}$$

According to the method of retarded potentials, its solution is

$$h_{\mu\nu}(x,y,z,t) = -4\int \frac{T_{\mu\nu}^{*}(x',y',z',t-r)}{r} dx'dy'dz' \tag{6.18}$$

where $r = \sqrt{x^2 + y^2 + z^2}$.

The Eqs. (6.16) and (6.17) are the linearized equations of general relativity for the case of a weak gravitational field. One can show that the solution Eq. (6.18), is consistent with the condition Eq. (6.16). These linearized equations show that the gravitational fields propagate with unit velocity (in gravitational units), *i.e.*, with *c*, the velocity of light in vacuum. When the energy–momentum tensor $T_{\mu\nu}$ is zero, the Eqs.(6.16) and (6.17), become

$$\frac{\partial}{\partial x^{\nu}}\psi_{\mu}^{\nu} = 0$$

$$\Box^2 h_{\mu\nu} = 0$$

It is obviously clear that $h_{\mu\nu}$ satisfy the wave equation and the gravitational waves travel with the velocity of light.

The Eqs. (6.16) and (6.17) can be simplified further as shown here. The condition, Eq. (6.16), does not fix the coordinates uniquely, *i.e.*, such class of coordinate transformations are allowable so that the new quantities still Ψ'^{ν}_{μ} and $h'_{\mu\nu}$ satisfy the Eqs. (6.16) and (6.17). In order to show this, let us consider an infinitesimal transformation

$$x'^{\mu} = x^{\mu} + \varepsilon^{\mu}(x) \tag{6.19}$$

where $\varepsilon^{\mu}(x)$ is an infinitesimal four-vector.

The new metric tensor

$$g'_{\mu\nu}(x') = \frac{\partial x^{\alpha}}{\partial x'^{\mu}} \frac{\partial x^{\beta}}{\partial x'^{\nu}} g_{\alpha\beta}(x)$$

$$= g_{\alpha\beta}\left(\delta^{\alpha}_{\mu} - \frac{\partial \varepsilon^{\alpha}}{\partial x^{\mu}}\right)\left(\partial^{\beta}_{\nu} - \frac{\partial \varepsilon^{\beta}}{\partial x^{\nu}}\right)$$

or

$$g'_{\mu\nu} = g_{\mu\nu} - g_{\mu\beta}\frac{\partial \varepsilon^{\beta}}{\partial x^{\nu}} - g_{\alpha\nu}\frac{\partial \varepsilon^{\alpha}}{\partial x^{\mu}} \tag{6.20}$$

where the arguments of $g'_{\mu\nu}$ and $g_{\mu\nu}$ and the higher order terms in ε^{μ} have been neglected. We also have from Eqs (6.4)

$$g'_{\mu\nu}(x') = \delta_{\mu\nu} + h'_{\mu\nu}(x') \tag{6.21}$$

$$g_{\mu\nu}(x) = \delta_{\mu\nu} + h_{\mu\nu}(x) \tag{6.22}$$

Thus, from Eqs. (6.21), (6.22) and (6.20)

$$h'_{\mu\nu}(x') = g'_{\mu\nu} - \delta_{\mu\nu}$$

$$= g_{\mu\nu} - g_{\mu\beta}\frac{\partial \varepsilon^{\beta}}{\partial x^{\nu}} - g_{\alpha\nu}\frac{\partial \varepsilon^{\alpha}}{\partial x^{\mu}} - \delta_{\mu\nu}$$

$$= h_{\mu\nu} - \frac{\partial \varepsilon_{\mu}}{\partial x^{\nu}} - \frac{\partial \varepsilon_{\nu}}{\partial x^{\mu}} \tag{6.23}$$

Rewriting Eq. (6.23), we have

$$h_{\mu\nu} = h'_{\mu\nu} + \partial_{\nu}\varepsilon_{\mu} + \partial_{\mu}\varepsilon_{\nu} \tag{6.24}$$

or raising the index ν, we get

$$h^{\nu}_{\mu} = h'^{\nu}_{\mu} + \partial^{\nu}\varepsilon_{\mu} + \partial_{\mu}\varepsilon^{\nu}$$

Now

$$h = h^{\mu}_{\mu} = h'^{\mu}_{\mu} + \partial^{\mu}\varepsilon_{\mu} + \partial_{\mu}\varepsilon^{\mu}$$

$$= h' + 2\partial^{\mu}\varepsilon_{\mu} \tag{6.25}$$

From Eq. (6.13), we have

$$\begin{aligned}\Psi^{\nu}_{\mu} &= h^{\nu}_{\mu} - \frac{1}{2}\delta^{\nu}_{\mu} h \\ &= h'^{\nu}_{\mu} + \partial^{\nu}\varepsilon_{\mu} + \partial_{\mu}\varepsilon^{\nu} - \frac{1}{2}\delta^{\nu}_{\mu}(h' + 2\partial_{\alpha}\varepsilon^{\alpha}) \\ &= h'^{\nu}_{\mu} - \frac{1}{2}\delta^{\nu}_{\mu}h' + \partial^{\nu}\varepsilon_{\mu} + \partial_{\mu}\varepsilon^{\nu} - (\partial_{\alpha}\varepsilon^{\alpha})\delta^{\nu}_{\mu}\end{aligned} \tag{6.26}$$

The condition, Eq. (6.16) requires that

$$\frac{\partial \psi^{\nu}_{\mu}}{\partial x^{\nu}} = 0$$

or

$$\partial_{\nu}\left(h'^{\nu}_{\mu} - \frac{1}{2}\delta^{\nu}_{\mu}h'\right) + \partial_{\nu}(\partial^{\nu}\varepsilon_{\mu} + \partial_{\mu}\varepsilon^{\nu} - \delta^{\nu}_{\mu}(\partial_{\alpha}\varepsilon^{\alpha})) = 0$$

or

$$\partial_{\nu}\left(h'^{\nu}_{\mu} - \frac{1}{2}\delta^{\nu}_{\mu}h'\right) + \square^{2}\varepsilon_{\mu} = 0 \tag{6.27}$$

since $\partial_{\nu}\partial_{\mu}\varepsilon^{\nu} = \partial_{\mu}(\partial_{\alpha}\varepsilon^{\alpha})$.

Thus, the coordinate transformation, Eq. (6.19), is acceptable provided our choice of satisfies the equation

$$\square^{2}\varepsilon_{\mu}(x) = 0 \tag{6.28}$$

When the energy–momentum tensor is zero, the Eqs. (6.16) and (6.17) become

$$\partial_{\nu}\Psi^{\nu}_{\mu} = 0 \tag{6.29}$$

$$\square^{2}h_{\mu\nu} = 0 \tag{6.30}$$

With the condition Eq. (6.28), Eq. (6.24) becomes

$$\begin{aligned}\square^{2}h_{\mu\nu} &= \square^{2}h'_{\mu\nu} + \partial_{\nu}\square^{2}\varepsilon_{\mu} + \partial_{\mu}\square^{2}\varepsilon_{\nu} = 0 \\ &= \square^{2}h'_{\mu\nu} = 0\end{aligned} \tag{6.31}$$

Finally, in the new coordinate system specified by Eqs. (6.19) and (6.28), the following equations hold

$$\partial_{\nu}\Psi'^{\nu}_{\mu} = 0 \tag{6.32}$$

$$\square^{2}h'_{\mu\nu} = 0 \tag{6.33}$$

and
$$\Psi'^{\nu}_{\mu} = h'^{\nu}_{\mu} - \frac{1}{2}\delta^{\nu}_{\mu} h' \tag{6.34}$$

Furthermore, we show that this coordinate transformation, will help us reduce the number of components of $h_{\mu\nu}$ from ten to two.

6.3 PLANE WAVE SOLUTIONS

Let us assume that a plane wave is propagating along x^1 direction and is given by

$$h^{\nu}_{\mu}(x) = \alpha^{\nu}_{\mu}(x) f(x^0 - x^1) \tag{6.35}$$

where $\alpha^{\nu}_{\mu}(x) = \alpha^{\nu}_{\mu}(x^0, x^1, x^2, x^3)$ and f is a function of $(x^0 - x^1)$.

In order to show which components of Ψ^{ν}_{μ} are zero, we proceed with the Eq. (6.32)

$$\partial_{\nu}\left(h^{\nu}_{\mu} - \frac{1}{2}\delta^{\nu}_{\mu} h\right) = 0$$

or
$$\frac{\partial h^{\nu}_{\mu}}{\partial x^{\nu}} - \frac{1}{2}\delta^{\nu}_{\mu}\frac{\partial h}{\partial x^{\nu}} = 0 \tag{6.36}$$

Writing Eq. (6.36) explicitly for $\mu = 0, 1, 2, 3$, we get

$$\begin{aligned}
\frac{\partial h^0_0}{\partial x^0} + \frac{\partial h^1_0}{\partial x^1} - \frac{1}{2}\frac{\partial h}{\partial x^0} &= 0 \\
\frac{\partial h^0_1}{\partial x^0} + \frac{\partial h^1_1}{\partial x^1} - \frac{1}{2}\frac{\partial h}{\partial x^1} &= 0 \\
\frac{\partial h^0_2}{\partial x^0} + \frac{\partial h^1_2}{\partial x^1} &= 0 \\
\frac{\partial h^0_3}{\partial x^0} + \frac{\partial h^1_3}{\partial x^1} &= 0
\end{aligned} \tag{6.37}$$

Making use of the relation

$$h^{\nu}_{\mu} = h^{\nu}_{\mu}(x^0 - x^1) = h^{\nu}_{\mu}(X) \tag{6.38}$$

where $X = (x^0 - x^1)$, we get

$$\frac{\partial h^{\nu}_{\mu}}{\partial x^0} = \frac{\partial h^{\nu}_{\mu}}{\partial X}\frac{\partial X}{\partial x^0} = \frac{\partial h^{\nu}_{\mu}}{\partial X} = \dot{h}^{\nu}_{\mu} \tag{6.39}$$

$$\frac{\partial h^{\nu}_{\mu}}{\partial x^1} = \frac{\partial h^{\nu}_{\mu}}{\partial X}\frac{\partial X}{\partial x^1} = \frac{\partial h^{\nu}_{\mu}}{\partial X} = -\dot{h}^{\nu}_{\mu} \tag{6.40}$$

where a dot stands for the differentiation with respect to $X = (x^0 - x^1)$. Writing the set of Eqs. (6.37), in view of Eqs. (6.39) and (6.40), we get

$$\begin{aligned} \dot{h}_0^0 - \dot{h}_0^1 - \frac{1}{2}\dot{h} &= 0 \\ \dot{h}_1^0 - \dot{h}_1^1 + \frac{1}{2}\dot{h} &= 0 \\ \dot{h}_2^0 - \dot{h}_2^0 &= 0 \\ \dot{h}_3^0 - \dot{h}_3^0 &= 0 \end{aligned} \tag{6.41}$$

Integrating these equations, we get

$$\begin{aligned} h_0^0 - h_0^1 - \frac{1}{2}h &= 0 \\ h_1^0 - h_1^1 + \frac{1}{2}h &= 0 \\ h_2^0 &= h_2^1 \\ h_3^0 &= h_3^1 \end{aligned} \tag{6.42}$$

From the first two equations, we get

$$h_0^0 = h_0^1 + \frac{1}{2}h$$

$$h_1^1 = h_1^0 + \frac{1}{2}h = -h_0^1 + \frac{1}{2}h$$

Adding these equations, one obtains

$$\begin{aligned} h_0^0 + h_1^1 &= h(= h_\mu^\mu) \\ &= h_0^0 + h_1^1 + h_2^2 + h_3^3 \end{aligned} \tag{6.43}$$

which gives

$$h_2^2 = -h_3^3 \tag{6.44}$$

Summarizing the results from Eqs. (6.42), (6.43) and (6.44), we get

$$\begin{aligned} h &= h_0^0 + h_1^1 \\ h_2^2 &= - h_3^3 \\ h_2^0 &= h_2^1 \\ h_3^0 &= h_3^1 \end{aligned} \tag{6.45}$$

On lowering the index *i.e.*, these become

$$\begin{aligned} h &= h_{00} - h_{11} \\ h_{22} &= - h_{33} \\ h_{20} &= - h_{21} \\ h_{30} &= - h_{31} \end{aligned} \tag{6.46}$$

From the first equation of the set (6.42), we have

$$\begin{aligned} h_0^1 &= h_0^0 - \frac{1}{2}h \\ &= h_0^0 - \frac{1}{2}(h_0^0 + h_1^1) \\ &= \frac{1}{2}(h_0^0 - h_1^1) \end{aligned} \tag{6.47}$$

or

$$-h_{01} = \frac{1}{2}(h_{00} + h_{11})$$

$$h_{01} = -\frac{1}{2}(h_{00} + h_{11}) = h_{10} \tag{6.48}$$

Writing $h_{\mu\nu}$ in the matrix form, we have

$$h_{\mu\nu} = \begin{pmatrix} h_{00} & h_{01} & h_{02} & h_{03} \\ h_{10} & h_{11} & h_{12} & h_{13} \\ h_{20} & h_{21} & h_{22} & h_{23} \\ h_{30} & h_{31} & h_{32} & h_{33} \end{pmatrix}$$

$$= \begin{pmatrix} h_{00} & -\frac{1}{2}(h_{00} + h_{11}) & -h_{12} & -h_{13} \\ -\frac{1}{2}(h_{00} + h_{11}) & h_{11} & h_{12} & h_{13} \\ -h_{12} & h_{12} & h_{22} & h_{23} \\ -h_{13} & h_{13} & h_{32} & -h_{22} \end{pmatrix} \tag{6.49}$$

The Eq. (6.49) represents the most general solution of the field equations in an arbitrary coordinate system. The solution can be further reduced so that all the terms of the matrix except the last block

$$\begin{pmatrix} h_{22} & h_{23} \\ h_{32} & -h_{22} \end{pmatrix}$$

are zero. This is achieved by transforming to a new coordinate system as described before and given by Eqs. (6.4) and (6.28).

We transform x^μ to x'^μ as

$$x'^\mu = x^\mu + \varepsilon^\mu$$

with the condition

$$\Box^2 \varepsilon^\mu = 0 \tag{6.50}$$

where $$\varepsilon^\mu = \varepsilon^\mu (x^0 - x^1)$$

According to these coordinate transformations, $h_{\mu\nu}(x) \to h'_{\mu\nu}(x')$ given by Eq. (6.23). We are to choose the four functions ε^μ in such a way that leads to the following new values of $h'_{\mu\nu}(x')$

$$\begin{aligned} h'_{00} &= 0 \\ h'_{11} &= 0 \\ h'_{12} &= 0 \\ h'_{13} &= 0 \end{aligned} \tag{6.51}$$

Writing Eq. (6.23), we have

$$h'_{\mu\nu}(x') = h_{\mu\nu}(x) - \frac{\partial \varepsilon_\mu}{\partial x^\nu} - \frac{\partial \varepsilon_\nu}{\partial x^\mu}$$

so, we get

$$\begin{aligned} h'_{00} &= h_{00} - \left(\frac{\partial \varepsilon_0}{\partial x^0} + \frac{\partial \varepsilon_0}{\partial x^0} \right) = h_{00} - 2\frac{\partial \varepsilon_0}{\partial x^0} \\ h'_{11} &= h_{11} - \left(\frac{\partial \varepsilon_1}{\partial x^1} + \frac{\partial \varepsilon_1}{\partial x^1} \right) = h_{11} - 2\frac{\partial \varepsilon_1}{\partial x^1} \\ h'_{12} &= h_{12} - \left(\frac{\partial \varepsilon_1}{\partial x^2} + \frac{\partial \varepsilon_2}{\partial x^1} \right) = h_{12} - \frac{\partial \varepsilon_2}{\partial x^1} \\ h'_{13} &= h_{13} - \left(\frac{\partial \varepsilon_1}{\partial x^3} + \frac{\partial \varepsilon_3}{\partial x^1} \right) = h_{13} - \frac{\partial \varepsilon_3}{\partial x^1} \end{aligned} \tag{6.52}$$

Further the four functions are chosen in a manner so that

$$\begin{aligned} \varepsilon_0 &= \frac{1}{2} A(x^0 - x^1) \\ \varepsilon_1 &= \frac{1}{2} B(x^0 - x^1) \\ \varepsilon_2 &= C(x^0 - x^1) \\ \varepsilon_3 &= D(x^0 - x^1) \end{aligned} \tag{6.53}$$

$A, B, C,$ and D are arbitrary functions of $(x^0 - x^1)$.

Evaluating the following functions, with the help of Eq. (6.53), we have

$$\begin{aligned}
\frac{\partial \varepsilon_0}{\partial x^0} &= \frac{1}{2}\frac{\partial A}{\partial x^0} = \frac{1}{2}\frac{\partial A}{\partial X}\frac{\partial X}{\partial x^0} = \frac{1}{2}\frac{\partial A}{\partial X} = \frac{1}{2}\dot{A} \\
\frac{\partial \varepsilon_1}{\partial x^1} &= \frac{1}{2}\frac{\partial B}{\partial x^1} = \frac{1}{2}\frac{\partial B}{\partial X}\frac{\partial X}{\partial x^1} = -\frac{1}{2}\frac{\partial B}{\partial X} = -\frac{1}{2}\dot{B} \\
\frac{\partial \varepsilon_2}{\partial x^1} &= \frac{\partial C}{\partial x^1} = \frac{\partial C}{\partial X}\frac{\partial X}{\partial x^1} = -\frac{\partial C}{\partial X} = -\dot{C} \\
\frac{\partial \varepsilon_3}{\partial x^1} &= \frac{\partial D}{\partial x^1} = \frac{\partial C}{\partial X}\frac{\partial X}{\partial x^1} = -\frac{\partial D}{\partial X} = -\dot{D}
\end{aligned} \tag{6.54}$$

Thus, if we choose the arbitrary functions $A, B, C,$ and D so that

$$\begin{aligned}
\dot{A} &= h_{00}(X) \\
\dot{B} &= h_{11}(X) \\
\dot{C} &= -h_{12}(X) \\
\dot{D} &= -h_{13}(X)
\end{aligned} \tag{6.55}$$

then the new tensor $h'_{\mu\nu}(x')$ becomes

$$h'_{\mu\nu}(x') = \begin{pmatrix} 0 & 0 & 0 & 0 \\ 0 & 0 & 0 & 0 \\ 0 & 0 & h'_{22} & h'_{23} \\ 0 & 0 & h'_{23} & -h'_{22} \end{pmatrix} \tag{6.56}$$

We can drop the primes and conclude that there exists a coordinate frame in which $h_{\mu\nu}$ has the form

$$h_{\mu\nu} = \begin{pmatrix} 0 & 0 & 0 & 0 \\ 0 & 0 & 0 & 0 \\ 0 & 0 & h_{22} & h_{23} \\ 0 & 0 & h_{23} & -h_{22} \end{pmatrix} \tag{6.57}$$

Thus, only two components of the tensor $h_{\mu\nu} = h_{\nu\mu}$ are independent.

Furthermore, the trace of $h_{\mu\nu}$ is

$$h = \sum h_{\mu\mu} = h_{00} + h_{11} + h_{22} + h_{33} = 0 \tag{6.58}$$

so $h_{\mu\nu}$ is a traceless tensor.

From Eq. (6.13), we have

$$\begin{aligned}
\psi^{\nu}_{\mu} &= h^{\nu}_{\mu} - \frac{1}{2}\delta^{\nu}_{\mu} h \\
&= h^{\nu}_{\mu}
\end{aligned} \tag{6.59}$$

Since, $h_{\mu\nu}$ is traceless, so is $\psi_{\mu\nu}$.

Finally, we have

$$\psi_{\mu\mu} = 0$$
$$\psi_{\mu\nu} = \psi_{\nu\mu} \tag{6.60}$$
$$\partial_\nu \psi_{\mu\nu} = 0$$

and
$$\Box^2 \psi_{\mu\nu} = 0 \tag{6.61}$$

Let us interpret this result with a view to extract the information regarding $\psi_{\mu\nu}$. The fact that the tensor $\psi_{\mu\nu}$ is symmetric, implies that it has ten independent components. Further, since $\psi_{\mu\nu}$ is traceless, the number of independent components reduces to nine. The condition $\partial_\nu \psi_{\mu\nu} = 0$, for each value of $\mu = 0,1,2,3$ puts four more conditions on $\psi_{\mu\nu}$, with the consequence that the number of independent components is reduced to five. This means that the spin of the gravitational field is two. Since $\psi_{\mu\nu}$ satisfies Klein-Gordon equation with mass term set equal to zero, Eq. (6.61), it is possible to make a further coordinate transformation to reduce the independent components to two, as has been demonstrated before. It may be remarked that for a field of tensorial rank S, for massless particles like photons, gravitons, the number of independent components is two instead of $(2S + 1)$.

The Eqs. (6.60) and (6.61) together are called transverse-traceless (T T) gauge conditions.

6.4 BEHAVIOUR OF A MASSIVE PARTICLE AS A GRAVITATIONAL WAVE PASSES

In order to analyze the behavior of a massive particle under the influence of a passing gravitational wave, we take up the case of two polarizations that are given by $h_{yy} \neq 0$, $h_{yz} = 0$, and $h_{yy} = 0$, $h_{yz} \neq 0$. Considering first the case corresponding to $h_{yy} \neq 0$, $h_{yz} = 0$, we use the expression for $h_{\mu\nu}$

$$h_{\mu\nu} = \begin{pmatrix} 0 & 0 & 0 & 0 \\ 0 & 0 & 0 & 0 \\ 0 & 0 & h_{yy} & h_{yz} \\ 0 & 0 & h_{yz} & -h_{yy} \end{pmatrix} \tag{6.62}$$

along with $g_{\mu\nu} = \eta_{\mu\nu} + \varepsilon h_{\mu\nu}$ where $\eta_{\mu\nu}$ is

$$\eta_{\mu\nu} = \begin{pmatrix} 1 & 0 & 0 & 0 \\ 0 & -1 & 0 & 0 \\ 0 & 0 & -1 & 0 \\ 0 & 0 & 0 & -1 \end{pmatrix} \tag{6.63}$$

and write out the line element

$$ds^2 = dt^2 - dx^2 - (1 - \varepsilon h_{yy})dy^2 - (1 + \varepsilon h_{yy})dz^2 \tag{6.64}$$

According to the metric, as the gravitational wave passes, the distances between the particles will change. The wave has oscillatory behavior and so we see the form of Eq. (6.64) depending upon the value of h_{yy}, as h_{yy} changes from $h_{yy} > 0$ to zero and then $h_{yy} < 0$.

For the sake of simplicity, we consider the particles lying in the *y*-*z* plane. In addition, imagine that the separation between the particles lies on a line that is parallel to the *z*-axis. Thus, *dy* vanishes and at a fixed time, we have

$$ds^2 = -(1+\varepsilon h_{yy})dz^2 \tag{6.65}$$

When $h_{yy)} > 0$, the distance between the particles on the *z*-axis, decreases since ds^2 becomes more negative, Fig. 6.1.

However, when $h_{yy} < 0$, the distance between the particles will increase, Fig. 6.2. Next we consider the separation of the particles along the *y*-axis. The proper distance is given as

$$ds^2 = -(1-\varepsilon h_{yy})dy^2 \tag{6.66}$$

When $h_{yy} > 0$, the line element becomes more positive, with the result that the relative separation between the particles will increase, Fig. 6.3.

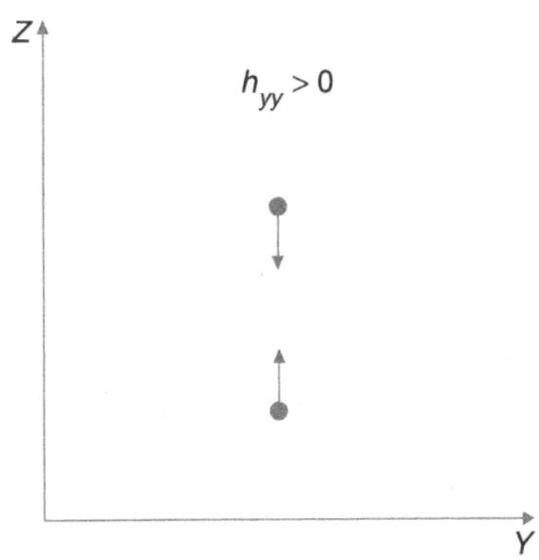

Figure 6.1 *When $h_{yy} > 0$, the relative distance between the two particles separated along the z-axis, will decrease.*

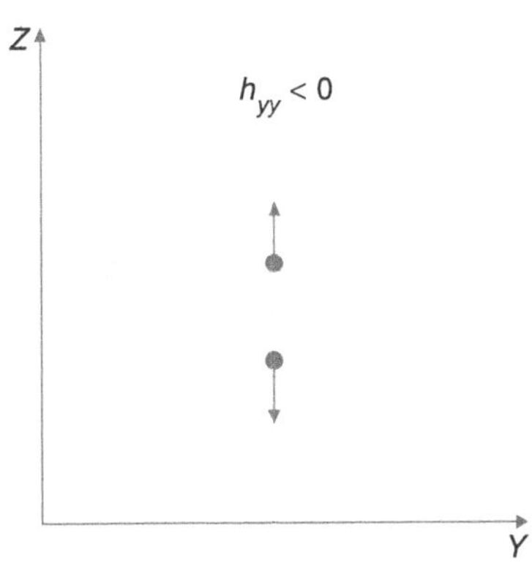

Figure 6.2 *With $h_{yy} < 0$, the relative distance between the particles on the z-axis will increase.*

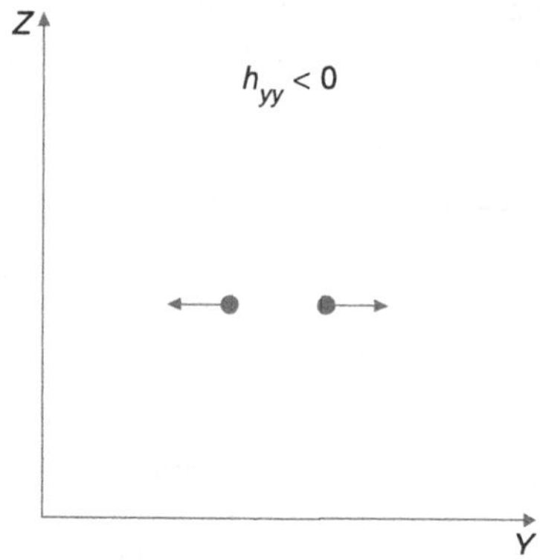

Figure 6.3 *With $h_{yy} > 0$, the relative distance between the particles increases.*

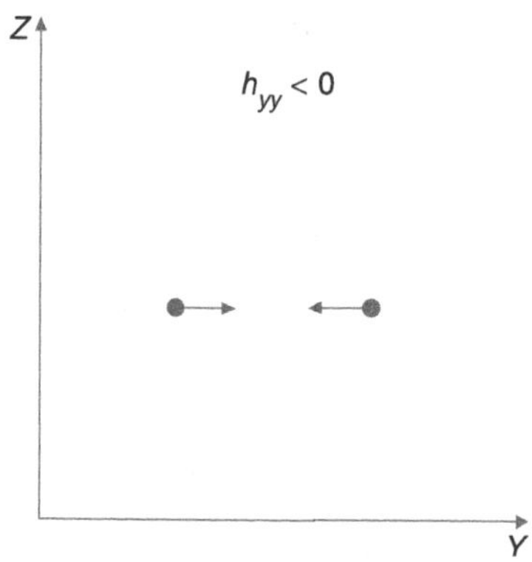

Figure 6.4 *When $h_{yy} < 0$, the relative distance between the particles along the y-axis, will decrease.*

However, when $h_{yy} < 0$, the relative distance between the particles will decrease, Fig. 6.4.We will now extrapolate to a more general situation. Consider a ring of particles lying in the yz-plane and examine how the ring is distorted by a passing gravitational wave. In particular, consider that the ring to begin with, is a perfect circle. As the wave passes, h_{yy} will oscillate between positive, zero and negative values, the relative distances between the particles will change as discussed earlier, Fig. 6.5. A transverse wave with $h_{yy} \neq 0$ and $h_{yz} = 0$ is described as a wave with +polarization.

Next we take up the case of other polarization with which is described as –polarization. The line element in this case becomes

$$ds^2 = dt^2 - dx^2 - dy^2 - dz^2 - 2\varepsilon h_{yz} dydz \tag{12.67}$$

Transforming the y-, z-axes through a rotation of $\pi/4$ around the origin

$$dy' = \frac{dy - dz}{\sqrt{2}} \tag{12.68}$$

$$dz' = \frac{dy + dz}{\sqrt{2}} \tag{6.69}$$

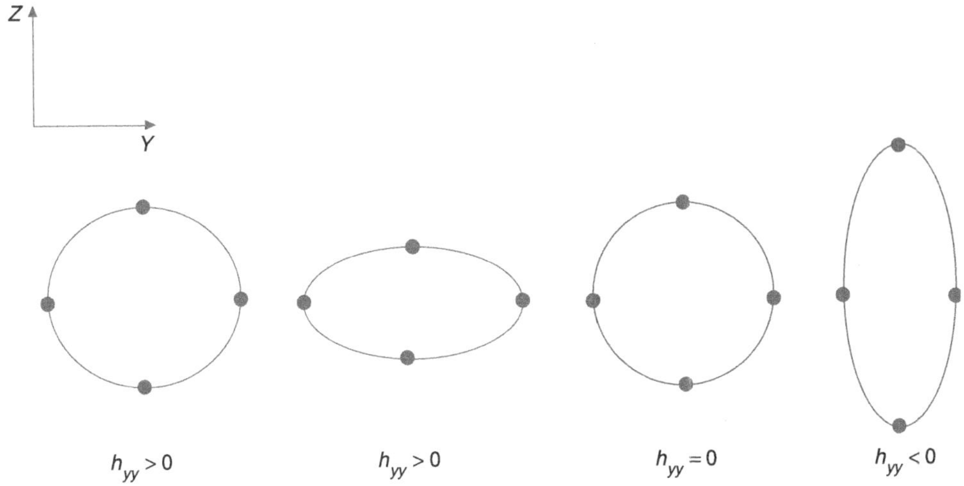

Figure 6.5 *The ring of particles as a gravitational wave, with +polarization passes.*

and expressing the line element, Eq. (6.67) in the coordinates x, y', z', we obtain

$$ds^2 = dt^2 - dx^2 - (1 - \varepsilon h_{yz})dy'^2 - (1 + \varepsilon h_{yz})dz'^2 \tag{6.70}$$

It is the same line element, Eq. (6.64), thus the behaviour induced by the wave is analogous to the behaviour discussed before, with the only difference that it is rotated through $\pi/4$.

6.5 THE DETECTION OF GRAVITATIONAL WAVES

The great strides that astronomy has made since 1960 is due to the fact that improved technology enabled the astronomers to observe, many different parts, other than the visible light, of the electromagnetic spectrum. These new diverse and exciting phenomena are giant radio galaxies, quasars, pulsars, compact x-ray binaries, molecular line masers in dense clouds and cosmic microwave background radiation. Still there are regions of the electromagnetic spectrum, that are greatly unexplored. However, there is another spectrum, which is completely untouched *i.e.*, gravitational wave spectra. All astrophysical phenomena emit gravitational waves and the violent ones like collision of black holes or stellar collapse, give off copious amount of gravitational waves.

In some cases, gravitational waves give the information that electromagnetic waves cannot. For instance, gravitational waves come directly from the heart of supernova explosions, the electromagnetic radiation from the same region will undergo scattering countless times and will lose most of the information, it might carry otherwise, whereas long wavelength gravitational waves can tell us about the formation of black holes. Furthermore, we may observe phenomena that we never expected. It is with these objectives that considerable effort is being directed towards the development of increasingly sensitive gravitational wave antennas.

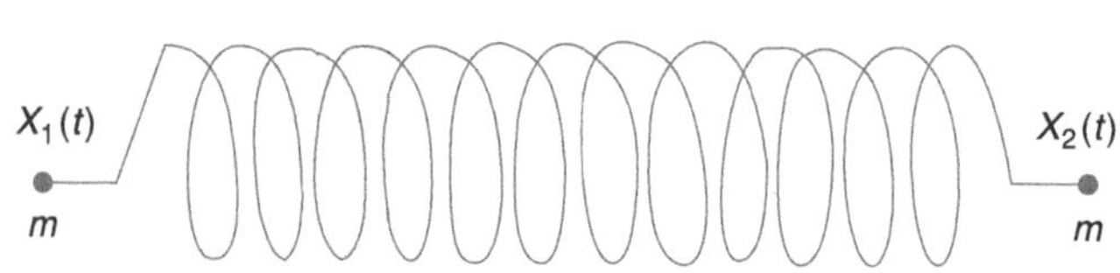

Figure 6.6 *A spring with two identical masses as a detector of gravitational waves.*

The technical difficulties encountered in the detection of gravitational waves are tremendous, since the amplitude of the metric perturbations $h_{\mu\nu}$, that can be expected from distant sources are exceedingly small. We discuss the case of resonant oscillator, pioneered by J. Weber (1961) for getting a feeling for what is involved in the detection of gravitational waves.

A Resonant Detector: An idealized detector, Fig. 6.6, consists of two point particles, each of mass m, damping constant ν and unstretched length l_0. The oscillator lies on the x-axis of T T coordinate system, the masses at positions x_1 and x_2.

In flat space–time, the masses are governed by the following equations

$$mx_{1,00} = -k(x_1 - x_2 + l_0) - \nu(x_1 - x_2)_{,0} \tag{6.71}$$

$$mx_{2,00} = -k(x_2 - x_1 - l_0) - \nu(x_2 - x_1)_{,0} \tag{6.72}$$

Defining

$$\xi = x_2 - x_1 - l_0$$

$$\omega_0^2 = 2k/m \tag{6.73}$$
$$\gamma = \nu/m$$

and taking the difference of Eq. (6.71) from Eq. (6.72), we get

$$m(x_2 - x_1)_{,00} = -2k(x_2 - x_1 - l_0) - 2\nu(x_2 - x_1)_{,0}$$

or

$$(x_2 - x_1)_{,00} = -\frac{2k}{m}(x_2 - x_1 - l_0) - \frac{2\nu}{m}(x_2 - x_1)_{,0}$$

$$\Rightarrow \xi_{,00} + 2\gamma\xi_{,0} + \omega_0^2\xi = 0 \tag{6.74}$$

which is a damped harmonic oscillator equation.

Further, we examine the effect of a passing gravitational wave. The analysis is split into three steps.

First Step: A free particle at rest in the T T coordinates: This implies that a local inertial frame at say, x_1, before the arrival of the wave, remains at rest after the impact of the wave. Let its coordinates be $\{x^{\alpha'}\}$. We assume that the only motions in the system are those caused by the wave, *i.e.*, $\xi = 0(l \mid h_{\mu\nu}) \ll l_0$. As the displacements are small, the resulting velocities will be small too and the motion can be treated by Newtonian dynamics in the local inertial frame.

$$mx^{j'}_{,0'0'} = F^{j'} \tag{6.75}$$

where $\{F^{j'}\}$ are the components of gravitational forces. The coordinates $\{x^{\alpha'}\}$ can differ from $\{x^{\alpha}\}$ by terms of the order of $h_{\mu\nu}$ and since $x_1, x_{1,0}$ and $x_{1,00}$ are of the order of $h_{\mu\nu}$, we are in a position to use T T coordinates in Eq. (6.75) with negligible error. Thus,

$$mx^j_{,00} = F^j + 0(|h_{\mu\nu}|^2) \tag{6.76}$$

Second Step: Any other non-gravitational force on each mass is due to the spring. Since all the displacements are small, the spring will exert a force according to Hook's law, *i.e.*, proportional to its instantaneous proper extension. If l is the proper length of the spring and the gravitational wave travels in the z-direction, then

$$l(t) = \int_{x_1(t)}^{x_2(t)} [1 + h^{TT}_{xx}(t)]^{1/2} dt \tag{6.77}$$

The Eq. (6.75) becomes

$$mx_{1,00} = -k(l_0 - l) - v(l_0 - l)_{,0} \tag{6.78}$$

$$mx_{2,00} = -k(l - l_0) - v(l - l_0)_{,0} \tag{6.79}$$

Third Step: Defining ω_0 and γ as in Eq. (6.73) and

$$\xi = l - l_0 \tag{6.80}$$

$$= x_2 - x_1 - l_0 + \frac{1}{2} h^{TT}_{xx}(x_2 - x_1) + 0(|h_{\mu\nu}|^2) \tag{6.81}$$

$$\Rightarrow x_2 - x_1 = l_0 + \xi - \frac{1}{2} h^{TT}_{xx}(x_2 - x_1) + 0(|h_{\mu\nu}|^2) \tag{6.82}$$

Taking the difference of Eqs. (6.79) and (6.78) and using Eq. (6.82), we get

$$\xi_{,00} + 2\gamma\xi_{,0} + \omega_0^2 \xi = \frac{1}{2} l_0 h^{TT}_{xx,00} \tag{6.83}$$

correct to the first order in h^{TT}_{xx}. This equation determines the response of the detector to the gravitational wave. It is the equation of a forced, damped simple harmonic oscillator.

A detector of this type may be used to detect gravitational waves as a resonant detector for gravitational radiation of a fixed frequency, which is the case for pulsars or close binary stars.

Suppose that

$$h^{TT}_{xx} = A \cos \Omega t \tag{6.84}$$

The steady state solution for ξ is

$$\xi = R \cos(\Omega t + \phi) \tag{6.85}$$

where

$$R = \frac{1}{2} l_0 \Omega^2 A / [(\omega_0 - \Omega)^2 + 4\Omega^2 \gamma^2]^{1/2} \tag{6.86}$$

$$\tan\phi = 2\gamma\Omega / (\omega_0^2 - \Omega^2) \tag{6.87}$$

However, the general initial-value solution for ξ, will have transients, but these will damp out on a time scale of $1/\gamma$.

The energy of oscillation of the detector is

$$E = \frac{1}{2} m(x_{1,0})^2 + \frac{1}{2} m(x_{2,0})^2 + \frac{1}{2} k\xi^2 \tag{6.88}$$

For a detector, which was at rest before the arrival of the gravitational wave, we have

$$x_{1,0} = -x_{2,0} = -\xi_{,0} / 2$$

Thus, the energy of oscillation of the detector becomes

$$\begin{aligned} E &= \frac{1}{4} m[(\xi_{,0})^2 + \omega_0^2 \xi^2] \\ &= \frac{1}{4} m R^2 [\Omega^2 \sin^2(\Omega t + \varphi) + \omega_0^2 \cos^2(\Omega t + \varphi)] \end{aligned} \tag{6.89}$$

The mean value of E is the average over one period, $2\pi/\Omega$

$$\langle E \rangle = \frac{1}{8} m R^2 (\omega_0^2 + \Omega^2) \tag{6.90}$$

If we know the frequency Ω of the specific source of radiation, we should adjust ω^0 to equal Ω for a resonant effect. Then the amplitude of the response will be

$$R_{\text{resonant}} = \frac{1}{4} l_0 A(\Omega/\gamma) \tag{6.91}$$

$$E_{\text{resonant}} = \frac{1}{64} m l_0^2 \Omega^2 A^2 (\Omega/\gamma)^2 \tag{6.92}$$

Defining the quality factor Q of the oscillator,

$$Q = \frac{\omega_0}{2\gamma} \tag{6.93}$$

Then, Eq. (6.92), expressed in terms of Q, becomes

$$E_{\text{resonant}} = \frac{1}{16} m l_0^2 \Omega^2 A^2 Q^2 \tag{6.94}$$

Most of such resonant detectors are massive cylindrical bars, in which the spring constant is provided by the elasticity of the bar, when it is stretched along its axis. The longitudinal modes of vibrations are executed when waves hit the bar broadside. The first detector built by J. Weber of the University of Maryland in the 1960s, had the following specifications. The aluminium bar had:

$$\text{mass} = 1\times 4\times 10^3 \text{ kg}$$
$$\text{length } l_0 = 1\times 5 \text{ m}$$
$$\omega_0 = 10^4 \text{s}^{-1}$$
$$Q = 10^5$$

A strong resonant gravitational wave of amplitude $A = 10^{-20}$, will give to the bar the resonant energy of 10^{-20} J. The resonant amplitude by Eq. (6.91) is only about 10^{-15} *m*. However, realistic gravitational waves will have amplitudes many orders of magnitude smaller and will exist for too short a time, with the consequence that the bar will not attain its full resonant amplitude.

There are difficulties in detecting such small excitations and there are random noises in the oscillator. For instance, at temperature T, the thermal noise in the detector induces random vibrations with energy $\sim kT$. If T is the room temperature $(\sim 300K)$, the thermal noise induced vibrations will be comparable to the energy of excitation. Other sources of noise include vibrations from the passing traffic and seismic disturbances, complicate the detection still further. It requires a very careful isolation from these disturbances.

The main problem is caused by noise and various steps are being taken to overcome these problems. These steps include, the growth of very large single crystals with high Q values, the cooling of massive bars to liquid helium temperatures, developing of stable high-power lasers for laser-interferometer detection and electronics for detecting small excitations of the bars. These high sensitivity technology developments inspire the hope that one day we will succeed in our main mission of detecting the gravitational waves.

We give a brief description of the current state of these detectors with their detection sensitivities.

Resonant Detectors: The first detector to reach astrophysically interesting sensitivity was the ultracold resonant bar at Stanford University, operating at 4 K. In 1980, it operated with a sensitivity to short bursts with strain amplitudes $(\Delta L / L)$ of around 1 part in 10^{18}. The first observations looking for coincident observations were performed between an Italian detector and the Allegro detector at Louisiana State University in 1991. This run determined the stringest upper limit yet on the flux of the gravitational waves.

A new generation of resonant detectors, was pioneered by the Nautilus and Auriga detectors in Italy. These use dilution refrigerators for cooling their several-ton resonators to temperatures of around 50 mK. These will attain a sensitivity of 1 part in 10^{20}.

Designs have been produced for resonant detectors that could reach sensitivities of parts in 10^{21}. These detectors will extend the technology already developed for aluminium cylinders to spheres as much as 10 times more massive.

Interferometric Detectors: For many years, work on interferometers planned at kilometre-scale devices, capable of achieving astrophysically motivated sensitivities, was directed mainly to proof-of-principle devices and engineering tests. Finally, in the early 1990s, large interferometer construction projects were approved in several countries around the world.

LIGO (Laser Interferometer Gravitational-Wave Observatory) at present is the most sensitive experiment designed to detect gravitational waves. It is looking for evidence of these waves by searching for small variations in the path of laser light as it bounces back and forth inside each 2.5 mile-long arm of the observatory's two detectors. So far (as of December 2007), no detections have been reported, but a second observatory, called 'Advanced LIGO' will have an event rate at 100 times that of LIGO, hopefully recording several events per year. But Advanced LIGO is still at the planning stage and is expected to go online in 2011, with completion scheduled for 2015.

Here are but some of the goals that LIGO is designed to accomplish.

1. Prove by direct measurement that gravitational waves exist.
2. Ascertain whether the gravitational waves propagate at the speed of light, as postulated by Einstein.
3. Verify that the gravitational waves cause the predicted displacements in the matter they travel through.
4. Confirm the presence in the universe of black holes and study their dynamics and evolution.
5. Observe cosmic cataclysms, from supernova to coalescing black holes to the big bang.

The LISA (Laser Interferometer Space Antenna) is a proposed joint project of NASA and the European Space Agency to build a laser interferometer gravitational wave detector. It will detect gravitational waves in a lower frequency band between 0.03 milli-Hertz and 1 Hertz whereas LIGO operates at frequencies between 10 Hertz and 1000 Hertz. LISA will be functional in 2016. It will be designed to detect the gravitational waves by employing laser interferometry to monitor the fluctuations in the relative distances between three spacecrafts, on the scale of picometers (10^{-12} metre). The three spacecrafts will be arranged in an equilateral triangle with 5-million-kilometre arms, and fly along an earth—like heliocentric orbit. Gravitational waves will create oscillations in the inner-spacecraft distances. Gravitational signals emanating from massive black holes and other sources of cosmological origin, such as the early universe, will be detected. Unlike LIGO, LISA will be free of interference from earth's motion, so it should be able to detect the subtle ripples in space–time caused by giant black holes that merge after their host galaxies collide.

6.6 QUADRUPOLAR NATURE OF GRAVITATIONAL WAVES

In an isolated nearly Newtonian system, masses move in slow motion about each other. It is desired to assess the type of radiation that they will emit.

One can apply the radiation formulae of electromagnetic theory, for an order of magnitude estimation, with the prescription of replacing, $e^2 \to -m^2$ which converts the static coulomb law into Newton's law of gravitation. This procedure has the disadvantage of treating gravity as a spin-one (vector) field, rather than a spin-two (tensor) field and as a consequence moderate errors get introduced in the numerical factors and changes angular distribution. But the estimate of the total radiated power is adequate.

In electrodynamics, the electric-dipole radiation is the major contribution and the total power output is

$$P_{\text{electric dipole}} = \frac{2}{3}e^2 a^2$$

for an accelerated particle with acceleration a and a dipole moment, changing as $\ddot{d} = e\ddot{r} = ea$. Thus,

$$P_{\text{electric dipole}} = \frac{2}{3}\ddot{d}^2$$

for an arbitrary system with electric dipole moment d.

In gravitation, the analogue of electric dipole moment is the mass dipole moment

$$d = \sum_{\text{particles } A} m_A x_A$$

The time rate of change of d is

$$\dot{d} = \sum_{\text{particles } A} m_A \dot{x}_A = p$$

The second time rate of change of the mass dipole moment has to vanish on account of the law of conservation of linear momentum

$$\ddot{d} = \dot{p} = 0$$

Therefore, there is no mass dipole radiation in gravitation physics.

The next types of electromagnetic radiation are magnetic-dipole and electric quadrupole. The magnetic-dipole radiation is generated by the second time-derivative of the magnetic moment, $\ddot{\mu}$. In this case also, the gravitational analogue, *i.e.*, angular momentum is a constant of motion. Thus,

$$\begin{aligned}\mu &= \sum_A (\text{position of } A) \times (\text{current due to } A) \\ &= \sum_A r_A \times (m v_A) = J\end{aligned}$$

Finally, taking the case of quadrupolar radiation, we get a non-zero emission as shown in the following.

The power emitted by a quadrupole, according to electromagnetic theory, is

$$P_{\text{electric quadrupole}} = \frac{1}{20}\dddot{Q}^2 = \frac{1}{20}\dddot{Q}_{jk}\dddot{Q}_{jk}$$

$$Q_{jk} = \sum_A e_A \left(x_{Aj} x_{AK} - \frac{1}{3}\delta_{jk} r_A^2 \right)$$

Its gravitational analogue is

$$\begin{aligned}P_{\text{mass quadrupole}} &= \frac{1}{5}\langle \dddot{I} \rangle^2 \\ &\equiv \frac{1}{5}\langle \dddot{I\!\!\!-}_{jk} \dddot{I\!\!\!-}_{jk} \rangle\end{aligned} \tag{6.95}$$

$$\begin{aligned}I\!\!\!-_{jk} &= \sum_A m_A \left(x_{Aj} x_{Ak} - \frac{1}{3}\delta_{jk} r_A^2 \right) \\ &= \int \rho \left(x_j x_k - \frac{1}{3}\delta_{jk} r_A^2 \right) d^3x\end{aligned} \tag{6.96}$$

Equation (6.95) contains the correct factor of $\frac{1}{5}$, which results from tensor calculations, instead of the incorrect factor $\frac{1}{20}$ in the electromagnetic analogue. The right-hand side of Eq. (6.96), has been averaged over several characteristic time periods of the source, since one is unable to locate the energy of the gravitational wave inside a wavelength.

Example 6.1

For an electric dipole and its radiation pattern, there are three independent orientations, corresponding to the three directions in which the dipole may point. How many independent orientations are there for a traceless quadrupole tensor?

Solution

The three independent orientations of the dipole correspond to three components of a three dimensional vector. The quadrupole is a second rank, three dimensional tensor (9 components), which is symmetric (3 constraints) and traceless (one constraint), therefore, there are five independent orientations. The five non-zero components of the quadrupole tensor are

$$(Q_{ij}) = \begin{pmatrix} -\frac{1}{2}Q_0 & I_{xy} & I_{xz} \\ I_{xy} & -\frac{1}{2}Q_0 & I_{yz} \\ I_{xz} & I_{yz} & Q_0 \end{pmatrix}$$

6.7 THE EMISSION OF GRAVITATIONAL WAVES

One can easily see that the amplitude of any gravitational wave reaching earth should be small. The perturbation resulting from a strong gravitational wave $h_{\mu\nu} = O(1)$, will be attained only near the source, where the Newtonian potential is of the order of 1. If the source has mass M, then such an amplitude will be attained at distances of order M from it. As is the case with electromagnetic wave, the amplitude of a gravitational wave will vary with distance as $1/R$ in the radiation zone. Thus, if earth is at a distance R from a source of mass M, the maximum amplitude waves, one can expect at earth are of the order of M/R. For instance, from a 10 solar mass black hole formed in a supernova explosion in a nearby galaxy $10^{23}\,m$ away, the amplitude will be about 10^{-17}. This obviously is the upper limit in such an event and less violent events will lead to much smaller amplitudes.

An approximate calculation of wave generation: It is intended to solve the wave equation, Eq. (6.15)

$$\left(\frac{\partial^2}{\partial t^2}, -\nabla^2\right)\bar{h}_{\mu\nu} = -16\pi T_{\mu\nu} \tag{6.97}$$

In order to facilitate the approximate solution, we make some simplifying but realistic assumptions. It is assumed that the time-dependent part of $T_{\mu\nu}$ is oscillating sinusoidally with frequency Ω, *i.e.*, it is the real part of

$$T_{\mu\nu} = S_{\mu\nu}(x^i)e^{-i\Omega t} \tag{6.98}$$

Further, we assume that the region of space in which the amplitude $S_{\mu\nu}$ is non-zero, is small as compared with $2\pi/\Omega$, the wavelength of the gravitational wave of frequency Ω. The first assumption is not restrictive since any general time dependence can be Fourier analyzed into sinusoidal motions. Moreover, many astrophysical sources are nearly periodic, such as pulsars, pulsating stars and binary systems. The second assumption is known as the slow-motion assumption, because it requires that the typical velocity inside the source region, which is Ω times the size of that region, should be much smaller than 1. Such a condition is satisfied by most of the sources of gravitational waves with the sole exception of the most powerful sources.

Let us assume that the solution is $\bar{h}_{\mu\nu}$ given by the real part of the wave

$$\bar{h}_{\mu\nu} = B_{\mu\nu}(x^i)e^{-i\Omega t} \tag{6.99}$$

Rewriting Eq. (6.97) in view of Eqs. (6.98) and (6.99), we get

$$(\nabla^2 + \Omega^2)B = 16\pi S_{\mu\nu} \tag{6.100}$$

It may be remarked that the indices $\bar{h}_{\mu\nu}$ in Eq. (6.97) play no role and $\bar{h}_{\mu\nu}$ is taken simply a function in Minkowski space, satisfying the wave equation.

In a region outside the source, $S_{\mu\nu} = 0$, and for finding the outgoing radiation, we define r to be the spherical polar radial coordinate with the origin inside the source. The solution outside the source is

$$B_{\mu\nu} = \frac{A_{\mu\nu}}{r}e^{i\Omega r} + \frac{Z_{\mu\nu}}{r}e^{-i\Omega r} \tag{6.101}$$

where $A_{\mu\nu}$ and $Z_{\mu\nu}$ are constants. The term $e^{-i\Omega r}$ represents a wave travelling towards the origin while the other term $e^{i\Omega r}$ is the outgoing wave. As we are to find the waves emitted by the source, we take $Z_{\mu\nu} = 0$.

Our job is to determine $A_{\mu\nu}$ in terms of the source. In order to do that we make the assumption that the source is confined inside a sphere of radius $\varepsilon \ll 2\pi/\Omega$. On integrating Eq. (6.100) over the interior of the sphere, the first term, we get

$$\int \Omega^2 B_{\mu\nu} dx^3 \ll \Omega^2 |B_{\mu\nu}|_{\max} 4\pi\varepsilon^3/3 \tag{6.102}$$

where $|B_{\mu\nu}|_{\max}$ is the maximum value of $B_{\mu\nu}$ inside the source.

The other term that we get by integrating the l.h.s. of Eq. (6.100), is

$$\int \nabla^2 B_{\mu\nu} d^3x = \oint n \cdot \nabla B_{\mu\nu} dS \tag{6.103}$$

by Gauss's theorem. As the surface integral is outside the source and $B_{\mu\nu}$ is given by Eq. (6.101), which is spherically symmetric, we get

$$\oint n \cdot \nabla B_{\mu\nu} dS = 4\pi\varepsilon^2 \left(\frac{d}{dr}B_{\mu\nu}\right)_{r=\varepsilon} \approx -4\pi A_{\mu\nu} \tag{6.104}$$

Lastly, we take the integral of the r.h.s. of Eq. (6.100)

$$J_{\mu\nu} = \int S_{\mu\nu} d^3x \tag{6.105}$$

Thus, we find the solution in the limit $\varepsilon \to 0$, is

$$\begin{aligned} A_{\mu\nu} &= -4J_{\mu\nu} \\ \bar{h}_{\mu\nu} &= -4J_{\mu\nu} e^{i\Omega(r-t)} / r \end{aligned} \tag{6.106}$$

The Eq. (6.106) are the gravitational waves generated by the source. As is obvious, we neglected terms of the order $1/r^2$ and any $1/r$ terms that are higher order in $\varepsilon\,\Omega$.

We proceed to simplify further. Rewriting Eq. (6.105), we get

$$\begin{aligned} J_{\mu\nu} e^{-i\Omega t} &= \int S_{\mu\nu} e^{-i\Omega t} d^3x \\ &= \int T_{\mu\nu} d^3x \end{aligned} \tag{6.107}$$

where use was made of Eq. (6.98). Making use of the conservation law for $T_{\mu\nu}$, we have

$$T^{\mu\nu}{}_{,\nu} = 0 \tag{6.108}$$

$$\Rightarrow T^{\mu 0}{}_{,0} = -T^{\mu k}{}_{,k} \tag{6.109}$$

Therefore,

$$\begin{aligned} i\Omega J^{\mu 0} e^{-i\Omega t} &= \int T^{\mu k}{}_{,k}\, d^3x \\ &= \oint T^{\mu k} n_k \, dS \end{aligned} \tag{6.110}$$

As $T^{\mu\nu} = 0$ on the surface around the volume, the r.h.s. of Eq. (6.110) vanishes. If $\Omega \neq 0$, we get

$$J^{\mu 0} = 0, \quad \bar{h}_{\eta\nu} = 0 \tag{6.111}$$

Now we have to write the expression for J_{ij}. We make use of the following result

$$\frac{d^2}{dt^2} \int T^{00} x^l x^m d^3x = 2 \int T^{lm} d^3x \tag{6.112}$$

For a source in slow motion, $T^{00} \approx \rho$, the Newtonian mass density. The integral on the l.h.s. of Eq. (6.112), is often referred to as the quadrupole moment tensor of the mass distribution

$$I^{lm} = \int T^{00} x^l x^m d^3x \tag{6.113a}$$

$$= D^{lm} e^{-i\Omega r} \tag{6.113b}$$

[It may be remarked that different conventions for defining the quadrupole moment are used in different texts. We followed Miser *et al.* (1973).]

Thus, we have

$$\bar{h}_{jk} = -2\Omega^2 D_{jk} e^{i\Omega(r-t)} / r \tag{6.114}$$

It may be remarked that Eq. (6.114) is an approximation, which ignores terms of order r^{-2} as well as r^{-1} terms that are not dominant in the slow motion approximation. In particular, $\bar{h}_{jk}{}^{,k}$ is of higher order and thus ensures that the T T gauge condition $\bar{h}^{\mu\nu}{}_{,\nu} = 0$, is satisfied by Eqs. (6.114) and (6.111) at the lowest order in r^{-1} and Ω. Due to Eq. (6.114), this approximation is normally called the *quadrupole* approximation for gravitational radiation.

6.8 EXPERIMENTAL SUPPORT FOR GRAVITATIONAL WAVES

In a manner analogous to atoms and molecules that emit electromagnetic radiation, a gravitating mass that is in quadrupole type or higher vibration, or is asymmetric and in rotation, can emit gravitational waves[6]. Orbiting objects and merging neutron stars and/or black holes are expected to emit gravitational radiation. As shown earlier, the gravitational waves travel at the speed of light. The planets orbiting the sun constantly lose energy via gravitational radiation, but this effect, being so small, is quite unlikely to be observed in the near future. (It is estimated that earth radiates about 300 watts of gravitational energy.). Gravitational waves have been indirectly detected from the Hulse–Taylor binary pulsar. However, there are no confirmed observations of gravitational radiation at this time.

6.8.1 Cosmological Tests

Cosmological tests of general relativity are not nearly so stringent as solar tests. Some cosmological tests search primordial gravity waves generated during cosmic inflation, which may be detected in the cosmic microwave background polarization or by a proposed space-based gravity wave interferometer called Big Bang observer.

The experimental support for the existence of gravitational radiation comes from observations of binary pulsar systems. In 1993, Hulse and Taylor[7-8] were awarded the Nobel Prize in physics for their 1974 discovery of a pulsar PSR 1913+16 (PSR stands for the pulsar and 1913+16 specifies the pulsar's position in the sky), in a binary system and for the subsequent confirmation of the existence of gravitational waves[3] as predicted by general relativity. They were able to determine that PSR 1913+16 is a part of binary system because the arrival times of its pulses—sometimes earlier and sometimes later—are consistent with the behaviour if it were in a binary system. That is, when the pulsar is on the part of its orbit closer to earth, its pulse arrives approximately 3 seconds earlier than it does when the pulsar is on the part of its orbit farthest away from earth.

Hulse and Taylor recorded the observed time of periastron (the minimum separation of an object to the centre of mass of a binary system in elliptical orbits) for the pulsar over a number of years. In 1983, they reported that there was a systematic shifting in periastron time—the pulse signal was arriving sooner and sooner than before. In 1982, PSR 1913+16 was arriving at its periastron over a second earlier than it had been in 1974.

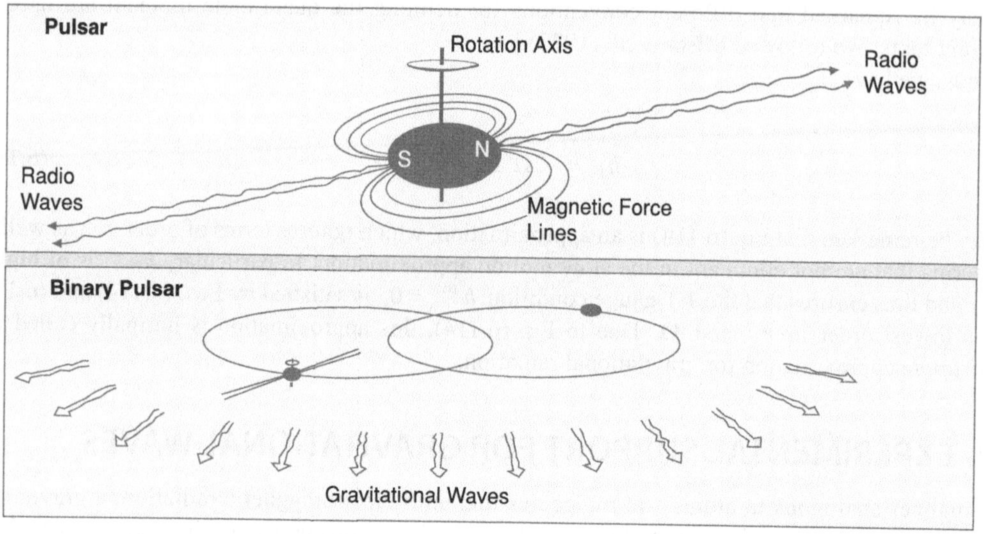

Figure 6.7 *The radio waves from a pulsar are emitted in two bunches which sweep across space at the same rate as the pulsar rotates (upper figure). Froma binary pulsar, gravitational waves are also emitted (lower figure)*

Hulse and Taylor found that shifts in periastron time were consistent with the general relativity prediction that, such a binary system will lose energy to gravitational radiation resulting in a shorter orbital period, Fig. 6.7. The orbit of PSR 1913+16 shrinks at a rate of approximately 3 mm per orbit. At this rate, the two stars should merge in around 300 million years. It is hoped that instruments, such as LIGO and LISA will be able to detect the gravitational radiation produced during such events.

References

1. Misner, C.W., Thorne, K.S., and Wheeler, J.A. (1973). *Gravitation*, New York.
2. Ciufolini, I. and Wheeler, J.A. (1995). *Gravitation and Inertia*, (Princeton, N. J), Princeton University Press.
3. Einstein, A. (1916). *Preuss. Akad. Wiss*, Berlin, Sitzber, 688.
4. Landau, L.D. and Lifshitz, E.M. (1975). *The Classical Theory of Fields*, Reading MA, Addison-Wesley.
5. Bondi, H.(1957). *Nature*, 179, 1072.
6. In general relativity, a perfectly spherical star (in vacuum) that expands or contracts while remaining perfectly spherical cannot emit any gravitational waves (similar to the lack of *e/m* radiation from a pulsating charge), as Birkoff's theorem says that the geometry remains the same exterior to the star. More generally, a radiating system will only emit gravitational waves if it lacks the axial symmetry with respect to the axis of rotation.
7. Hulse, R.A. and Taylor, J.H. (1975). *Astrophys. J. Lett*, 195 151–153.
8. Taylor J.H., Fowler, L.A, and McCulloch, P.M.(1979). *Nature*, 277, 437–440.

Problems

6.1 Two bodies attracting each other according to Newton's law, move in circular orbits around their common centre of inertia. Determine the average (over a rotation period) of the intensity of gravitational waves and its distribution in polarization and direction.

6.2 Determine the time-averaged rate of loss of angular momentum from a system of bodies in stationary motion and emitting gravitational waves.

6.3 Verify that the retarded solution

$$\phi^{\mu\nu}(t,\mathbf{x}) = -\frac{k}{4\pi}\int \frac{T^{\mu\nu}(t-|\mathbf{x}-\mathbf{x}'|,\mathbf{x}')}{|\mathbf{x}-\mathbf{x}'|}d^3x'$$

is a solution of the linear field equation

$$\partial_\lambda \partial^\lambda \phi^{\mu\nu}(t,\mathbf{x}') = -kT^{\mu\nu}(t,\mathbf{x}')$$

where $T^{\mu\nu}$ is the energy–momentum tensor of the matter.

6.4 Show that gravitational waves, just like electromagnetic waves, carry not only energy but also momentum. Show that the flux of momentum is equal to the energy flux.

6.5 Show that the distance between two particles on the y-axis, at , $y = \pm y_1$ exposed to the gravitational wave

$$\phi^{\mu\nu} = A\varepsilon^{\mu\nu}\cos k_\alpha x^\alpha$$

varies as follows

$$\Delta l = \left(1 + \frac{k}{2}A\cos\omega t(2y_0)\right)$$

6.6 Show that the pulsation frequencies of an incompressible fluid sphere held together by gravitation are of the order of $\omega \sim \sqrt{G\rho}$, where ρ is the density.
Hint: The velocity of sound waves in a fluid is $v \sim \sqrt{p/\rho}$ where p is the pressure. The typical pressure for a sphere of radius R is Gm^2/R^4 and $\omega \sim v/R$.

6.7 Show that the radiated power by a quadrupole rotator consisting of two spherical masses moving in circular orbits about their common centre of mass is

$$-\frac{dE}{dt} = \frac{32G}{5c^5}\left(\frac{m_1 m_2}{m_1+m_2}\right)^2 r^4\omega^6$$

where m_1 and m_2 are the masses, r is the distance between them and 2ω is the frequency of rotation. The gravitational waves emitted by the system have a frequency 2ω, this is because the quadrupole moment repeats when the masses move through one-half of the orbit.

6.8 Two stars of equal mass M orbit a common centre. The radius of the orbit of each star is R. Assume that each star has a mass equal to 1.5 solar masses (solar mass $= 2\times10^{30}$ kg) and the initial separation of the stars is 2.0×10^9 m. The period of revolution of each star is 7.8 years. Two stars on opposite sides of a circular orbit and a distance of two stars separating them,

$$\text{Force on each star} = \frac{GM^2}{4R^2}$$

$$\text{Period of revolution of each star } T^2 = \frac{16\pi^2 R^3}{GM}$$

The two stars lose energy as a result of emitting gravitational radiation, so that the stars will move closer to each other.

(i) Explain why the fractional loss of energy per unit time may be calculated from the expression

$$\frac{\Delta E}{E}\Delta t = \frac{2}{3}\frac{\Delta T}{T}\Delta t$$

where T is the period of revolution of each star and $\frac{\Delta T}{T}\Delta t$ is the fractional decrease in period per unit time.

(ii) The orbital period decreases at a rate of $\frac{\Delta T}{\Delta t} = 72\,\mu s\,\text{yr}^{-1}$. Estimate the fractional energy loss per year. ($1.7 \times 10^{-9}\,\text{J yr}^{-1}$)

(iii) The two stars will collapse into each other when $\Delta E \sim E$. Estimate the lifetime in years of this binary star system. (6×10^8 yr)

7

Black Holes

7.1 INTRODUCTION

A black hole is a region of space–time that has so much mass concentrated in it that gravity is so strong that not even light signal can escape its gravitational pull. It is believed that black holes form at the end of a star's lifetime when a massive star runs short of fuel and collapses. Incidentally, the name black hole was suggested by John Archibald Wheeler and being much catchier, seems to have stuck. Before Wheeler, these objects were referred to as 'frozen stars.'

In the collapse of a dying star to a black hole, all the information about the collapsing star is lost, such as its shape and the stuff it was made of. However, the information, which is not wiped out, is its mass, charge and angular momentum. This fact, the *no-hair* theorem, implies that the most general black hole metric will be a function of these three numbers only. Hair is just a metaphor for all other information about the constituent matter of the black hole or the matter falling into the black hole, disappears beyond the event horizon of the black hole and is not accessible to external observers. However, the real world black holes will probably be spherically neutral, so we will not discuss the case of a charged black hole.

Thus, a black hole is just characterized by just three parameters—mass, charge and angular momentum. This characterization results in three general classes of black holes. These are as follows:

- Static black holes with no charge, described by Schwarzschild solution.
- Rotating black holes described by Kerr solution.
- Black holes with electrical charge, described by Reissner–Nordström solution.

We will be discussing the Schwarzschild black holes, Kerr black holes and charged black holes in some detail.

7.2 SCHWARZSCHILD BLACK HOLES—SINGULARITIES

Almost immediately after Einstein's general relativity, Karl Schwarzschild discovered a mathematical solution to the field equations that described such an object. Let us consider such a metric

$$ds^2 = \left(1 - \frac{2m}{r}\right)dt^2 - \frac{dr^2}{(1-2m/r)} - r^2(d\theta^2 + \sin^2\theta d\phi^2) \tag{7.1}$$

It exhibits a singularity at $r = 2m$. At $r > 2m$, $g_{00} > 0$ and $g_{11} < 0$. However, for $r < 2m$, $g_{00} < 0$ and $g_{11} > 0$, the signs of these components of metric reverse. This implies that a world line along the t-axis has $ds^2 < 0$ and so describes a space–like curve. However, a world line along the r-axis has $ds^2 > 0$ and as such describes a time–like curve. The space and time character of the coordinates gets reversed and a massive object inside the Schwarzschild radius at a constant value of *r*, cannot remain stationary. At $r = 2m$, $g_{00} = 1 - 2m/2m = 0$. Although, it is well-behaved mathematically but the vanishing of g_{00} means that a surface $r = 2m$ is a surface of infinite red-shift.

Let us examine as to what happens to g_{11}.

$$g_{11} = -\frac{1}{1-2m/r} \to \infty \quad \text{as}\, r \to 2m.$$

Obviously, it means that the point $r = 2m$, constitutes a singularity. The next question that arises is to analyse whether the singularity is physically real or is due to the choice of coordinates. It will turn out that the $r = 2m$ surface has unusual properties and is the coordinate singularity. By using a different set of coordinates, it is possible to get rid of the singularity. Furthermore, the singularity at $r = 0$, is due to infinite curvature and cannot be removed by the change of coordinates.

One way to see it, is through the invariants, as invariants being scalars do not depend on a particular choice of coordinates. Thus,

$$R_{\mu\nu\rho\sigma} R^{\mu\nu\rho\sigma} = \frac{48m^2}{r^6} \tag{7.2}$$

The expression, Eq. (7.2) shows that the curvature tensor blows apart at $r = 0$ and nothing happens at $r = 2m$. This tantamounts to the fact that the singularity at $r = 2m$, can be removed by the choice of a particular coordinate system. The $r = 0$ singularity is a curvature singularity and cannot be removed by a coordinate system change.

We investigate the problem further by examining the behaviour of light cones near $t = \pm r + C$. Consider the paths along the radial lines for the sake of simplicity by putting $d\theta = d\phi = 0$. As a result, the Schwarzschild metric becomes

$$ds^2 = (1 - 2m/r)dt^2 - \frac{dr^2}{(1 - 2m/r)} \tag{7.3}$$

The path of the light rays is given by putting $ds^2 = 0$, so the slope of the light cone is given by

$$\frac{dt}{dr} = \pm(1 - 2m/r)^{-1} \tag{7.4}$$

Far from $r = 2m$, that is, as $r \to \infty$, Eq. (7.4) becomes

$$\frac{dt}{dr} = \pm 1 \tag{7.5}$$

An integration gives $t = \pm r + C$, where C is a constant of integration. It gives the motion of light rays in flat space, just the result for light cones in Minkowski space.

Further, we examine the behaviour as we approach smaller values of *r* tending towards $r = 2m$. Taking the positive sign, corresponding to outgoing radial null curves, we get from Eq. (7.4)

$$\frac{dt}{dr} = \frac{r}{r - 2m} \tag{7.6}$$

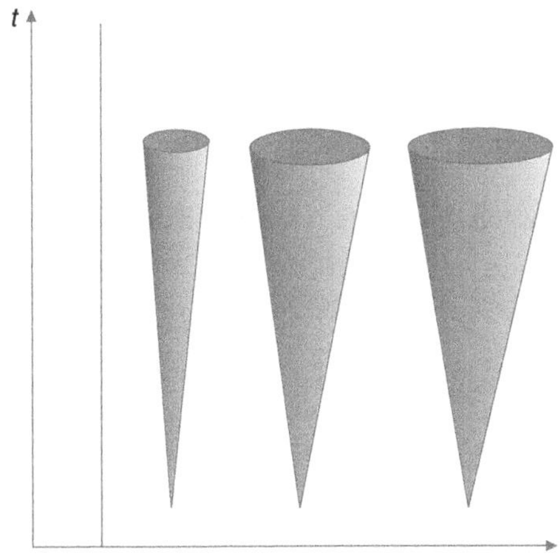

Figure 7.1 *With Schwarzschild coordinates, light cones become narrower as they approach r = 2m.*

As $r \to 2m, dt/dr \to \infty$, meaning that light cones are becoming progressively narrower as we approach $r = 2m$. At $r = 2m$, the world lines are vertical, Fig. (7.1). The surface $r = 2m$, is a surface of infinite red-shift.

The singularity can be got rid off by integrating Eq. (7.4) and taking the positive sign, so we get (ignoring the integration constant)

$$t = r + 2m\ln|r - 2m| \tag{7.7}$$

It suggests a coordinate transformation, one can use. Such a coordinate transformation, called *tortoise* transformation, was suggested by Eddington–Finkelstein for removing the coordinate singularity. Thus, introducing a new coordinate

$$r^* = r + 2m\ln\left(\frac{r}{2m} - 1\right) \tag{7.8}$$

alongwith two null coordinates

$$u = t - r^* \tag{7.9a}$$

$$v = t + r^* \tag{7.9b}$$

Differentiating Eq. (7.8), we get

$$\begin{aligned} dr^* &= dr + \frac{2m}{(r/2m - 1)}\left(\frac{1}{2m}\right)dr \\ &= \frac{dr}{1 - 2m/r} \end{aligned} \tag{7.10}$$

From Eq. (7.9b). We obtain

$$\begin{aligned} dt &= dv - dr^* \\ &= dv - \frac{dr}{(1 - 2m/r)} \end{aligned} \tag{7.11}$$

Squaring Eq. (7.11), we obtain

$$dt^2 = dv^2 - 2\frac{drdv}{(1-2m/r)} + \frac{dr^2}{(1-2m/r)^2} \tag{7.12}$$

Substituting Eq. (7.12) into Eq. (7.3), we get the Eddington–Finkelstein form of metric

$$ds^2 = (1-2m/r)dv^2 - 2dvdr - r^2(d\theta^2 + \sin^2\theta d\phi^2) \tag{7.13}$$

The coordinate singularity at $r = 2m$ is eliminated, but the singularity at $r = 0$ is very much there.

Let us again consider the radial light rays by putting $d\theta = d\phi = 0$ and $ds^2 = 0$ in Eq. (7.13). Thus,

$$(1-2m/r)dv^2 - 2dvdr = 0$$

or

$$(1-2m/r) - 2dr/dv = 0 \tag{7.14}$$

Putting $r = 2m$ in Eq. (7.14), we get

$$dr/dv = 0 \tag{7.15}$$

The radial coordinate velocity is eliminated. On integrating Eq. (7.15), one gets

$$r(v) = \text{constant} \tag{7.16}$$

The light ray stays where it is. Rewriting Eq. (7.14), one gets

$$\frac{dv}{dr} = \frac{2}{(1-2m/r)}$$

which on integrating, gives

$$v(r) = 2(r + 2m\ln|r-2m|) + \text{constant} \tag{7.17}$$

This equation gives us the paths that radial rays will follow on using (v, r) coordinates. When $r > 2m$, then as r increases, v increases. This will hold for light rays that are outgoing. However, when $r < 2m$, as r decreases, v decreases and this will pertain to the light rays that are ingoing.

In the (v, r) coordinates, light cones do not progressively become narrow and these cross-over the line $r = 2m$; but since the character of the time and radial coordinates reverses inside $r = 2m$; it implies that light cones tilt over in this region, Fig. (7.2).

One can draw the following references from the discussion above.

- The surface $r = 2m$, is a coordinate singularity. Through a choice of a particular coordinate system, the singularity can be eliminated.

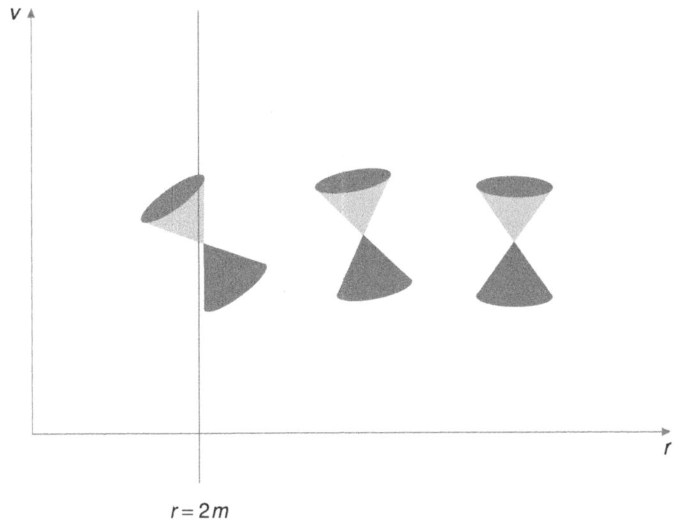

Figure 7.2 *The coordinate singularity at $r = 2m$ is removed with Eddington–Finkelstein coordinates (v, r). With decreasing r, light cones trip over.*

- The surface $r = 2m$, is called the surface horizon and is a one-way membrane. The light-like and time-like curves do cross from $r > 2m$ to $r < 2m$, but the reverse is not tenable. The events inside the event horizon cannot be seen by external observers. The surface $r = 2m$, is a surface of infinite red-shift.
- As r decreases, the light cones start tipping over and at $r = 2m$, the outward travelling photons become stationary.
- For $r = 2m$, future directed light-like and time-like curves move towards $r = 0$.
- The Schwarzschild coordinates are adequate for describing the geometry over the region $2m < r < \infty$ and $-\infty < t < \infty$; but another coordinate system have to be used for describing the point $r = 2m$ and the region inside the event horizon.

7.3 KRUSKAL COORDINATES

To describe the geometry in the region $r < 2m$, Kruskal–Szekeres coordinates are adequate. These are labelled u and v and are related to Schwarzschild coordinates t, r in the following manner, whether $r < 2m$ or $r > 2m$.

$r > 2m$:

$$u = e^{\frac{r}{4m}}\sqrt{\frac{r}{2m} - 1}\cosh\frac{t}{4m} \tag{7.18a}$$

$$v = e^{\frac{r}{4m}}\sqrt{\frac{r}{2m} - 1}\sinh\frac{t}{4m} \tag{7.18b}$$

$r < 2m$:

$$u = e^{\frac{r}{4m}} \sqrt{\frac{r}{2m} - 1} \sinh \frac{t}{4m} \quad (7.19a)$$

$$v = e^{\frac{r}{4m}} \sqrt{\frac{r}{2m} - 1} \cosh \frac{t}{4m} \quad (7.19b)$$

The Kruskal form of metric is

$$ds^2 = \frac{32m^3}{r} e^{-r/2m} (dv^2 - du^2) - r^2 (d\theta^2 + \sin^2 \theta d\phi^2) \quad (7.20)$$

The coordinates, Fig. (7.3) have the following characteristics:

- The outside world is labelled by O in the region $r \geq 2m$, which corresponds to $u \geq v$.
- The line $u = v$ relates to the Schwarzschild coordinate $t \to \infty$ whereas $u = -v$ corresponds to $t = -\infty$.
- The region inside the event horizon, $r < 2m$, corresponds to $v > |u|$.
- In these coordinates, the light cones are at 45°

In addition, the following relationships hold

$$u^2 - v^2 = \left(\frac{r}{2m} - 1\right) e^{\frac{r}{2m}} \quad (7.21a)$$

and

$$\frac{v}{u} = \tanh \frac{t}{4m} \quad (7.21b)$$

The coordinate singularity at $r = 2m$, corresponds to $u^2 - v^2 = 0$. The physical curvature singularity is a hyperbola $v^2 - u^2 = 1$.

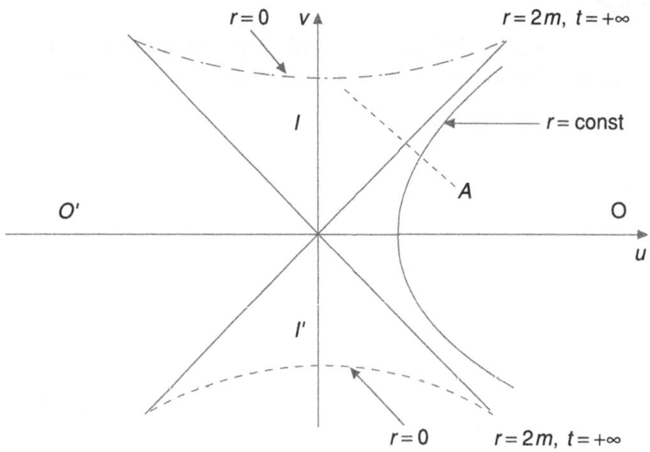

Figure 7.3 *Kruskal coordinates. Regions O and O′ are outside the event horizon and as such relate to region r > 2m. Regions I and I′ correspond to region r < 2m.*

Let us again study the path of light rays by putting $ds^2 = 0$ and consider the paths along radial lines, *i.e.*, putting $d\theta = d\phi = 0$, in Eq. (7.20). We get

$$ds^2 = 0 = \frac{32m^3}{r} e^{-\frac{r}{2m}} (du^2 - dv^2) \tag{7.22}$$

Which gives $(du / dv)^2 = 1$. This shows that the velocity of light is 1 everywhere. Thus, there is no barrier to light propagation in these coordinates. One can summarize the conclusions arrived at, as follows:

u is a global radial coordinate.
v is a global time coordinate.
The metric is analogous to the Schwarzschild solution, but does not correspond to the spherical coordinates at large distances.
The coordinate singularity at $r = 2m$, is removed.
The curvature singularity at $r = 0$ still remains.

7.4 THE KERR METRIC IN BOYER–LINDQUIST COORDINATES

The Kerr metric describes the geometry of space-time around a rotating massive body and according to this metric, the rotating bodies exhibit unusual new phenomena such as frame dragging. An object placed near a spinning black hole will rotate alongwith the black hole, irrespective of what state of motion we give to the object. Measurement of this frame dragging effect was a primary goal of the Gravity Probe B experiment. The object in the close vicinity of the rotating mass will rotate not because of any applied force or torque that can be felt, but rather due to the curvature of space-time associated with the rotating bodies. In fact, the rotation even affects light.

As we will see below, a rotating black hole has two horizons and the region in between these horizons is called the *ergosphere*. It is in this region that frame dragging holds. This feature can be used to extract energy from a rotating black hole, upto its invariant mass energy mc^2, according to a process called *Penrose process*.

The Kerr metric is an exact solution of Einstein's field equations of general relativity and is a generalization of the Schwarzschild metric. The effects of spinning of the space-time around the black hole, will be observed by the presence of angular momentum in the metric alongwith mixed or cross terms like $d\phi dt$. These terms show the change of angle with time. Angular momentum per unit mass, $a = J / m$ where J is the angular momentum and m, the gravitational mass. The units of a then become meters.

The Kerr metric, describing the space-time around a spinning black hole is relatively complicated and is merely stated as

$$\begin{aligned} ds^2 &= \left(1 - \frac{2mr}{\Sigma}\right) dt^2 + 2\frac{2amr\sin^2\theta}{\Sigma} dt d\phi - \frac{\Sigma}{\Delta} dr^2 - \Sigma d\theta^2 \\ &\quad - \left(r^2 + a^2 + \frac{2a^2 mr \sin^2\theta}{\Sigma}\right) \sin^2\theta d\phi^2 \end{aligned} \tag{7.23}$$

where the length scale symbols have the following connotation,

$$\Delta = r^2 - 2mr + a^2$$

$$\sum = r^2 + a^2 \cos^2 \theta$$

and $$a = J / m$$

The coordinates r, θ, ϕ are standard spherical coordinate system. In the non-relativistic limit where m (or equivalently the Schwarzschild radius $r_s = 2m$) goes to zero, the Kerr metric becomes the orthogonal metric for the oblate spherical coordinates

$$ds^2 = dt^2 - \frac{\Sigma}{\Delta} dr^2 - \sum d\theta^2 - (r^2 + a^2)\sin^2 \theta d\phi^2 \tag{7.24}$$

which are equivalent to the Boyer–Lindquist coordinates

$$x = \sqrt{r^2 + a^2} \sin\theta \cos\phi$$
$$y = \sqrt{r^2 + a^2} \sin\theta \sin\phi$$
$$z = r\cos\theta$$

The components of the metric tensor are as follows:

$$g_{00} = (1 - 2mr / \Sigma)$$
$$g_{03} = g_{30} = 2mar \sin^2 \theta / \Sigma$$
$$g_{11} = -\Sigma / \Delta$$
$$g_{22} = -\Sigma$$
$$g_{33} = -\left(r^2 + a^2 + 2ma^2 r \sin^2 \theta / \Sigma\right) \sin^2 \theta \tag{7.25}$$

As is evident, the metric components are independent of t and ϕ and this means that two suitable killing vectors for the space–time are ∂_t and ∂_ϕ. The off-diagonal terms involve g_{03}. Thus, we can invert the terms g_{11} and g_{22}, which are given as

$$g^{11} = \frac{1}{g_{11}} = -\frac{\Delta}{\Sigma}$$
$$g^{22} = \frac{1}{g_{22}} = -\frac{1}{\Sigma} \tag{7.26}$$

To find the remaining components, we consider the matrix

$$\begin{pmatrix} g_{00} & g_{03} \\ g_{30} & g_{33} \end{pmatrix} = \begin{pmatrix} 1 - 2mr / \Sigma & 2mar \sin^2 \theta / \Sigma \\ 2mar \sin^2 \theta / \Sigma & -(r^2 + a^2 + 2ma^2 r \sin^2 \theta / \Sigma) \sin^2 \theta \end{pmatrix} \tag{7.27}$$

It is a cumbersome effort to invert this matrix but the easier path is by using computer, one gets

$$
\begin{aligned}
g^{00} &= \frac{(r^2+a^2)^2 - \Delta a^2 \sin^2\theta}{\Sigma\Delta} \\
g^{03} &= \frac{2mar}{\Sigma\Delta} \\
g^{33} &= -\frac{\Delta - a^2\sin^2\theta}{\Sigma\Delta}
\end{aligned}
\tag{7.28}
$$

The presence of mixed components of the metric tensor leads to interesting results. Consider a particle with four-momentum $p_\mu = (p_0, p_1, p_2, p_3)$. Thus,

$$
\begin{aligned}
p^0 &= \sum_{k=0,3} g^{0k} p_k = g^{00} p_0 + g^{03} p_3 \\
p^3 &= \sum_{k=0,3} g^{3k} p_k = g^{30} p_0 + g^{33} p_3
\end{aligned}
\tag{7.29}
$$

Therefore, a particle with $p_3 = 0$ will still have p^3.

Next we simplify the matters a little and learn the important features of the Kerr metric. We consider a plane through the centre of the black hole and at right angles to its spin axis. In this plane, $\theta = \pi/2$, the Kerr metric is simplified considerably and becomes

$$
\begin{aligned}
ds^2 = {} & (1 - 2m/r)dt^2 + 2\frac{2am}{r}dt d\phi - \frac{1}{(1 - 2m/r + a^2/r^2)}dr^2 \\
& - (1 + a^2/r^2 + 2a^2 m/r^3)r^2 d\phi^2
\end{aligned}
\tag{7.30}
$$

It may be remarked that the time coordinate used in the metric, is the time recorded by a distant observer as was the case of Schwarzschild metric. The g_{00} coefficient is the same in Eq. (7.30), as in the case of Schwarzschild metric. This goes to zero at

$$
r_s = 2m \tag{7.31}
$$

This is called the static limit. Next let us consider the g_{11} term, which is

$$
g_{11} = -\frac{1}{(1 - 2m/r + a^2/r^2)} \tag{7.32}
$$

This term will blow up when

$$
1 - 2m/r + a^2/r^2 = 0
$$

or

$$r^2 - 2mr + a^2 = 0 \tag{7.33}$$

with solutions

$$r_\pm = \frac{2m \pm \sqrt{4m^2 - 4a^2}}{2} \tag{7.34}$$

Thus, we learn that in the case of a rotating black hole, it has two horizons. By equating $a = 0$, we get the Schwarzschild result

$$r_\pm = m \pm m$$

which gives

$$r_+ = 2m \quad \text{or} \quad r_- = 0 \tag{7.35}$$

In the case of the rotating black holes, r_+ gives the outer horizon and r_-, the inner horizon. The inner horizon is called the *Cauchy horizon*. It is obvious from Eq. (7.34) that inner horizon will have maximum value when $m = a$, actually under this condition $r_\pm = m$ and the inner and outer horizons coincide. Both these horizons are real horizons, *i.e.*, these are one-way membranes, which allow the going in but not the coming out.

The outer horizon given by

$$r_+ = m + \sqrt{m^2 - a^2} \tag{7.36}$$

marks the boundary between the black hole and the outside universe. This is similar to the Schwarzschild horizon if we set $a = 0$, that gives the earlier result $r = 2m$. The inner horizon lies inside the outer horizon, as such cannot be accessed by an outside observer.

We marked that $g_{00} = 0$ at $r_s = 2m$, Eq. (7.31), thus, the solution $r_s = 2m$ can be taken as the outer infinite redshift surface that lies outside the outer horizon $r_+ = m + \sqrt{m^2 - a^2}$. Light signals and particles can go in or out of the outer surface, but the inner horizon is one-way membrane that allows the particles and light to go in but not coming out. But at $\theta = 0, \pi$ the horizon and the surface of infinite redshift coincide and at these points if light and massive particles cross, these cannot escape and get entrapped.

The volume contained between the surface of static limit and the outer horizon, *i.e.*, the region $r_+ < r < r_s$, is called the *ergosphere*. Inside the ergosphere, the frame dragging effect comes into play, which states that any object inside this region is dragged along irrespective of its energy or state of motion. In other words, one can state that all time-like geodesics rotate with the mass that is source of the gravitational field. However, in between the two horizons, *i.e.*, $r_- < r < r_+$, *r* becomes a time-like coordinate. Anything there will be pulled inexorably to the *Cauchy horizon*, $r = r_-$.

7.5 FRAME DRAGGING. LENSE-THIRRING EFFECT

General relativity predicts that massive rotating objects should drag space-time around themselves as they rotate. Frame dragging is what happens if a bowling ball spins in a thick fluid such as molasses. As the ball spins, it pulls the molasses around itself. Anything stuck in the molasses will also move around the ball. Similarly, as the earth rotates, it pulls spacetime in its vicinity around itself. This will shift the orbits of the satellites near the earth.

An international team of NASA and university researchers has detected the effect by precisely measuring shifts in the orbits of two earth-orbiting laser-ranging satellites, the Laser Geodynamics Satellite I (LAGEOS I), a NASA spacecraft, and LAGEOS II, a joint NASA/Italian Space Agency (ASI) spacecraft. It is the first direct measurement of a bizarre effect called '*frame dragging*'. The effect has recently been observed around distant celestial objects with intense gravitational fields such as black holes and neutron stars.

For describing this effect, we consider the momentum four-vector and the components of the metric tensor. From Eq. (7.28), we get

$$\frac{g^{03}}{g^{00}} = \frac{2mar}{(r^2+a^2)^2 - \Delta a^2 \sin^2\theta} \tag{7.37}$$

Consider a massive particle with zero angular momentum dropped in. The angular velocity is

$$\frac{d\phi}{dt} = \frac{d\phi/d\tau}{dt/d\tau} = \frac{p^3}{p^0} \tag{7.38}$$

With $p_3 = 0$, we get from Eq. (7.29)

$$p^0 = g^{00} p_0 \tag{7.39}$$

and

$$p^3 = g^{30} p_0 \tag{7.40}$$

Thus, Eq. (7.38), becomes in view of Eqs. (7.37), (7.39) and (7.40)

$$\frac{p^3}{p^0} = \frac{g^{30}}{g^{00}} = \omega = \frac{2mar}{(r^2+a^2)^2 - \Delta a^2 \sin^2\theta} \tag{7.41}$$

As the angular velocity ω is proportional to the components of the metric, so it is due to the gravitational field. Thus, a particle dropped from infinity with zero angular momentum, picks up an angular velocity from the gravitational field. This effect is the frame dragging that causes a gyroscopic precession effect, called the *Lense–Thirring effect*. To get a simplified expression for angular velocity, we consider the equatorial plane perpendicular to the spin axis of the black hole and set $\theta = 0$, so that

$$\omega = \frac{2mar}{(r^2+a^2)^2} \tag{7.42}$$

Next we consider the case of light near a Kerr black hole. Consider light that in the beginning is moving on a tangential path, *i.e.*, $dr = 0$ and for a null ray $ds^2 = 0$. Once again restricting to the equatorial plane $\theta = \pi/2$, we get from Eq. (7.30)

$$0 = \left(1 - \frac{2m}{r}\right)dt^2 + 2\frac{2ma}{r}dt\,d\phi - \left(1 + \frac{a^2}{r^2} + \frac{2ma^2}{r^3}\right)r^2 d\phi^2 \tag{7.43}$$

Introducing the reduced circumference following Taylor and Wheeler (2000), we get

$$R^2 = r^2 + m^2 + \frac{2m^3}{r} \tag{7.44}$$

Thus, rewriting Eq. (7.43) in the light of Eq. (7.44), one gets

$$0 = \left(1 - \frac{2m}{r}\right) dt^2 + 2\frac{2ma}{r} dt d\phi - R^2 d\phi^2 \tag{7.45}$$

Rearranging the terms, we get

$$\left(\frac{d\phi}{dt}\right)^2 - \frac{4ma}{rR^2}\frac{d\phi}{dt} - \frac{1}{R^2}(1 - 2m/r) = 0 \tag{7.46}$$

Considering the static limit, we put $r = r_s = 2m$, we get from Eq. (7.44)

$$R^2 = 6m^2 \tag{7.47}$$

Thus, the angular velocity, Eq. (7.46), at the static limit, becomes

$$\left(\frac{d\phi}{dt}\right)^2 - \frac{a}{3m^2}\frac{d\phi}{dt} = 0 \tag{7.48}$$

This equation has the solutions

$$\frac{d\phi}{dt} = \frac{a}{3m^2} \tag{7.49}$$

and

$$\frac{d\phi}{dt} = 0 \tag{7.50}$$

According to the solution $d\phi / dt = a / (3m^2)$, the Eq. (7.49), the emitted light has the same direction as the rotating black hole and is governed by the angular momentum *a* of the black hole. The solution $d\phi / dt = 0$, Eq. (7.50), is all the more astonishing since if the light is emitted in the direction opposite to the spinning black hole's rotation, the light is stationary. As no material particle can move at a velocity higher than the velocity of light, it implies that no rocket ship, probe or any particle, can move in the direction opposite to the direction of rotation of the black hole.

Moving beyond the coordinate singularity, consider any singularity, one can find from invariant quantity formed from the Riemannian tensor $R^{\mu\nu\rho\sigma} R_{\mu\nu\rho\sigma}$. This leads to a real singularity described by

$$r^2 + a^2 \cos^2\theta = 0 \tag{7.51}$$

In the equatorial plane, we have $\theta = 0$ and the singularity is described by the equation $r^2 + a^2 = 0$. This equation describes a ring of radius *a*, lying in the *x*-*y* plane at $z = 0$. Thus, for a rotating black hole, the curvature singularity is not given by $r = 0$ but by a ring of radius *a*.

7.6 ENERGY EXTRACTION FROM A ROTATING BLACK HOLE. PENROSE PROCESS

In 1969, Roger Penrose[1] discovered a way, atleast in theory, that can be used to extract energy from a rotating black hole. Before we discuss the details of the process, we treat another concept about *ergospace*. As a Kerr black hole rotates, it drags inertial frames alongwith it and as a result, there is a region of space that is dragged sufficiently enough to make it impossible for an observer to remain static, as we expressed before. We thus have a surface in spacetime that forms the limit of where an observer can remain static and we call this ergospace, after the Greek word $\varepsilon\rho\nu o\nu$ that means work, since within this region energy extraction can be carried out. The ergospace is given in Boyer–Lindquist coordinates, where g_{00} is equal to zero. Thus,

$$r_{s\pm} = m \pm \sqrt{m^2 - a^2 \cos^2\theta} \tag{7.52}$$

We take the larger of the two solutions, r_{s+} to be the ergosurface.

This is outside the outer event horizon for all possible values of θ and is coincident at the axis of rotation $(\theta = 0, \pi)$. The ergosurface and the event horizon are shown diagrammatically in Fig. (7.4). Penrose process is as follows:

- Let a particle (A) fall towards a Kerr black hole from infinity.
- Once the particle enters the ergosphere, it is made to break up into two pieces, (B) and (C).
- One fragment, say (B), escapes to infinity and the other is absorbed by the black hole, Fig. (7.5).

If we ensure that (C) has angular momentum that is counter-rotational to the black hole, then it will have a negative energy with respect to the observer at infinity. Therefore since $E(\mathrm{C}) < 0$, we have by conservation of energy

$$\begin{aligned} E_{(\mathrm{A})} &= E_{(\mathrm{B})} + E_{(\mathrm{C})} \\ \Rightarrow E_{(\mathrm{B})} &> E_{(\mathrm{A})} \end{aligned} \tag{7.53}$$

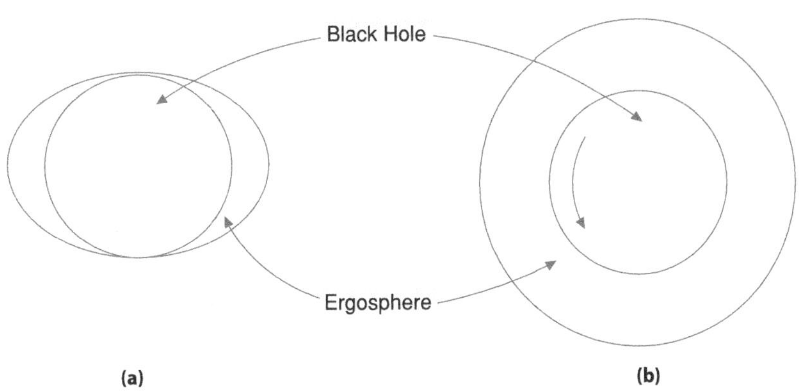

Figure 7.4 *The ergosurface of a Kerr black hole. (a) a side view (b) a top view.*

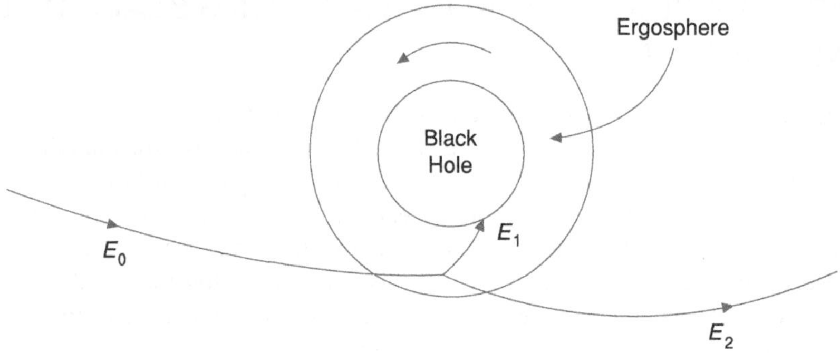

Figure 7.5 *Penrose process.*

Thus, as the energy of the ejected particle is greater than the energy of the original particle, the gain of the energy must be the energy extracted from the black hole.

Next the question arises, can we extract all the energy from a black hole by this process. The Penrose process requires the particle (B) to have a counter rotational angular momentum to ensure that it has a negative energy. The angular momentum of the black hole will get reduced on capture of (B) and consequently the black hole will rotate slower. Eventually, all the angular momentum will be extracted and the black hole will stop rotating. As the black hole stops spinning, it will no longer have an ergosphere, as it is due to rotation. As there is no ergosphere there is no more extraction of energy. We are left with an element of the mass-energy called the irreducible mass[2], defined as Schwarzschild black hole, which is the spherically symmetric static solution, $r = 2m$.

The Penrose process extracts only the rotational part of the mass-energy. There is, in fact, an element of the mass-energy called the irreducible mass, defined as

$$m_{irr}^2 = \frac{1}{2} m r_+ \tag{7.54}$$

We rearrange it to get an expression for the total mass-energy in terms of the irreducible mass and the angular momentum

$$\begin{aligned} m^2 &= m_{irr}^2 + \frac{J^2}{4m_{irr}^2} \\ &\geq m_{irr}^2 \end{aligned} \tag{7.55}$$

The irreducible mass cannot, by definition, be reduced by any continuous process. The universal nature of the limit on energy extraction implied by Eq. (7.55) obtained above only for the specific process proposed by Penrose, can be seen from the area theorem. The area of the event horizon of a Kerr black hole is given by

$$\begin{aligned} A &= \int_{r=r_S} \sqrt{g_{22} g_{33}}\, d\theta d\phi \\ &= \int (r_+^2 + a^2) \sin\theta d\theta d\phi \\ &= 4\pi (r_+^2 + a^2) \\ &= 16\pi m_{irr}^2 \end{aligned} \tag{7.56}$$

Thus, from the area theorem we obtain the general result that m_{irr} can never decrease, from which follows the above energy extraction limit for all possible processes.

If a Kerr black hole has arbitrarily large angular momentum, then one could extract an arbitrarily large amount of rotational energy. It is the extremel black hole (where $a = m$) that will give the maximum energy. For a maximally rotating black hole, $J = m^2$, the energy extracted represents $(1-1/\sqrt{2}) \approx 29\%$ of the mass-energy of the black hole[2].

7.7 THE REISSNER–NORDSTRÖM SOLUTION

Hans Reissner, a German aeronautical engineer in 1916 and Gunnar Nordström, a Finnish theoretical physicist in 1918, derived independently a modified version of Schwarzschild solution of Einstein's equations of general relativity for a radially symmetric body of mass M and having an electric charge Q. It is written as follows:

$$ds^2 = f(r)c^2dt^2 - \frac{dr^2}{f(r)} - r^2(d\theta^2 + \sin^2\theta d\phi) \tag{7.57}$$

where
$$f(r) = 1 - \frac{2MG}{rc^2} + \frac{Q^2G}{r^2c^4} \tag{7.58}$$

If the body were to be free of its electrical charge, the Eq. (7.57) will reduce to the metric Eq. (4.40) describing the Schwarzschild solution.

To deduce the Reissner–Nordström solution, we imagine two points A and B stationary relative to each other in a gravitational field and they will occupy two different warped inertial frames, according to the equivalence principle. Imagine that point B lies deep inside the gravitational field and the point A lying far enough above the point B, so as to be effectively at infinity. As such, point A is close to a set of inertial frames characterized by freely falling bodies and passing point B at speed determined by the difference of gravitational potential between the points A and B, thus given by

$$\frac{1}{2}v^2 = \frac{MG}{r} \tag{7.59}$$

Substituting this value of velocity into special Lorentz transformations, one obtains transformation equations equivalent to the Schwarzschild metric equation.

We propose an imaginary experiment in order to know how the warpage of space and time, is brought about. Let a mass M be at an extremely far away distance from the origin of the coordinate system where our gravitating body will eventually be situated. Imagine that the mass M exists in the form of a spherical shell that surrounds the origin and one by one the component shells, each of mass dM, leave the main shell and collapse on to the origin. Each of the collapsing shells warps the spacetime into a gravitational field. The next subsequent shell to peel off marks an inertial frame deformed by the gravitational field and, thus warping the spacetime still further. The work done by the gravitational field on each collapsing shell, increases its mass-energy but that work gets converted into heat when the shell collapses on the body at the origin and the heat energy radiates into space. Ultimately, the body has a total mass $M = \sum dM$, which then surrounds itself with the Schwarzschild metric of Eq. (7.57) with $f(r) = 1 - \frac{2MG}{rc^2}$.

Further imagine that a charge Q is added to the above imaginary experiment. The electric force, does not contribute to the warping of the spacetime, since it does not conform to the equivalence principle. But the process of shell by shell build-up of the charged body, requires the gravitational field to do work in order to overcome the repulsion between the growing body and the shell coming to collapse on it. The work done has to be accounted for out of the final mass of the fully grown body. Assuming that we add an electric charge Q to the gravitating body, then each shell brings mass dM and charge dQ on to the growing body, the work done to overcome the electric repulsion is

$$\begin{aligned} W &= \int\int \frac{Q}{r^2} dQdr \\ &= \frac{Q^2}{2r} \end{aligned} \tag{7.60}$$

where r is distance from the centre of mass of the body and that much work has been done by the gravitational field. As a consequence, the effective mass of the body generating the gravitational field becomes

$$M' = M - \frac{Q^2}{2rc^2} \tag{7.61}$$

Substituting Eq. (7.61) into the Schwarzschild metric, Eq. (4.40), we get the Reissner–Nordström solution for a radially symmetric body having mass M and an electric charge Q

$$ds^2 = \left[1 - \frac{2MG}{rc^2} + \frac{Q^2G}{r^2c^4}\right]c^2dt^2 - \left[1 - \frac{2MG}{rc^2} + \frac{Q^2G}{r^2c^4}\right]^{-1} dr^2 - r^2d\theta^2 - r^2\sin^2\theta d\phi^2 \tag{7.62}$$

We rewrite the Reissner–Nordström metric Eq. (7.62), through the definitions of Schwarzschild radius r_s of the massive body, which is related to its mass M by

$$r_s = \frac{2GM}{c^2} \tag{7.63}$$

and r_Q, a length scale corresponding to the electric charge Q of the mass

$$r_Q = \frac{Q^2G}{c^4} \tag{7.64}$$

Therefore, the Reissner–Nordström solution, Eq. (7.62) becomes

$$ds^2 = \left[1 - \frac{r_s}{r} + \frac{r_Q^2}{r^2}\right]c^2dt^2 - \left[1 - \frac{r_s}{r} + \frac{r_Q^2}{r^2}\right]^{-1} dr^2 - r^2d\theta^2 - r^2\sin^2\theta d\phi^2 \tag{7.65}$$

Although charged black holes with $r_Q \ll r_s$ are similar to Schwarschild black holes; these have two horizons, the event horizon and an internal Cauchy horizon. These horizons are located where g^{rr} diverges, *i.e.*,

$$(g^{rr})^{-1} = 1 - \frac{r_s}{r} + \frac{r_Q^2}{r^2} = \frac{1}{r^2}(r^2 - r_s r + r_Q^2) = 0 \tag{7.66}$$

Thus, the event horizons are given by

$$r_{\pm} = \frac{1}{2}\left(r \pm \sqrt{r_s^2 - 4r_Q^2}\right) \tag{7.67}$$

These concentric event horizons become degenerate when $2r_Q = r_s$, that corresponds to extremel black hole. The black holes with $2r_Q > r_s$ are conjectored not to exist in nature, since they would contain a naked singularity.

7.8 EVIDENCE FOR THE EXISTENCE OF BLACK HOLES

NASA's Hubble space telescope and the Chandra x-ray observatory have independently provided the best direct evidence yet for the existence of an event horizon, the defining feature of a black hole. An event horizon is the theoretical 'one-way ticket' boundary around a black hole, from which nothing, not even light can escape. As no other object, other than a black hole, can have an event horizon, so evidence for its existence is the resounding proof of the existence of black holes in space.

The photographs, Fig. (7.6), released by NASA show gravity drawing gas from a companion star into a black hole in a swirling pattern. As the gas nears the event horizon, a strong gravitational red-shift makes it appear redder and dimmer. When the gas finally crosses the event horizon, it disappears from view, the region within the event horizon appears black.

Astronomers have found convincing evidence for the supermassive black hole in the centre of the giant elliptical galaxy M87, as well as several other galaxies and quasars. The discovery is based on velocity measurements of a whirlpool of hot gas orbiting the black hole. In 1994, Hubble telescope data produced an unprecedented measurement of the mass of an unseen object at the centre of M87. Based on the kinetic energy of the material whirling about the centre, the object is about 3 billion times the mass of our sun and appears to be concentrated into a space smaller than our solar system. A supermassive black hole, known as '*Sagittarius A*' exists at the centre of our own galaxy, the Milky Way. Its existence was proven by tracking the motions of 28 stars circulating around it.

In the case of quasars, other explanations of the output of energies are not as good as the explanation using a supermassive black hole. This can be understood easily. When matter falls in a gravitational field, its speed and therefore, its energy increases. If great amount of matter is falling at the same time, and swirling around in the black hole, then the friction between the various pieces of matter will

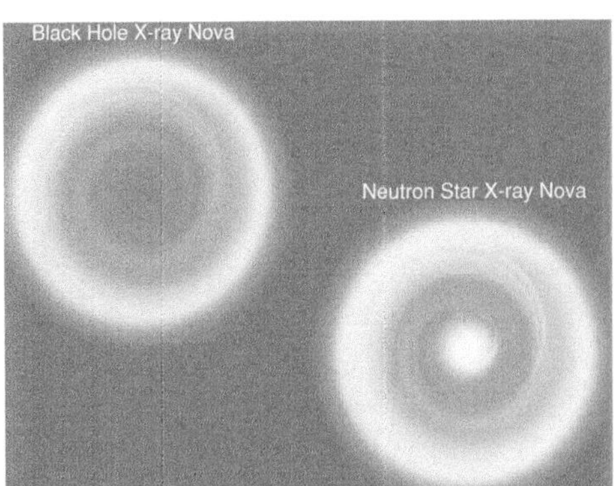

Figure 7.6 *Existence of the event horizon of a black hole* (**Source:** NASA).
(See the colour plate on next page)

turn much of that speed energy picked up during the fall into heat, which then gets radiated away. This way matter surrounding a supermassive black hole, can radiate more energy per gram than can be released by any other mechanism, we know including nuclear fusion.

Furthermore, astronomers have found a half-dozen or so binary star systems (the stars orbiting each other) where one of the stars is invisible, yet must be there since it pulls with enough gravitational force on the other visible star, to make that star orbit around their common centre of gravity and the mass of the invisible star is considerably greater than 3 to 5 solar masses. Therefore, these invisible stars are thought to be good candidates for black holes.

References

1. Penrose, R., *Riv. Nuovo.* (1969). *Cimento Soc. Ital. Fi.,* 1, 252.
2. Christodoulou, D. and Ruffini, R. (1971). *Phys. Rev.,* D4, 3552.

Problems

7.1 (a) For a Kerr black hole, prove that for fixed *m*, the largest area is obtained for $a = 0$ (Schwarzschild).
(b) Conversely prove that for fixed area, the smallest mass is obtained for $a = 0$.

7.2 Show that the Penrose process decreases the angular momentum of the black hole.

7.3 Consider two fixed observers located nearby a Schwarzschild black hole. One observer, located at $r_1 = 3m$, emits a pulse of ultraviolet light to a second observer located at $r_1 = 8m$. Show that light appears to be red-shifted to orange to the second observer.

7.4 Show that space–time is locally flat at the centre of the Kruskel–Ezekeres coordinate system, $u = v = 0$.

7.5 Given the area of a Kerr hole, $A = 4\pi(r_+^2 + a^2)$ where $r_+ = m + \sqrt{m^2 - a^2}$, show that any two holes with masses m_1 and m_2 and angular momenta m_1a_1 and m_2a_2, respectively have total area less than that of a single hole of mass $m_1 + m_2$ and angular momentum $m_1a_1 + m_2a_2$.

8

Cosmology

8.1 INTRODUCTION

Cosmology is the study of the dynamics of the entire universe. The mathematical formulation of cosmology is relatively simpler due to two reasons. The first reason is that gravity dominates on large scales of space–time, so the local complexity arising from electromagnetic and nuclear forces, can be ignored. The second reason for simplicity is that the universe, on a large scale and to a good approximation, is homogeneous and isotropic. The incorporation of the characteristics of homogeneity and isotropy into the spatial part of the metric, enables us to consider spaces of constant curvature.

Consider an n-dimensional space R^n and let k designate its curvature. According to Schurs theorem from differential geometry, if all points in some neighbourhood N about a point, are isotropic and the dimensions of the space $n \geq 3$, then the curvature k is constant throughout N.

The three space $d\sigma^2 = g_{ij}dx^i dx^j$ is homogenous and isotropic. With these properties, we are able to construct a maximally symmetric space with the Riemann tensor in terms of the curvature and the metric, as

$$R_{\mu\nu\rho\sigma} = k(g_{\mu\rho}g_{\nu\sigma} - g_{\mu\sigma}g_{\nu\rho}) \tag{8.1}$$

where $R_{\mu\nu\rho\sigma}$ is the Riemann tensor of the space and k, the curvature of space. In Riemannian geometry, a space with this feature is called a constant curvature space.

8.2 THE COSMOLOGICAL PRINCIPLE AND WEYL'S POSTULATE

The modern cosmology is based on three assumptions, namely cosmological principle, Weyl's postulate and Einstein's general relativity. Application of the theory of general relativity to the large scale structure of the universe, leads to various cosmological theories. The starting point of all these theories is the cosmological principle. Viewed on sufficiently large distance scales, there are no preferred directions or preferred places in the universe. In simple words, the principle implies that averaged over large enough distances, one part of the universe looks approximately like any other part. The principle is to be applied on very large scales, where local variations in density are averaged over. A number of different observations like number count of galaxies and observations of diffuse x-ray and γ-ray backgrounds and more remarkably the 3 K microwave background radiation, lend credence to the validity of the principle. Although the microwave background radiation is not perfectly smooth but the anisotropies are of the order of 10^{-5} or less and thus it constitutes an adequate basis for an approximate description of space–time on large scales. Thus on a large enough scale, *i.e.*, on the level of cluster of galaxies, we assume tacitly that the universe is homogeneous and isotropic in the spatial part of the metric.

Homogeneity means indifference to origin and implies that the nature of universe is invariant under spatial translation. One cannot distinguish one space from the other in the universe. We are not considering local variations such as those in the vicinity of a black hole.

Isotropy means indifference to direction and implies that the spatial view of the universe is rotationally invariant about a point in space, *i.e.*, all spatial directions are equivalent.

It is natural to ask whether the Einstein's field equations based on these two assumptions about the universe,

$$R_{\mu\nu} - \frac{1}{2} g_{\mu\nu} R = -8\pi T_{\mu\nu} \tag{8.2}$$

are consistent with the cosmological observations.

It turned out that these assumptions are not consistent with the field equations and Einstein introduced a modification of his original equations of relativity, by the addition of a positive constant Λ, called cosmological constant, that does render general relativity consistent with the above equations. The modified equations are

$$R_{\mu\nu} - \frac{1}{2} g_{\mu\nu} R + \Lambda g_{\mu\nu} = -8\pi T_{\mu\nu} \tag{8.3}$$

The constant Λ, is such that its effect is negligible in our solar system or even our galaxy. However, it becomes important when we consider universe as a whole.

A second assumption of the standard cosmological model is Weyl's postulate-sometimes denoted principle, assumption or hypothesis. In a version close to Robertson's[1] (1933), it states

> *The world lines of galaxies, or 'fundamental particles', form (on average) a spacetime-filling family of non-intersecting geodesics converging towards the past.*

The importance of the postulate emanates from the fact that it provides a reference frame based on an expanding 'substratum of fundamental particles.' If the geodesic world lines are required to be orthogonal to a series of space-like hypersurfaces, we define a co-moving reference frame in which constant spatial coordinates 'are carried' by the fundamental particles. The time coordinate is a cosmic time which labels the series of hypersurfaces and which is taken as the proper time along any of the particle world line.

The postulate means that the substratum can be represented by a perfect fluid with energy momentum tensor

$$T_\mu^\nu = (\rho + p) u_\mu u^\nu - p\delta_\mu^\nu \tag{8.4}$$

where the energy density ρ and the pressure p depend only on time and related by the equation of state

$$p = p(\rho) \tag{8.5}$$

In addition, the fluid is assumed to be at rest in the co-moving frame in which the spacelike coordinates of each particle are constant along its geodesic. Thus in the synchronous gauge, $u^\mu = (1, 0, 0, 0)$ and T_μ^ν become diagonal

$$T_0^0 = \rho(t), \qquad T_i^j = -p(t)\delta_i^j \tag{8.6}$$

Weyl postulate is not explicitly mentioned in modern texts on cosmology, nonetheless, it is implicitly implied in these texts. In some of the recent literature, Weyl's postulate has been replaced by cosmological principle. However Roberson (1933) clearly states that he uses two assumptions amounting to both Weyl's hypothesis and cosmological principle. Furthermore, Robertson notes that the cosmic time implied by Weyl's postulate 'allows us to give a relatively precise formulation of the assumption that our ideal approximation to the actual world is spatially uniform' and thus suggests that Weyl postulate is actually a precondition for the cosmological principle.

8.3 A SPATIAL METRIC INCORPORATING HOMOGENEITY AND ISOTROPY

The properties of homogeneity and isotropy apply only to the spatial part of the metric. Since we know that the universe is expanding, so these properties cannot be extended to the space–time manifold. Under these conditions, the metric can be written in the general form as

$$ds^2 = dt^2 - a^2(t)\,d\sigma^2 \tag{8.4}$$

where $d\sigma^2$ is the spatial part of the metric and $a(t)$ is the scale factor which describes the expansion of the spatial part of metric. $a(t) > 0$, implies that the universe is expanding.

The universe incorporating the above properties of homogeneity and isotropy and expanding in time, can be visualized by representing it mathematically by a set of three-dimensional space-like hypersurfaces or slices S, stacked on top of each other. An observer who sits at a fixed point in space continues to be there but moves forward in time along a geodesic that is parallel to the time-axis.

Let S represent the spatial hypersurface at time t_1 and S' is the spatial hypersurface at a later time t_2. P and Q are two points on these hypersurfaces and a geodesic moves between these points, Fig. (8.1). On moving from S to S', the spatial coordinates of the points P and Q do not change since the observer is sitting at the same point. The arc length of the geodesic between S and S', is given by the time coordinate, *i.e.*, arc length $= t_2 - t_1$.

Thus,

$$ds^2 = dt^2 \tag{8.5}$$

and the component of the metric is

$$g_{00} = 1 \tag{8.6}$$

In order to derive the spatial components of the metric, we exploit the Schwarzschild metric, since it has the property of spherical symmetry. Re-writing the general form of the Schwarzschild metric, we have

$$ds^2 = e^{\nu(r)}dt^2 - e^{\lambda(r)}dr^2 - r^2(d\theta^2 + \sin^2\theta\, d\phi^2) \tag{8.7}$$

In view of its form, the spatial component of the metric in the present case, can be written as

$$d\sigma^2 = e^{f(r)}dr^2 + r^2 d\theta^2 + r^2\sin^2\theta\, d\phi^2 \tag{8.8}$$

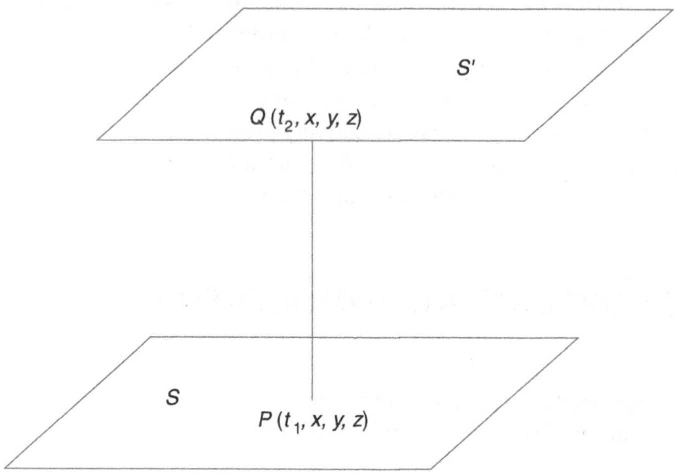

Figure 8.1 *A geodesic at the same spatial point, moving along the time-axis.*

We evaluate the function $e^{f(r)}$ by using our earlier results for the components of the Ricci tensor derived in the case of Schwarzschild metric, Eq. (4.27).

$$\begin{aligned} R_{00} &= -\frac{1}{2}e^{\nu-\lambda}\left(\nu''+\frac{1}{2}\nu'^2-\frac{1}{2}\nu'\lambda'+\frac{2}{r}\nu'\right) \\ R_{11} &= \frac{1}{2}\nu''+\frac{1}{4}\nu'^2-\frac{1}{4}\nu'\lambda'-\frac{\lambda'}{r} \\ R_{22} &= e^{-\lambda}\left(1+\frac{1}{2}r\nu'-\frac{1}{2}r\lambda'\right)-1 \\ R_{33} &= \sin^2\theta\, R_{22} \end{aligned} \tag{8.9}$$

Let us solve the problem term by term. Taking the term R_{11} while allowing for the change of sign for the 3-space vs 4-space and working in the coordinate basis, we get from Eq. (8.9)

$$R_{11} = -\left[\frac{1}{2}\nu''+\frac{1}{4}\nu'^2-\frac{1}{4}\nu'\lambda'-\frac{\lambda'}{r}\right] \tag{8.10}$$

To obtain the equations in the present case, we set $\nu \to 0$ and $\lambda \to f$, therefore

$$R_{11} = \frac{1}{r}\frac{df}{dr} \tag{8.11}$$

Next using Eq. (8.1), we obtain the Ricci tensor in terms of the scalar curvature k and the metric g_{ij}. We use the Latin indices i, j, k, l as we are considering only the spatial part of the metric.

The Ricci tensor is

$$R_{jl} = R^k_{jkl} \tag{8.12}$$

Thus,

$$\begin{aligned} R^k_{jkl} &= g^{ki} R_{ijkl} \\ &= g^{ki} k(g_{ik} g_{jl} - g_{il} g_{jk}) \\ &= k(g^{ki} g_{ik} g_{jl} - g^{ki} g_{il} g_{jk}) \\ &= k(\delta^k_k g_{jl} - \delta^k_l g_{jk}) \\ &= k(\delta^k_k g_{jl} - g_{jl}) \end{aligned} \tag{8.13}$$

As we are working in three dimensions, we have

$$\delta^k_k = 1+1+1 = 3 \tag{8.14}$$

Therefore, the Ricci tensor in the constant curvature case is given by

$$\begin{aligned} R_{jl} &= k(3g_{jl} - g_{jl}) \\ &= 2kg_{jl} \end{aligned} \tag{8.15}$$

Since $d\sigma^2 = e^{f(r)} dr^2 + r^2 d\theta^2 + r^2 \sin^2\theta d\phi^2$, we get the metric as

$$g_{ij} = \begin{pmatrix} e^f & 0 & 0 \\ 0 & r^2 & 0 \\ 0 & 0 & r^2 \sin^2\theta \end{pmatrix} \tag{8.16}$$

Employing $g_{11} = e^f$ and $R_{jl} = 2kg_{jl}$, alongwith Eq. (8.11), we get the differential equation

$$\frac{1}{r}\frac{df}{dr} = 2ke^f$$

or

$$e^{-f} df = 2kr\, dr \tag{8.17}$$

Integrating, we obtain

$$-e^{-f} + C = kr^2$$

which leads to

$$e^f = \frac{1}{C - kr^2} \tag{8.18}$$

The constant of integration C is determined by using the other components of the Ricci tensor. In the coordinate basis, R_{22} for the Schwarzschild metric is, Eq. (8.9)

$$R_{22} = 1 - e^{-\lambda} - \frac{1}{2} r\nu' e^{-\lambda} + \frac{1}{2} r\lambda' e^{-\lambda} \tag{8.19}$$

Using $v \to 0$ and $\lambda \to f$, we get

$$R_{22} = 1 - e^{-f} + \tfrac{1}{2} r \frac{df}{dr} e^{-f} \tag{8.20}$$

Making use of the previous result $\frac{1}{r}\frac{df}{dr} = 2k\, e^{f}$, we get

$$\frac{1}{2} r \frac{df}{dr} e^{-f} = kr^2 \tag{8.21}$$

Therefore,

$$\begin{aligned} R_{22} &= 1 - e^{-f} + kr^2 \\ &= 1 + kr^2 - C + kr^2 \end{aligned} \tag{8.22}$$

Next using the result $R_{jl} = 2kg_{jl}$ alongwith $g_{22} = r^2$, we obtain

$$R_{22} = 2kg_{22} = 2r^2 k \tag{8.23}$$

Hence

$$1 + 2r^2 k - C = 2r^2 k$$

or

$$C = 1 \tag{8.24}$$

The spatial part of the metric becomes

$$d\sigma^2 = \frac{dr^2}{1 - kr^2} + rd\theta^2 + r^2 \sin^2\theta d\phi^2 \tag{8.25}$$

We assume that the curvature scalar is normalized and denoted by k. There are three possible cases corresponding to the values of k. If $k = +1$, it corresponds to positive curvature, when $k = -1$, it corresponds to negative curvature and finally if $k = 0$, it relates to flat space.

8.4 SPACES OF POSITIVE, NEGATIVE, AND ZERO CURVATURE

Let us consider the three cases that arise with the normalized curvature scalar k.

1. **Case I** $k = -1$ The spatial metric can be written as

$$d\sigma^2 = \frac{dr^2}{1 - r^2} + r^2 d\theta^2 + r^2 \sin^2\theta d\phi^2 \tag{8.26}$$

which in the more general form is

$$d\sigma^2 = \frac{dr^2}{1 - r^2(\chi)} + r^2(\chi) d\theta^2 + r^2(\chi) \sin^2\theta d\phi^2 \tag{8.27}$$

Putting $r(\chi) = \sin\chi$, we get

$$d\sigma^2 = d\chi^2 + \sin^2\chi d\theta^2 + \sin^2\chi \sin^2\theta d\phi^2 \tag{8.28}$$

This space can be understood by examining the surface that results by setting θ equal to some constant value, say $\theta = \pi/2$. The line element, then is

$$d\sigma^2 = d\chi^2 + \sin^2\chi d\phi^2 \tag{8.29}$$

This surface is a two sphere, Fig. (8.2).

A three-dimensional space with constant curvature has two similarities with the surface of a sphere. Starting at some point and travelling in a straight line on the sphere, we will end up at the same point eventually. This will also hold in the three-dimensional space of a universe with positive curvature. Secondly, a triangle drawn on the surface of a sphere will have the sum of its angles greater than 180°

2. **Case II** $k = -1$ It corresponds to negative curvature. The spatial metric becomes

$$d\sigma^2 = \frac{dr^2}{1+r^2} + r^2 d\theta^2 + r^2 \sin^2\theta d\phi^2 \tag{8.30}$$

which in the more general form is

$$d\sigma^2 = \frac{dr^2}{1+r^2(\chi)} + r^2(\chi) d\theta^2 + r^2(\chi)\sin^2\theta d\phi^2 \tag{8.31}$$

Putting $r(\chi) = \sin h\chi$, the line element becomes

$$d\sigma^2 = d\chi^2 + \sinh^2\chi d\theta^2 + \sinh^2\chi \sin^2\theta d\phi^2 \tag{8.32}$$

For the metric $ds^2 = dt^2 - a^2(t) d\sigma^2$, the spatial slices have the property that they have infinite volume. The sum of the angles of the triangle add up to less than 180°.

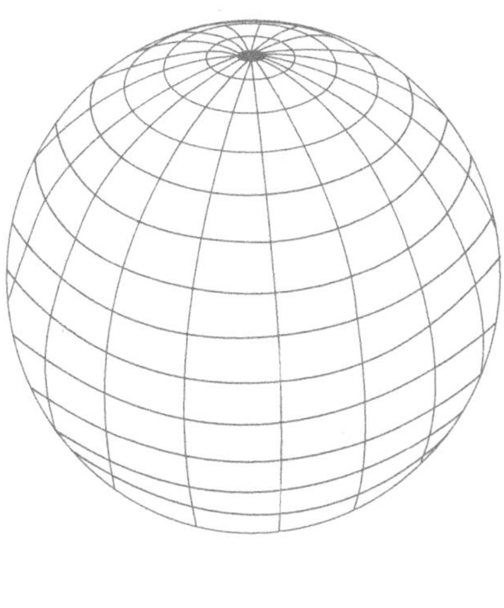

Figure 8.2 *For a space of positive curvature k=+1, with $\theta = \pi/2$, the two surface is a sphere.*

Putting $\theta = \pi / 2$, for obtaining an embedding diagram, we get

$$d\sigma^2 = d\chi^2 + \sinh^2 \chi d\phi^2 \tag{8.33}$$

The embedding diagram for a surface with negative curvature is a saddle, Fig. (8.3). This is an example of an open universe. This can extend to infinity.

3. **Case III** $k = 0$. It corresponds to the case of zero curvature. The line element for $r = \chi$, becomes

$$d\sigma^2 = d\chi^2 + \chi^2 d\theta^2 + \chi^2 \sin^2 \theta d\phi^2 \tag{8.34}$$

which is a flat Euclidean space. Putting $\theta = \pi / 2$, we get

$$d\sigma^2 = d\chi^2 + \chi^2 d\phi^2 \tag{8.35}$$

which is a flat plane for the embedding diagram, Fig. (8.4).

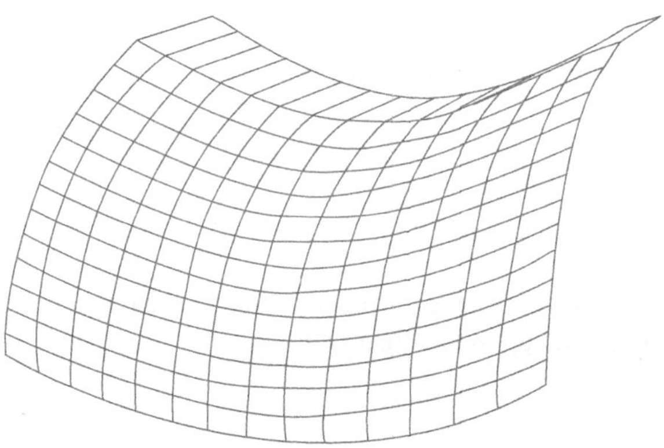

Figure 8.3 *A surface of negative curvature, $k = -1$, is a saddle.*

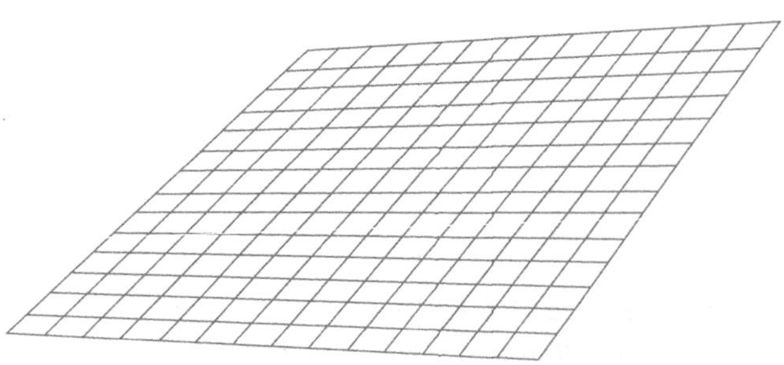

Figure 8.4 *With $k = 0$, the space is perfectly flat.*

8.5 STATIC COSMOLOGICAL MODELS

The models of the universe based on the following three postulates lead to static cosmological models. These are:

- The pressure and proper density, as measured by a local observer, will be the same everywhere.
- Universe is isotropic
- Universe is homogeneous

The following three elements

- Einstein line-element
- de-Sitter line-element
- Special relativity line-element

cover all the possible line-elements admitted by a static, isotropic and homogeneous universe.

The line-element, satisfying the requirement of spatial symmetry, is given by

$$ds^2 = e^{\nu} dt^2 - e^{\lambda} dr^2 - r^2 (d\theta^2 + \sin^2\theta d\phi^2) \tag{8.36}$$

where $\nu = \nu(r)$ and $\lambda = \lambda(r)$ are to be determined.

The pressure p_0 and density ρ_0 determined by the field equations

$$R_{\mu\nu} - \frac{1}{2} g_{\mu\nu} R + \Lambda g_{\mu\nu} = -8\pi T_{\mu\nu} \tag{8.37}$$

are the Schwarzschild's interior solution, Eqs. (5.27)–(5.29)

$$8\pi p_0 = e^{-\lambda}\left(\frac{\nu'}{r} + \frac{1}{r^2}\right) - \frac{1}{r^2} + \Lambda$$

$$8\pi p_0 = e^{-\lambda}\left(\frac{\nu'}{r} - \frac{1}{r^2}\right) + \frac{1}{r^2} - \Lambda$$

$$\frac{dp_0}{dr} = -\frac{1}{2}(p_0 + \rho_0)\nu' \tag{8.38}$$

The solution of Eqs. (8.38) has to be in compliance with the postulates and hence must fulfill the following requirements.

- The pressure p_0 and proper density ρ_0 as measured by a local observer, will be the same everywhere
- The line-element for small values of *r*, should reduce to the flat space–time of special relativity, since we know that local gravitational fields in small space–time regions, can be neglected and hence special relativity is applicable there.

To satisfy these requirements, the last equation of the set (8.38) requires that

$$\frac{dp_0}{dr} = -\frac{1}{2}(p_0 + \rho_0)\nu' = 0 \tag{8.39}$$

Equation (8.39) is satisfied with any of the three situations:

i. $\nu' = 0$

ii. $p_0 + \rho_0 = 0$

iii. $\nu'(p_0 + \rho_0) = 0,$ (8.40)

These three solutions lead respectively to the Einstein, de-Sitter and the special relativity line-elements of the universe-a universe, which is static, isotropic and homogeneous. These three cases exhaust all the possibilities of such a universe.

8.5.1 Einstein Universe

The Einstein Line-Element Let us take the first possibility, *i.e.*,

$$\nu' = 0 \tag{8.41}$$

On integration, we get

$$\nu = C \tag{8.42}$$

From (iii) condition, $\lambda = 0 = \nu$ at $r = 0$. Thus, from Eq. (8.42), we get

$$C = 0$$

And the first equation in Eqs. (8.38), becomes

$$e^{-\lambda} = 1 - (\Lambda - 8\pi p_0) r^2 = 1 - \frac{r^2}{R_0^2} \tag{8.43}$$

where $\frac{1}{R_0^2} = (\Lambda - 8\pi p_0)$.

The Eq (8.36) for line-element becomes

$$ds^2 = dt^2 - \frac{dr^2}{(1 - r^2 / R_0^2)} - r^2 (d\theta^2 + \sin^2\theta d\phi^2) \tag{8.44}$$

This is known as Einstein's line-element for static homogeneous universe.

Geometry of Einstein Universe The Einstein line-element can be expressed in different forms. Let us consider the transformation

$$r = \frac{\rho}{\left(1 + \frac{\rho^2}{4R_0^2}\right)}$$

Then

$$r\left(1 + \frac{\rho^2}{4R_0^2}\right) = \rho$$

Differentiating, we get

$$dr\left(1+\frac{\rho^2}{4R_0^2}\right)+\frac{2\rho r}{4R_0^2}d\rho = d\rho$$

which gives

$$dr = \frac{1-\dfrac{\rho^2}{2R_0^2}}{1+\dfrac{\rho^2}{4R_0^2}}d\rho$$

and

$$\frac{dr^2}{1-\dfrac{r^2}{R_0^2}} = \frac{1}{1-\dfrac{r^2}{R_0^2}}\left(\frac{1-\dfrac{\rho^2}{2R_0^2}}{1+\dfrac{\rho^2}{4R_0^2}}\right)^2 d\rho^2$$

or after simplification

$$\frac{dr^2}{1-\dfrac{r^2}{R_0^2}} = \frac{d\rho^2}{\left(1+\dfrac{\rho^2}{4R_0^2}\right)} \tag{8.45}$$

Einstein line-element, Eq. (8.44) becomes

$$ds^2 = dt^2 - \frac{1}{(1+\rho^2/4R_0^2)^2}[d\rho^2+\rho^2(d\theta^2+\sin^2\theta d\phi^2)] \tag{8.46}$$

Consider a second transformation

$$z_1 = R_0\left(-\sqrt{1-r^2/R_0^2}\right)$$
$$z_2 = r\sin\theta\cos\phi$$
$$z_3 = r\sin\theta\sin\phi$$
$$z_4 = r\cos\theta$$

with

$$z_1^2+z_2^2+z_3^2+z_4^2 = R_0^2 \tag{8.47}$$

which is the equation of a hypersphere of radius R_0.

The Einstein line-element takes the form

$$ds^2 = dt^2 - (dz_1^2+dz_2^2+dz_3^2+dz_4^2) \tag{8.48}$$

This form of Einstein's line-element suggests that the spatial extent of the universe holds on the surface of a hypersphere in Euclidean space (z_1, z_2, z_3, z_4) of four-dimensions. It shows the isotropic and homogeneous nature of the universe.

Let us consider the transformation

$$r = R\sin\chi \tag{8.49}$$

Substituting it in Eq (8.36), we get

$$ds^2 = dt^2 - R_0^2(d\chi^2 + \sin^2\chi d\theta^2 + \sin^2\chi \sin^2\theta d\phi^2) \tag{8.50}$$

We know that $0 \le \theta \le \pi, 0 \le \phi \le 2\pi$. Regarding χ, we find that line-element Eq. (8.50) is the same at $\chi = 0$ and $\chi + \pi$, other variables θ and ϕ being arbitrary. This implies that an event at $\chi = 0$, has a similar event at $\chi = \pi$. The existence of a mirror event for any event means that Einstein universe may be taken as spherical.

The proper distance around the universe (major circle) is

$$l_0 = 2\int_0^{\pi} R_0 d\chi = 2\pi R_0 \tag{8.51}$$

and the proper volume V_0 of the spherical universe is

$$\begin{aligned} V_0 &= \int_{\chi=0}^{\pi}\int_{\theta=0}^{\pi}\int_{\phi=0}^{2\pi} (R_0 d\chi)(R_0 \sin\chi d\theta)(R_0 \sin\theta\, \chi d\phi) \\ &= 4\pi R_0^3 \int_0^{\pi} \frac{1}{2}(1-\cos 2\chi)\, d\chi = 2\pi R_0^2 \left[1 - \frac{1}{2}\sin 2\chi\right]_0^{\pi} \\ &= 2\pi^2 R_0^3 \end{aligned} \tag{8.52}$$

Thus the proper volume of the so called spherical universe is $2\pi^2 R_0^3$.

Alternatively, if we take the two events at $\chi = 0$ and $\chi = \pi$ and one and the same, then the Einstein universe is taken to be elliptical. The proper volume of this universe is

$$\begin{aligned} &\int_{r=0}^{R_0}\int_{\theta=0}^{\pi} \frac{drrd\theta\, r\sin\theta d\phi}{\sqrt{\left(1-\frac{r^2}{R_0^2}\right)}} = 4\pi\int_0^{R_0} \frac{r^2\, dr}{\sqrt{\left(1-\frac{r^2}{R_0^2}\right)}} \\ &= 2\pi R_0^3 \int_0^{\pi/2} (1-\cos 2\eta)\, d\eta = 2\pi R_0^3 \left[\eta - \frac{1}{2}\sin 2\eta\right]_0^{\pi/2} \\ &= \pi^2 R_0^3 \end{aligned} \tag{8.53}$$

The proper volume of the so called elliptical universe is one half of the spherical universe.

The proper distance in case of elliptical universe is

$$\begin{aligned} l_0 &= 2\int_0^{R_0} \frac{dr}{\sqrt{1 - r^2 / R_0^2}} \\ &= 2\int_0^{\pi/2} \frac{R_0 \cos\eta d\eta}{\cos\eta} \quad \text{where } \frac{r}{R_0} = \sin\eta \\ &= \pi R_0 \end{aligned} \tag{8.54}$$

Density and Pressure of the Matter in Einstein's Universe We have from Eqs. (8.38)

$$8\pi p_0 = e^{-\lambda}\left(\frac{\nu'}{r}+\frac{1}{r^2}\right)-\frac{1}{r^2}+\Lambda \tag{8.55}$$

$$8\pi\rho_0 = e^{-\lambda}\left(\frac{\lambda'}{r}-\frac{1}{r^2}\right)+\frac{1}{r^2}-\Lambda \tag{8.56}$$

where

$$\nu' = 0 = \nu$$

$$e^{-\lambda} = 1-\frac{r^2}{R_0^2}$$

or

$$e^{-\lambda}\frac{\lambda'}{r} = \frac{2}{R_0^2} \tag{8.57}$$

Equation (8.55), in the light of Eq. (8.57) becomes

$$8\pi p_0 = \left(1-\frac{r^2}{R_0^2}\right)\frac{1}{r^2}-\frac{1}{r^2}+\Lambda$$

$$= -\frac{1}{R_0^2}+\Lambda \tag{8.58}$$

Rewriting Eq (8.56), in the light of Eq. (8.57), we get

$$8\pi\rho_0 = \frac{2}{R_0^2}-\left(\frac{1}{r^2}-\frac{1}{R_0^2}\right)+\frac{1}{r^2}-\Lambda$$

$$= \frac{3}{R_0^2}-\Lambda \tag{8.59}$$

From Eqs. (8.58) and (8.59), we get

$$8\pi(p_0+\rho_0) = \frac{2}{R_0^2} \tag{8.60}$$

We consider the following three cases.

(i) **Case I.** Suppose that the universe is filled with incoherent matter, such as free particles (stars), exerting no pressure. Then $p_0 = 0$ and Eq. (8.60) becomes

$$8\pi\rho_0 = \frac{2}{R_0^2}$$

or

$$\rho_0 = \frac{1}{4\pi R_0^2} \tag{8.61}$$

which gives

$$\begin{aligned}\text{Mass of the spherical universe} &= \rho_0 V_0 \\ &= \frac{1}{4\pi R_0^2} 2\pi^2 R_0^3 \\ &= \frac{\pi R_0}{2}\end{aligned} \tag{8.62}$$

$$\begin{aligned}\text{Analogously, the mass of elliptical universe} &= \frac{1}{4\pi R_0^2} \pi^2 R_0^3 \\ &= \frac{\pi R_0}{4}\end{aligned} \tag{8.63}$$

(ii) **Case II** Suppose the universe is completely filled with radiation for which $\rho_0 = 3p_0$. Then Eq. (8.60) becomes

$$p_0 = \frac{1}{16\pi R_0^2} = \frac{\rho_0}{3} \tag{8.64}$$

$$\begin{aligned}\text{Mass of the spherical universe} &= V_0 \rho_0 \\ &= 2\pi^2 R_0^3 \times \frac{3}{16\pi R_0^2} \\ &= \frac{3\pi R_0}{8}\end{aligned} \tag{8.65}$$

(iii) **Case III** Suppose that the universe is completely empty. Then

$$P_0 = \rho_0 = 0 \tag{8.66}$$

The Eqs. (8.58) and (8.59) become

$$\Lambda = \frac{1}{R_0^2}$$

and

$$\Lambda = \frac{3}{R_0^2} \tag{8.67}$$

This leads to

$$\Lambda = \frac{1}{R_0^2} = 0$$

Thus, the line-element, Eq. (8.46) becomes

$$ds^2 = dt^2 - dr^2 - r^2 d\theta^2 - r^2 \sin^2 \theta d\phi^2 \tag{8.68}$$

which is the line-element of special relativity for flat space time.

Motion of a Particle in the Einstein Universe The geodesic equations governing the motion of a particle are

$$\frac{d^2 x^\alpha}{ds^2} + \Gamma^\alpha_{\mu\nu} \frac{dx^\mu}{ds} \frac{dx^\nu}{ds} = 0 \tag{8.69}$$

Due to the static nature of the universe, we make the assumption that the particle is initially at rest, so that the velocity components become

$$\frac{dx^0}{ds} = 1, \frac{dx^1}{ds} = \frac{dx^2}{ds} = \frac{dx^3}{ds} = 0 \tag{8.70}$$

In view of Eq. (8.70), the Eq. (8.69) becomes

$$\frac{d^2 x^\alpha}{ds^2} + \Gamma^\alpha_{00} \frac{dx^0}{ds} \frac{dx^0}{ds} = 0$$

But for the Einstein line-element

$$ds^2 = dt^2 - \frac{dr^2}{1 - r^2 / R_0^2} - r^2 d\theta^2 - r^2 \sin^2 \theta d\phi^2 \tag{8.71}$$

We get from Eq. (4.20) that

$$\Gamma^\alpha_{00} = 0 \tag{8.72}$$

for all values of α.Thus we have

$$\frac{d^2 r}{ds^2} = \frac{d^2 \theta}{ds^2} = \frac{d^2 \phi}{ds^2} = 0 \tag{8.73}$$

and the particle will have no acceleration.Therefore, in Einstein universe, matter is without motion.

Velocity of Light in Einstein Universe For light $ds = 0$, so we have from Eq. (8.71)

$$0 = dt^2 - \frac{dr^2}{1 - r^2 / R_0^2} - r^2 d\theta^2 - r^2 \sin^2 \theta d\phi^2 \tag{8.74}$$

The radial velocity is

$$\left(\frac{dr}{dt}\right)^2 = \left(1 - \frac{r^2}{R_0^2}\right)$$

or
$$\frac{dr}{dt} = \sqrt{1 - \frac{r^2}{R_0^2}} \tag{8.75}$$

Thus the time to go around the universe once is

$$t = 4\int_0^{R_0} \frac{dr}{\sqrt{1 - \frac{r^2}{R_0^2}}}$$

$$= 4R_0 \sin^{-1} \frac{r}{R_0}\Bigg|_0^{R_0}$$

$$= 2\pi R_0 \quad \text{(in gravitationalunits)}$$

$$= \frac{2\pi R_0}{c} \quad \text{(in c.g.units)} \tag{8.76}$$

for spherical space. One can easily show that for elliptical space, the expression for time to go around the universe once is $\frac{\pi R_0}{c}$.

Doppler Shift in Spectral Lines Let us consider that an observer is situated at $r = 0$ and the light source, say a star at $r = r_1$, both being permanently at rest. For a ray of light emanating from the star and traveling along the radial direction towards the origin, we have

$$d\theta, d\phi = 0$$

so we get from Eq. (8.74)

$$0 = dt^2 - \frac{dr^2}{1 - r^2/R_0^2}$$

or
$$dt = \pm \frac{dr}{\sqrt{1 - r^2/R_0^2}} \tag{8.77}$$

A light pulse leaves the star at time t_1 and reaches the observer at time t_2, which is given by

$$\int_{t_2}^{t_2} dt = \int_0^{r_1} \frac{1}{\sqrt{(1 - r^2/R_0^2)}}$$

or
$$t_2 - t_1 = \int_0^{r_1} \frac{1}{\sqrt{(1 - r^2/R_0^2)}}$$

$$= R_0 \sin^{-1} \frac{r_1}{R_0} \tag{8.78}$$

As in Einstein universe, a particle at rest remains at rest, the distance r_1 is constant. Taking differentials in Eq. (8.78), we get

$$dt_2 - dt_1 = 0$$

or

$$dt_2 / dt_1 = 1 \tag{8.79}$$

But in Einstein universe the coordinate time t is identical to the proper time as measured by the local observer at rest w.r.t. the coordinate time. Thus the equality, Eq (8.79) also implies the equality of the proper periods of the emitted and observed light by the observer at rest w.r.t. original source at rest at the origin. There is no change in the wavelength of light at emission and at reception. Therefore there is no Doppler shift in the light received from distant stars. Thus the nebulae do not recede from each other and the actual universe is not expanding according to the Einstein universe.

Summarily, it may be stated that according to Einstein static universe, it has the following conclusions.

- It contains matter without motion.
- There is no Doppler shift in the spectral lines. It contradicts the observation that the nebulae are receding from each other and that the universe is expanding.

In our concluding remarks on Einstein universe, let us see how it compares with the actual universe.

The most satisfactory feature of Einstein universe, is its prediction of containing a finite concentration of uniformly distributed matter. In this respect, it is superior to the de-Sitter universe. This advantage is achieved at the expense of introducing the extra cosmological term $\Lambda g_{\mu\nu}$ into the original field equations derived by Einstein. However, the most unsatisfactory feature, is its prediction of no Doppler shift in the wavelength of light from distant stars. This goes contrary to the actual observations of Hubble and Humason, which show a definite red shift in the light from nebulae which increases with distance. This limitation of Einstein universe is the main consideration for presence of non-static models of universe as a basis for actual cosmology.

8.5.2 De-Sitter Universe

De-Sitter Line-Element We now consider the second possibility according to which

$$\rho_0 + p_0 = 0 \tag{8.80}$$

Adding the first two equations of the set Eqs. (8.38), we get

$$8\pi(\rho + p_0) = e^{-\lambda}\left(\frac{\nu' + \lambda}{r}\right)$$

which in view of Eq. (8.80) yields

$$e^{-\lambda}\left(\frac{\nu' + \lambda'}{r}\right) = 0$$

or

$$\nu' + \lambda' = 0 \tag{8.81}$$

On integration, it yields

$$\nu + \lambda = C$$

where C is the constant of integration.

According to the boundary condition that at $r = 0$, the line-element should approach the special relativity, we get $C = 0$. Thus

$$\nu = -\lambda \tag{8.82}$$

From Eq. (8.38), we get

$$8\pi\rho_0 = e^{-\lambda}\left(\frac{\lambda'}{r} - \frac{1}{r^2}\right) + \frac{1}{r^2} - \Lambda$$

or

$$(8\pi\rho_0 + \Lambda)r^2 - 1 = e^{-\lambda}(\lambda' r - 1)$$

or

$$1 - (8\pi\rho_0 + \Lambda)r^2 = \frac{d}{dr}(re^{-\lambda}) \tag{8.83}$$

Integrating it, we obtain

$$re^{-\lambda} = r - \frac{r^3}{3}(8\pi\rho_0 + \Lambda) + C_1$$

The condition that as $r \to 0$, both λ and $\nu \to 0$, yields $C_1 = 0$. Therefore,

$$e^{-\lambda} = e^{\nu} = 1 - \frac{r^2(8\pi\rho_0 + \Lambda)}{3} \tag{8.84}$$

Calling

$$\frac{1}{R_0^2} = \frac{8\pi\rho_0 + \Lambda}{3}$$

one gets

$$ds^2 = (1 - r^2 / R_0^2)dt^2 - \frac{dr^2}{1 - r^2 / R_0^2} - r^2 d\theta^2 - r^2 \sin^2\theta d\phi^2 \tag{8.85}$$

This line-element is called de-sitter line-element for a static homogeneous universe, since is was discovered by de-sitter.

Geometry of De-Sitter Universe By coordinate transformations, the line-element Eq. (8.85) can be put in different forms.

Putting

$$r = R_0 \sin\chi \tag{8.86}$$

In Eq (8.85), we get

$$ds^2 = \cos^{2\chi} dt^2 - R_0^2\left[d\chi^2 + \sin^2\chi(d\theta^2 + \sin^2\theta d\phi^2)\right] \tag{8.87}$$

The line-element may be simplified further by introducing a large number of variables

$$\begin{aligned} \alpha &= r\sin\theta\cos\phi \\ \beta &= r\sin\theta\sin\phi \\ \gamma &= r\cos\theta \end{aligned} \tag{8.88}$$

$$\begin{aligned} \delta + \varepsilon &= R_0 e^{t/R_0}\sqrt{1 - r^2/R_0^2} \\ \delta - \varepsilon &= R_0 e^{-t/R_0}\sqrt{1 - r^2/R_0^2} \end{aligned}$$

Noting that

$$d\alpha^2 + d\beta^2 + d\lambda^2 = dr^2 + r^2 d\theta^2 + r^2\sin^2\theta d\phi^2 \tag{8.89}$$

$$d(\delta^2 - \varepsilon^2) = d(\delta + \varepsilon)d(\delta - \varepsilon)$$

Now $$d(\delta + \varepsilon) = e^{t/R_0}\left[(1 - r^2/R_0^2)^{1/2} dt - \frac{rdr}{R_0\sqrt{1 - r^2/R_0^2}}\right]$$

$$d(\delta - \varepsilon) = e^{-t/R_0}\left[-(1 - r^2/R_0^2)^{1/2} dt - \frac{rdr}{R_0\sqrt{1 - r^2/R_0^2}}\right]$$

We obtain $$d(\delta + \varepsilon)d(\delta - \varepsilon) = -\left(1 - \frac{r^2}{R_0^2}\right)dt^2 + \frac{r^2}{R_0^2\left(1 - \frac{r^2}{R_0^2}\right)^{dr^2}} \tag{8.90}$$

Combining Eqs. (8.89) and (8.90), we get

$$\begin{aligned} -d\alpha^2 - d\beta^2 - d\gamma^2 - d\delta^2 + d\varepsilon^2 &= -dr^2 - r^2 d\theta^2 - r^2\sin\theta d\phi^2 + \left(1 - \frac{r^2}{R_0^2}\right)dt^2 - \frac{r^2}{R_0^2}\left(1 - \frac{r^2}{R_0^2}\right)^{-1} dr^2 \\ &= (1 - r^2/R_0^2)dt^2 - \frac{dr^2}{1 - r^2/R_0^2} - r^2 d\theta^2 - r^2\sin^2\theta d\phi^2 \end{aligned}$$

Therefore, $$ds^2 = -d\alpha^2 - d\beta^2 - d\gamma^2 - d\delta^2 + d\varepsilon^2 \tag{8.91}$$

Furthermore, we introduce

$$z_1 = i\alpha,\ z_2 = i\beta,\ z_3 = i\gamma,\ z_4 = i\delta,\ z_5 = \varepsilon$$

and get $$ds^2 = dz_1^2 + dz_2^2 + dz_3^2 + dz_4^2 + dz_5^2 \tag{8.92}$$

with $$z_1^2 + z_2^2 + z_3^2 + z_4^2 + z_5^2 = (iR_0)^2 \tag{8.93}$$

Eq. (8.93) suggests that the geometry of the de-Sitter universe is one which holds on the surface of a sphere embedded in a Euclidean space of five dimensions. The symmetric form of the line-element shows that the universe is isotropic and homogeneous. But this simplicity at the cost of losing the distinction of space-like and time-like intervals.

Another useful form of the de-Sitter line-element is obtained with the help of Lemaitre-Robinson transformations

$$r' = \frac{re^{-t/R_0}}{\sqrt{\left(1-\dfrac{r^2}{R_0^2}\right)}} \tag{8.94}$$

$$t' = t + \frac{1}{2}R_0 \log\sqrt{\left(1-\frac{r^2}{R_0^2}\right)}$$

With these transformations, Eq. (8.87) becomes

$$ds^2 = -e^{2t'/R_0}\left[dr'^2 + r'^2(d\theta^2 + \sin^2\theta d\phi^2)\right] + dt'^2 \tag{8.95}$$

Dropping dashes, one obtains

$$ds^2 = -e^{2t/R_0}\left[dr^2 + r^2(d\theta^2 + \sin^2\theta d\phi^2)\right] + dt^2$$

and taking, $k = \frac{1}{R_0}$, we get

$$ds^2 = -e^{2kt}\left\lfloor dr^2 + r^2(d\theta^2 + \sin^2\theta d\phi^2)\right\rfloor + dt^2 \tag{8.96}$$

Expressed in Cartesian coordinates, it becomes

$$ds^2 = -e^{2kt}[dx^2 + dy + dz^2] + dt^2 \tag{8.97}$$

As is obvious, with this transformation, the static line-element has been converted into a non-static line-element.

Pressure and Density of Matter in de-Sitter Universe The de-Sitter line-element is based on the assumption that

$$\rho_0 + p_0 = 0 \tag{8.98}$$

On physical considerations, the proper density ρ_0 can be either zero or greater than zero, *i.e.*, $\rho_0 \geq 0$. Now if we imagine the universe to be filled by some idealized fluid which exhibits cohesive forces, the negative pressure equal to the density, cannot be achieved by any known material. The Eq (8.98), can be reconciled if we assume that

$$\rho_0 = 0 = p_0 \tag{8.99}$$

This equation implies that the de-sitter universe is completely empty and is devoid of any matter or radiation.

In view of Eq (8.99), the de-Sitter line-element becomes

$$ds^2 = (1 - r^2/R_0^2)dt^2 - \frac{dr^2}{1 - r^2/R_0^2} - r^2 d\theta^2 - r^2 \sin^2\theta d\phi^2 \tag{8.100}$$

where

$$\frac{1}{R_0^2} = \frac{\Lambda}{3} \tag{8.101}$$

The Eq. (8.101) does not fix up the sign of the cosmological constant Λ. Depending whether Λ is positive, zero or negative, the de-sitter line-element has the form

$$\begin{aligned}
&\text{(a)}\quad ds^2 = (1 - \Lambda r^2/3)dt^2 - \frac{dr^2}{1 - \Lambda r^2/3} - r^2 d\theta^2 - r^2 \sin^2\theta d\phi^2 \qquad \Lambda > 0 \\
&\text{(b)}\quad ds^2 = dt^2 - dr^2 - r^2 d\theta^2 - r^2 \sin^2\theta d\phi^2 \qquad \Lambda = 0 \\
&\text{(c)}\quad ds^2 = (1 + \Lambda r^2/3)dt^2 - \frac{dr^2}{1 + \Lambda r^2/3} - r^2 d\theta^2 - r^2 \sin^2\theta d\phi^2 \qquad \Lambda < 0
\end{aligned} \tag{8.102}$$

The models represented by the line-elements a, b,and c respectively are called closed, flat and spatially open but curved. However, we will assume that the cosmological constant Λ is positive.

Motion of a Test Particle and Light Rays in De-Sitter Universe The geodesic equation describing the motion of a test particle is

$$\frac{d^2 x^\alpha}{ds^2} + \Gamma^\alpha_{\mu\nu} \frac{dx^\mu}{ds}\frac{dx^\nu}{ds} = 0 \tag{8.103}$$

Consider the de-Sitter line-element in the general form

$$ds^2 = e^\nu dt^2 - e^\lambda dr^2 - r^2(d\theta^2 + \sin^2\theta d\phi^2) \tag{8.104}$$

with

$$e^{-\lambda} = e^\nu = 1 - \frac{r^2}{R_0^2}$$

$$\begin{aligned}
&\Gamma^1_{11} = \lambda'/2 \\
&\Gamma^1_{22} = -re^{-\lambda} \\
&\Gamma^1_{00} = \nu' e^{\nu-\lambda}/2 \qquad \Gamma^1_{33} = -r\sin^2\theta\, e^{-\lambda} \\
&\Gamma^2_{12} = \frac{1}{r} = \Gamma^3_{13} \\
&\Gamma^3_{23} = \cot\theta \\
&\Gamma^0_{10} = \frac{\nu'}{2} \qquad \Gamma^2_{33} = -r\sin\theta\cos\theta
\end{aligned} \tag{8.105}$$

For $\alpha = 2$, we get

$$\frac{d^2x^2}{ds^2} + \Gamma^2_{\mu\nu}\frac{dx^\mu}{ds}\frac{dx^\nu}{ds} = 0$$

or

$$\frac{d^2x^2}{ds^2} + \Gamma^2_{12}\frac{dx^1}{ds}\frac{dx^2}{ds} + \Gamma^2_{21}\frac{dx^2}{ds}\frac{dx^1}{ds} + \Gamma^2_{33}\frac{dx^3}{ds}\frac{dx^3}{ds} = 0$$

or

$$\frac{d^2\theta}{ds^2} + \frac{2}{r}\frac{dr}{ds}\frac{d\theta}{ds} - r\sin\theta\cos\theta\left(\frac{d\phi}{ds}\right)^2 = 0 \tag{8.106}$$

For $\alpha = 3$, we get

$$\frac{d^2x^3}{ds^2} + \Gamma^3_{\mu\nu}\frac{dx^\mu}{ds}\frac{dx^\nu}{ds} = 0$$

or

$$\frac{d^2x^2}{ds^2} + 2\Gamma^3_{13}\frac{dx^1}{ds}\frac{dx^3}{ds} + 2\Gamma^3_{23}\frac{dx^2}{ds}\frac{dx^3}{ds} = 0$$

or

$$\frac{d^2\phi}{ds^2} + \frac{2}{r}\frac{dr}{ds}\frac{d\phi}{ds} + 2\cot\theta\left(\frac{d\theta}{ds}\right)\left(\frac{d\phi}{ds}\right) = 0 \tag{8.107}$$

Further $\alpha = 0$, for we obtain

$$\frac{d^2x^0}{ds^2} + \Gamma^0_{\mu\nu}\frac{dx^\mu}{ds}\frac{dx^\nu}{ds} = 0$$

or

$$\frac{d^2x^0}{ds^2} + \Gamma^0_{10}\frac{dx^1}{ds}\frac{dx^0}{ds} + \Gamma^0_{01}\frac{dx^0}{ds}\frac{dx^1}{ds} =$$

or

$$\frac{d^2t}{ds^2} + v'\frac{dr}{ds}\frac{dt}{ds} = 0$$

or

$$\frac{d^2t}{ds^2} + \frac{dv}{ds}\frac{dt}{ds} = 0 \tag{8.108}$$

To integrate the Eqs. (8.106), (8.107) and (8.108), we choose the initial motion to be in the plane $\theta = \pi/2$. There we have $\sin\theta = 1$, $\cos\theta = 0 = \frac{d\theta}{ds}$. Substituting these values in the Eqs. (8.106), (8.107) and (8.108), we get

$$\frac{d^2\theta}{ds^2} = 0 \tag{8.109}$$

$$\frac{d^2\phi}{ds^2} + \frac{2}{r}\frac{dr}{ds}\frac{d\phi}{ds} = 0 \tag{8.110}$$

and

$$\frac{d^2t}{ds^2} + \frac{dv}{ds}\frac{dt}{ds} = 0 \tag{8.111}$$

The Eq. (8.109) implies that the particle which starts moving in the plane $\theta = \pi/2$, continues to move in the same plane.

Rewriting Eq (8.110), we get

$$r^2 \frac{d^2\phi}{ds} + 2r \frac{dr}{ds}\frac{d\phi}{ds} = 0$$

or

$$\frac{d}{ds}\left(r^2 \frac{d\phi}{ds}\right) = 0$$

which on integration yields

$$r^2 \frac{d\phi}{ds} = h \tag{8.112}$$

where h is a constant of integration.

Further rewriting Eq (8.111), we obtain

$$e^{\nu} \frac{d^2t}{ds^2} + \nu' e^{\nu} \frac{dr}{ds}\frac{dt}{ds} = 0$$

or

$$\frac{d}{ds}\left(e^{\nu} \frac{dt}{ds}\right) = 0$$

which on integration, gives

$$e^{\nu} \frac{dt}{ds} = k \tag{8.112}$$

where k is a constant of integration.

In lieu of taking the equation with $\alpha = 1$, we take the line-element (8.102) which for $\theta = \pi/2$, becomes

$$ds^2 = -e^{\lambda} dr^2 - r^2 d\phi^2 + e^{\nu} dt^2$$

or

$$e^{\lambda}\left(\frac{dr}{ds}\right)^2 + r^2\left(\frac{d\phi}{ds}\right)^2 - e^{\nu}\left(\frac{dt}{ds}\right)^2 + 1 = 0 \tag{8.113}$$

Substituting the expressions for $\frac{d\phi}{ds}$ and $\frac{dt}{ds}$, in the above equations, we get

$$e^{\lambda}\left(\frac{dr}{ds}\right)^2 + r^2 \frac{h^2}{r^4} - e^{\nu} k^2 e^{-2\nu} + 1 = 0$$

or

$$e^{\nu+\lambda}\left(\frac{dr}{ds}\right)^2 + \frac{h^2 e^{\nu}}{r^2} - k^2 + e^{\nu} = 0$$

or

$$\left(\frac{dr}{ds}\right)^2 + \frac{h^2}{r^2}\left(1 - \frac{r^2}{R_0^2}\right) - k^2 + \left(1 - \frac{r^2}{R_0^2}\right) = 0$$

or

$$\left(\frac{dr}{ds}\right)^2 = k^2 - 1 + \frac{r^2}{R_0^2} - \frac{h^2}{r^2} + \frac{h^2}{R_0^2}$$

or

$$\left(\frac{dr}{ds}\right) = \pm\sqrt{\left(k^2 - 1 + \frac{r^2}{R_0^2} - \frac{h^2}{r^2} + \frac{h^2}{R_0^2}\right)} \tag{8.114}$$

The parameter h can be positive or negative, depending on the direction of motion; k is positive for $r < R_0$, since t is related to the proper time ds.

$$\frac{dr}{ds} = \frac{dr}{d\phi}\frac{d\phi}{ds}$$

$$= \left(k^2 - 1 + \frac{r^2}{R_0^2} - \frac{h^2}{r^2} + \frac{h^2}{R_0^2}\right)^{1/2}$$

or

$$\frac{dr}{d\phi}\frac{h}{r^2} = \left(k^2 - 1 + \frac{r^2}{R_0^2} - \frac{h^2}{r^2} + \frac{h^2}{R_0^2}\right)^{1/2} \tag{8.115}$$

Substituting $u = 1/r$, we get

$$-h\frac{du}{d\phi} = \left(k^2 - 1 + \frac{r^2}{R_0^2} - \frac{h^2}{r^2} + \frac{h^2}{R_0^2}\right)^{1/2}$$

which on squaring gives

$$h^2\left(\frac{du}{d\phi}\right)^2 = k^2 - 1 + \frac{r^2}{R_0^2} - \frac{h^2}{r^2} + \frac{h^2}{R_0^2}$$

Differentiating w.r.t. ϕ and simplifying, we get

$$\frac{d^2u}{d\phi^2} + u = -\frac{1}{h^2 R_0^2 u^3} = -\frac{1}{3}\frac{\Lambda}{h^2 u^3} \tag{8.116}$$

This is the equation of the orbit of a particle in the de-Sitter universe and in Newtonian mechanics corresponds to the motion of a particle with a central repulsive force proportional to the distance r when Λ is positive.

Velocity of a Particle in de-Sitter Universe

$$\frac{dr}{dt} = \frac{dr}{ds}\frac{ds}{dt} = \frac{e^{\nu}}{k}\frac{dr}{ds}$$

$$= \frac{1}{k}\left(1 - \frac{r^2}{R_0^2}\right)\frac{dr}{ds}$$

or
$$\frac{dr}{dt}=\frac{1}{k}\left(1-\frac{r^2}{R_0^2}\right)\left(k^2-1+\frac{r^2}{R_0^2}-\frac{h^2}{r^2}+\frac{h^2}{R_0^2}\right)^{1/2} \tag{8.117}$$

and
$$\frac{d\phi}{dt}=\frac{d\phi/ds}{dt/ds}=\frac{h}{k}\left(\frac{1-r^2/R_0^2}{r^2}\right) \tag{8.118}$$

The Eqs. (8.117) and (8.118) give the two components in terms of ordinary time as measured at the origin. The radial velocity dr/dt is zero when

$$k^2-1+\frac{r^2}{R_0^2}-\frac{h^2}{r^2}+\frac{h^2}{R_0^2}=0 \tag{8.119}$$

Further $\frac{d\phi}{dt}$ is zero when

$$r=R_0 \tag{8.120}$$

Equation (8.119) determines the value of r at perihelion when the particle approaches the origin most closely. The Eq. (8.120) shows that all motion ceases at R_0. This value of r is called the apparent horizon of the universe.

Acceleration of a Particle in de-Sitter Universe Differentiating Eq. (8.117) w.r.t. t, we get

$$\frac{d^2r}{dt^2}=\frac{1}{k}\left(-\frac{2r}{R_0^2}\right)\frac{dr}{dt}\left(k^2-1+\frac{r^2}{R_0^2}-\frac{h^2}{r^2}+\frac{h^2}{R_0^2}\right)^{1/2}$$
$$+\frac{1}{k}\left(1-\frac{r^2}{R_0^2}\right)\frac{1}{2}\left(k^2-1+\frac{r^2}{R_0^2}-\frac{h^2}{r^2}+\frac{h^2}{R_0^2}\right)^{-1/2}\left(\frac{2r}{R_0^2}+\frac{2h^2}{r^3}\right)\frac{dr}{dt}$$
$$=\left(-\frac{2r}{R_0^2}\right)\left(\frac{dr}{dt}\right)^2\frac{1}{(1-r^2/R_0^2)}$$
$$+\frac{1}{k^2}\left(1-\frac{r^2}{R_0^2}\right)^2\frac{1}{dr/dt}\left(\frac{r}{R_0^2}+\frac{h^2}{r^3}\right)\frac{dr}{dt}$$

or
$$\frac{d^2r}{dt^2}=\left(-\frac{2r}{R_0^2}\right)\left(\frac{dr}{dt}\right)^2\frac{1}{(1-r^2/R_0^2)}+\frac{1}{k^2}\left(1-\frac{r^2}{R_0^2}\right)^2\left(\frac{r}{R_0^2}+\frac{h^2}{r^3}\right) \tag{8.121}$$

Diffferentiating Eq. (8.118) w.r.t. t, we get

$$\frac{d^2\phi}{dt^2}=-\frac{2h}{kr^3}\frac{dr}{dt} \tag{8.122}$$

Equation (8.121) shows that

$$\frac{d^2r}{dt^2} > 0 \qquad \text{for } \frac{dr}{dt} = 0$$

It implies that a particle once reaches perihelion, starts to move away from the perihelion and would never return. Furthermore, for a particle at the origin $r = 0$, $h = 0$, the acceleration is zero, so the particle will remain at rest for ever.

Lastly let us consider the behaviour of light rays in de-Sitter universe. Putting $ds = 0$ in Eq. (8.100), we get

$$\left(\frac{dr}{dt}\right)^2 + \left(1 - \frac{r^2}{R_0^2}\right)\left[r^2\left(\frac{d\theta}{dt}\right)^2 + r^2 \sin^2\theta\left(\frac{d\phi}{dt}\right)^2\right] = \left(1 - \frac{r^2}{R_0^2}\right)^2 \tag{8.123}$$

which for a pure radial motion gives

$$\frac{dr}{dt} = \pm\left(1 - \frac{r^2}{R_0^2}\right) \tag{8.124}$$

Integrating it, we get

$$\int_0^{R_0} \frac{dr}{(1 - r^2/R_0^2)} = \frac{1}{R_0}\tan^{-1}\frac{r}{R_0}\Bigg|_0^{R_0}$$
$$= \int dt = \infty \tag{8.125}$$

The light signal will take infinite time to travel from the origin to $r = R_0$. Thus the observer situated at the origin will not be able to know as what happens at distances $r \geq R_0$. R_0 is the horizon to the universe to him. Another observer located at a different place, will locate his horizon differently.

Shift in Spectral Lines in de-Sitter Universe The question to be answered is how does the wavelength of light emitted by a freely moving particle at $r = r$, appears to the observer located at the origin of de-Sitter universe, $r = 0$. Let the source located at $r = r$, emit a light pulse, at time t and it will reach $r = 0$ at a later time t', according to Eq. (8.125) as

$$t' = t + \int_0^r \frac{dr}{1 - \frac{r^2}{R_0^2}} \tag{8.126}$$

If $\delta t'$ is the interval between the reception of two successive wave crests at the origin and δt, the interval between the emission of these waves by the distant star, we get by differentiating the Eq. (8.126)

$$dt' = dt + \frac{1}{1 - r^2/R_0^2} \frac{dr}{dt} dt \tag{8.127}$$

or

$$dt' = \left[1 + \frac{1}{1 - r^2/R_0^2} \frac{dr}{dt}\right] dt \tag{8.128}$$

where $\frac{dr}{dt}$ is the radial velocity of the source at the time of emission.

The proper time interval ds for an observer on the distant star, corresponding to the coordinate time dt; the motion being in the plane $\theta = \pi/2$. plane, is from Eq. (8.112)

$$ds = \frac{e^{\nu}}{k} dt$$

The proper time interval between the crests at the origin is dt'. Assuming the frequencies at emission and reception to be ν_e and ν_0 respectively, we obtain

$$\begin{aligned} \nu_o dt' &= \nu_e ds \\ \nu_o &= \nu_e \frac{ds}{dt'} = \nu_e \frac{ds}{dt} \frac{dt}{dt'} \\ &= \nu_e \frac{1 - r^2/R_0^2}{k} \frac{1}{1 + \frac{1}{1 - r^2/R_0^2} \frac{dr}{dt}} \end{aligned} \tag{8.129}$$

As is obvious, the frequency ν_0 as observed by the observer at the origin, is different from the frequency ν_e, as emitted by the distant star and thus there is a Doppler shift

$$\frac{\delta \nu}{\nu} = \frac{\nu_o - \nu_e}{\nu_e} \tag{8.130}$$

Whether the shift is towards the red or violet, depends on the sign and magnitude of the radial velocity of the radiating atom at the time of emission.

When $dr/dt > 0$, the frequency ν_o observed by the observer at the origin decreases, *i.e.*, there is red shift in the spectral light. However, when $dr/dt < 0$, we will get $\nu_o > \nu_e$.

Rewriting Eq (8.129) after simplification, we get

$$\frac{\nu_o}{\nu_e} = \frac{1 - r^2/R_0^2 - dr/dt}{k} \tag{8.131}$$

If λ is the wavelength at emission and $\lambda + d\lambda$ at reception, we get

$$\frac{\lambda + d\lambda}{\lambda} = \frac{k}{1 - r^2/R_0^2 - dr/dt} \tag{8.132}$$

At the perihelion, $dr/dt = 0$, so it becomes

$$\frac{\lambda + d\lambda}{\lambda} = \frac{k}{(1-r^2/R_0^2)} = \frac{\sqrt{\left(1-\frac{r^2}{R_0^2}+\frac{h^2}{r^2}-\frac{h^2}{R_0^2}\right)}}{(1-r^2/R_0^2)} \tag{8.133}$$

Thus we see that there is a possibility of both red and violet shifts, but the possibility of red shift is more prominent.

Lastly let us examine how the de-Sitter universe compares with the actual universe. The de-Sitter universe is completely devoid of matter and radiation in any form. This is the most unsatisfactory feature of de-Sitter model as a basis for the cosmology of the actual universe. However, unlike Einstein model, the de-Sitter model allows a red shift in the light from distant stars but does not allow for the observed finite concentration of matter in the actual universe. Furthermore, the cosmological constant turns out to be much greater in the de-Sitter universe than the Einstein universe.

8.5.3 Red Shift of Light and Weyl's Hypothesis

In order to apply Eq. (8.133) for the red-shift, one has to have the exact value of the velocity and the orbit of the emitting star. We will derive an alternative result for the red-shift on the basis of Weyl's hypothesis as discussed by Weyl himself and Robertson. Weyl's hypothesis states that, 'Nebulae in the actual universe lie on a coherent pencil of geodesics deriving from a common point in the past.' We will use the de-Sitter line-element as derived by Robertson, Eq. (8.97)

$$ds^2 = -e^{2kt}(dr^2 + r^2 d\theta^2 + r^2 \sin^2\theta d\phi^2) + dt^2 \tag{8.134}$$

It has been shown earlier, while discussing the acceleration of a particle in the de-Sitter universe, that for a particle at rest at the origin, the acceleration will be zero and the particle will remain at rest for ever.

The Weyl hypothesis supposes that the nebulae are uniformly distributed in the actual universe and except for small motion, are at rest w.r.t. the spatial coordinates r, θ, ϕ. The time t is the proper time. Although the particle is at rest with respect to the spatial coordinates system, the proper distance between any two such particles will change with time since $g_{\mu\nu}$ are now time-dependent. The light pulse emitted by one particle will be Doppler shifted on reception by the other particle.

The radial velocity of light is obtained from Eq. (8.134) by putting $ds = 0$, as

$$\frac{dr}{dt} = e^{-kt} \tag{8.135}$$

We assume that the particle is permanently located at r and emits a pulse of radiation at t_1, which is received by the observer at the origin at time t_2. Then

$$\int_{t_1}^{t_2} e^{-kt} dt = \int_0^r dr = \text{const} \tag{8.136}$$

Differentiating Eq. (8.136), we get

$$e^{-kt_2}\delta t_2 - e^{-kt_1}\delta t_1 = 0$$

or

$$\delta t_2 = e^{k(t_2 - t_1)}\delta t_1 \tag{8.137}$$

This equation relates the times of reception of two pulses to the time of emission δt_1. Thus,

$$\frac{\delta t_2}{\delta t_1} = e^{k(t_2 - t_1)} \tag{8.138}$$

As δt_1 and δt_2 are the proper times and if λ is the wavelength at emission and $\lambda + d\lambda$ at reception, then

$$\frac{\lambda + d\lambda}{\lambda} = \frac{\delta t_2}{\delta t_1} = e^{k(t_2 - t_1)} \tag{8.139}$$

Assuming that r is not too large, we get from Eq. (8.139)

$$\frac{d\lambda}{\lambda} = k(t_2 - t_1) \approx kr \tag{8.140}$$

As k is positive, there results a red-shift in the spectral lines and the shift is approximately linearly proportional to r.

8.6 THE ROBERTSON-WALKER METRIC. FRIEDMANN EQUATIONS

In order to model the large scale behaviour of the universe, so that Einstein equations are satisfied, we model the energy and matter in the universe by a perfect fluid. A perfect fluid has no heat conduction or viscosity and is characterized by its mass density ρ and pressure p. The particles in the fluid are galaxy clusters. In co-moving coordinates, $\dot{t} = 1$ and $\dot{x}_i = 0$, the four-velocity $u^\mu = (1, 0, 0, 0)$, the stress-energy tensor becomes

$$T_{\mu\nu} = (\rho + p)\frac{dx^\mu}{ds}\frac{dx^\nu}{ds} - g_{\mu\nu}p \tag{8.141}$$

$$= \begin{pmatrix} \rho & 0 & 0 & 0 \\ 0 & -p & 0 & 0 \\ 0 & 0 & -p & 0 \\ 0 & 0 & 0 & -p \end{pmatrix} \tag{8.142}$$

The Robertson-Walker metric, that describes a homogeneous, isotropic and expanding universe, is

$$\begin{aligned} ds^2 &= g_{\mu\nu} dx^\mu dx^\nu \\ &= dt^2 - \frac{a^2(t)}{1-kr^2} dr^2 - a^2(t) r^2 d\theta^2 - a^2(t) r^2 \sin^2\theta d\phi^2 \end{aligned} \tag{8.143}$$

where $a(t)$ is the scale factor and $g_{\mu\nu}$ is the metric for a three-dimensionally symmetric space with $(x^0, x^1, x^2, x^3) = (t, r, \theta, \phi)$ and

$$\begin{aligned} g_{0i} &= 0 \\ g^{ii} &= \frac{1}{g_{ii}} \end{aligned} \tag{8.144}$$

where $i = 1, 2, 3$. Thus,

$$g_{00} = 1 \qquad g_{11} = -\frac{a^2}{1-kr^2} \tag{8.145a}$$

$$g_{22} = -r^2 a^2 \qquad g_{33} = -r^2 \sin^2\theta\, a^2 \tag{8.145b}$$

$$g^{00} = 1 \qquad g^{11} = -\frac{1-kr^2}{a^2} \tag{8.146a}$$

$$g^{22} = -\frac{1}{r^2 a^2} \qquad g^{33} = -\frac{1}{r^2 \sin^2\theta a^2} \tag{8.146b}$$

$$g^{ij} = g_{ij} = 0 \qquad \text{if} \qquad i \neq j. \tag{8.146c}$$

To obtain, the 00- and 11- components of the Einstein field equations,

$$R_{\mu\nu} - \frac{1}{2} g_{\mu\nu} R = 8\pi T_{\mu\nu} \tag{8.147}$$

we calculate the R_{00} component of the Ricci tensor

$$R_{00} = \Gamma^\lambda_{00,\lambda} - \Gamma^\lambda_{0\lambda,0} + \Gamma^\lambda_{00}\Gamma^\sigma_{\lambda\sigma} - \Gamma^\sigma_{0\lambda}\Gamma^\lambda_{0\sigma} \tag{8.148}$$

As the signatures of Ricci tensor are different from the one, we have been using before, there is a change of sign of the term on the r.h.s. of Einstein field equation.

Ist and 3rd terms are zero because $\Gamma^\lambda_{00} = 0$. Now

$$\begin{aligned} \Gamma^\lambda_{0\lambda} &= \frac{1}{2} g^{\lambda\sigma} (g_{\sigma 0,\lambda} + g_{\sigma\lambda,0} - g_{0\lambda,\sigma}) \\ &= \frac{1}{2} g^{\lambda\sigma} g_{\sigma\lambda,0} \\ &= \frac{1}{2} g^{\lambda\sigma} \frac{\partial g_{\sigma\lambda}}{\partial t} \end{aligned} \tag{8.149}$$

Expanding Eq. (8.149) according to Einstein summation convention and discarding zero terms, we obtain

$$\Gamma^{\lambda}_{0\lambda} = 3\frac{\dot{a}}{a} \tag{8.150}$$

$$\begin{aligned}\Gamma^{\lambda}_{0\lambda,0} &= 3\left(\frac{a\ddot{a}-\dot{a}^2}{a^2}\right)\\ &= 3\left[\frac{\ddot{a}}{a}-\left(\frac{\dot{a}}{a}\right)^2\right]\end{aligned} \tag{8.151}$$

Therefore,

$$\begin{aligned}R_{00} &= -3\left[\frac{\ddot{a}}{a}-\left(\frac{\dot{a}}{a}\right)^2\right]-\Gamma^1_{01}\Gamma^1_{01}-\Gamma^2_{02}\Gamma^2_{02}-\Gamma^3_{03}\Gamma^3_{03}\\ &= -3\frac{\ddot{a}}{a}+3\left(\frac{\dot{a}}{a}\right)^2-3\left(\frac{\dot{a}}{a}\right)^2\\ &= -3\frac{\ddot{a}}{a}\end{aligned} \tag{8.152}$$

Calling

$$g_{ij} = a^2(t)\,\tilde{g}_{ij}(x) \tag{8.153}$$

and

$$g^{ij} = \frac{\tilde{g}^{ij}}{a^2(t)} \tag{8.154}$$

then

$$\begin{aligned}&\tilde{g} = -(1-kr^2)^{-1}\\ &\tilde{g}_{22} = -r^2\\ &\tilde{g}_{33} = -r^2\sin^2\theta\\ &\tilde{g}^{11} = -(1-kr^2)\\ &\tilde{g}^{22} = -\frac{1}{r^2}\\ &\tilde{g}^{33} = -\frac{1}{r^2\sin^2\theta}\\ &\tilde{g}^{ij} = 0\end{aligned} \tag{8.155}$$

Thus,

$$\begin{aligned}\Gamma^0_{ij} &= \frac{1}{2} g^{0\sigma}(g_{\sigma i,j} + g_{\sigma j,i} - g_{ij,\sigma}) \\ &= \frac{1}{2} g^{00}(g_{0i,j} + g_{0j,i} - g_{ij,0}) \\ &= -\frac{1}{2} g_{ij,0} \\ &= -\frac{1}{2}\frac{\partial}{\partial t}(a^2 \tilde{g}_{ij}) \\ &= -\tilde{g}_{ij}\, a\dot{a} \end{aligned} \tag{8.156}$$

Also

$$\begin{aligned}\Gamma^i_{0j} &= \frac{1}{2} g^{i\sigma}(g_{\sigma 0,j} + g_{\sigma j,0} - g_{0j,\sigma}) \\ &= \frac{1}{2} g^{ii}(g_{i0,j} + g_{ij,0} - g_{0j,i}) \\ &= \frac{1}{2} g^{ii}(g_{ij,0}) \\ &= \frac{1}{2}\frac{\tilde{g}^{ii}}{a^2}\frac{\partial}{\partial t}(a^2\,\tilde{g}_{ij}) \\ &= \frac{1}{2}\frac{\tilde{g}^{ii}}{a^2}\,\tilde{g}_{ij}\; 2a\dot{a} \\ &= \frac{\dot{a}}{a}\tilde{g}^{ii}\,\tilde{g}_{ij} \\ &= \frac{\dot{a}}{a}\delta^i_j \end{aligned} \tag{8.157}$$

Now

$$\begin{aligned}\Gamma^i_{jk} &= \frac{1}{2} g^{i\sigma}(g_{\sigma j,k} + g_{\sigma k,j} - g_{jk,\sigma}) \\ &= \frac{1}{2} g^{ii}(g_{ij,k} + g_{ik,j} - g_{jk,i}) \\ &= \frac{1}{2}\tilde{g}^{ii}\left(\frac{\partial \tilde{g}_{ij}}{\partial x^k} + \frac{\partial \tilde{g}_{ik}}{\partial x^j} - \frac{\partial \tilde{g}_{jk}}{\partial x^i}\right) \end{aligned} \tag{8.158}$$

Thus,

$$\begin{aligned}R_{0i} &= \Gamma^{\lambda}_{oi,\lambda} - \Gamma^{\lambda}_{o\lambda,i} + \Gamma^{\lambda}_{0i}\Gamma^{\sigma}_{\lambda\sigma} - \Gamma^{\sigma}_{0\lambda}\Gamma^{\lambda}_{i\sigma} \\ &= -\frac{\partial}{\partial x^i}\left(\frac{\dot{a}}{a}\right) = 0 \end{aligned} \tag{8.159}$$

Finally we calculate R_{11}, which is

$$\begin{aligned}R_{11} &= -\Gamma^{\alpha}_{11,\alpha} + \Gamma^{\alpha}_{1\alpha,1} + \Gamma^{\alpha}_{\sigma 1}\Gamma^{\sigma}_{1\alpha} - \Gamma^{\alpha}_{\sigma\alpha}\Gamma^{\sigma}_{11} \\ &= -\Gamma^{0}_{11,0} + \Gamma^{0}_{10,1} + \Gamma^{0}_{\sigma 1}\Gamma^{\sigma}_{10} - \Gamma^{0}_{\sigma 0}\Gamma^{\sigma}_{11} \\ &\quad -\left(\Gamma^{1}_{11,1} + \Gamma^{2}_{11,2} + \Gamma^{3}_{11,3}\right) \\ &\quad +\left(\Gamma^{1}_{11,1} + \Gamma^{2}_{12,1} + \Gamma^{3}_{13,1}\right) + \Gamma^{i}_{\sigma 1}\Gamma^{\sigma}_{1i} - \Gamma^{i}_{\sigma i}\Gamma^{\sigma}_{11}\end{aligned} \tag{8.160}$$

Collecting the terms involving time

$$\begin{aligned}&= -\Gamma^{0}_{11,0} + \Gamma^{0}_{10,1} + \Gamma^{0}_{\sigma 1}\Gamma^{\sigma}_{10} - \Gamma^{0}_{\sigma 0}\Gamma^{\sigma}_{11} + \Gamma^{1}_{01}\Gamma^{0}_{11} - \left(\Gamma^{1}_{01} + \Gamma^{2}_{02} + \Gamma^{3}_{03}\right)\Gamma^{0}_{11} \\ &= -\Gamma^{0}_{11,0} + 0 + \Gamma^{0}_{11}\Gamma^{1}_{10} - 0 + \Gamma^{1}_{01}\Gamma^{0}_{11} - \left(3\Gamma^{1}_{01}\right)\Gamma^{0}_{11} \\ &= -\Gamma^{0}_{11,0} - \Gamma^{0}_{11}\Gamma^{1}_{10} \\ &= -\frac{\partial}{\partial t}(\dot{a}a\,\tilde{g}_{11}) - (\dot{a}a\,\tilde{g}_{11})\left(\frac{\dot{a}}{a}\right) \\ &= -(\ddot{a}a + 2\dot{a}^2)\tilde{g}_{11}\end{aligned} \tag{8.161}$$

Then

$$R_{11} = -(\ddot{a}a + 2\dot{a}^2)\tilde{g}_{11} + \tilde{R}_{11} \tag{8.162}$$

where

$$\tilde{R}_{11} = -\Gamma^{i}_{11,i} + \Gamma^{i}_{1i,1} + \Gamma^{i}_{j1}\Gamma^{j}_{1i} - \Gamma^{i}_{ji}\Gamma^{j}_{11} \tag{8.163}$$

Now compute $\tilde{R}_{11}$.

The terms with derivatives

$$-\Gamma^{i}_{11,i} + \Gamma^{i}_{1i,1} = -\Gamma^{1}_{11,1} + \left(\Gamma^{1}_{11,1} + \Gamma^{2}_{12,1} + \Gamma^{3}_{13,1}\right) = 2\frac{d}{dr}r^{-1} = -\frac{2}{r^2} \tag{8.164}$$

The other terms

$$\begin{aligned}\Gamma^{i}_{j1}\Gamma^{j}_{1i} - \Gamma^{i}_{ji}\Gamma^{j}_{11} &= \Gamma^{1}_{11}\Gamma^{1}_{11} + \Gamma^{2}_{12}\Gamma^{2}_{12} + \Gamma^{3}_{13}\Gamma^{3}_{13} - \left(\Gamma^{1}_{11} + \Gamma^{2}_{21} + \Gamma^{3}_{31}\right)\Gamma^{1}_{11} \\ &= \frac{2}{r^2} + \frac{2k}{1-kr^2}\end{aligned} \tag{8.165}$$

Therefore,

$$\tilde{R}_{11} = -\frac{2}{r^2} + \frac{2}{r^2} + \frac{2k}{1-kr^2} \tag{8.166}$$

Since this is true for all diagonal elements, not only for the 11-term, we have

$$R_{ii} = -(\ddot{a}a + 2\dot{a}^2 + 2k)\tilde{g}_{ii} \tag{8.167}$$

The Ricci scalar is given by

$$R = g^{\mu\nu} R_{\mu\nu} \tag{8.168}$$

Expanding Eq. (8.168), we find that only the diagonal terms are non-zero, so

$$\begin{aligned} R &= -\frac{3\ddot{a}}{a} - \frac{(1-kr^2)}{a^2}(a\ddot{a} + 2\dot{a}^2 + 2k)\frac{1}{(1-kr^2)} \\ &\quad - \frac{1}{r^2a^2}(a\ddot{a} + 2\dot{a}^2 + 2k)r^2 \\ &\quad - \frac{1}{r^2a^2\sin^2\theta}(a\ddot{a} + 2\dot{a}^2 + 2k)r^2\sin^2\theta \\ &= -\frac{6}{a^2}(a\ddot{a} + \dot{a}^2 + k) \end{aligned} \tag{8.169}$$

Further,

$$\begin{aligned} T_{00} &= \rho \\ T_{11} &= (\rho + p)\frac{dx^1}{ds}\frac{dx^1}{ds} - pg_{11} \\ &= -p\left(-\frac{a^2}{1-kr^2}\right) \\ &= \frac{pa^2}{1-kr^2} \end{aligned} \tag{8.170}$$

The 00- and 11-components of $R_{\mu\nu} - \frac{1}{2}g_{\mu\nu}R = 8\pi T_{\mu\nu}$ are calculated as follows:

The 00-component becomes

$$-\frac{3\ddot{a}}{a} + \frac{1}{2}\frac{6}{a^2}\left(\ddot{a}a + \dot{a}^2 + k\right) = 8\pi\rho$$

or

$$3\left(\frac{\dot{a}}{a}\right)^2 + \frac{3k}{a^2} = 8\pi\rho$$

or

$$3(\dot{a}^2 + k) = 8\pi\rho a^2 \tag{8.171}$$

Lastly the 11-component is

$$\frac{\ddot{a}a+2\dot{a}^2+2k}{1-kr^2}+\frac{1}{2}\frac{a^2}{1-kr^2}\left[-\frac{6(\ddot{a}a+\dot{a}^2+k)}{a^2}\right]=8\pi\frac{pa^2}{1-kr^2}$$

$$\Rightarrow \quad \frac{1}{1-kr^2}[-2\ddot{a}a-\dot{a}^2-k]=\frac{8\pi\, pa^2}{1-kr^2}$$

$$\Rightarrow \quad 2\ddot{a}a+\dot{a}^2+k=-8\pi\, pa^2 \tag{8.172}$$

The Eqs. (8.171) and (8.172) are known as Friedmann equations and the metrices of the form Eq. (8.143), which obey these equations define Friedmann-Robertson-Walker (FRW) universes.

Multiplying Eq. (8.172) by 3 and subtracting the resulting equation from Eq. (8.178), we obtain

$$6\ddot{a}a=8\pi(-3pa^2-\rho a^2)$$

$$\Rightarrow \quad \ddot{a}=-\frac{4\pi}{3}(\rho+3p)a \tag{8.173}$$

We supplement these equations by the equation of the conservation of energy using the stress energy-momentum tensor

$$T^{\mu\nu}_{;\mu}=0 \tag{8.174}$$

Considering the conservation of energy equation using the energy-momentum tensor, we get

$$T^{\mu}_{0;\mu}=\partial_\mu T^\mu_0+\Gamma^\mu_{\mu\nu}T^\nu_0-\Gamma^\nu_{\mu 0}T^\mu_\nu=0 \tag{8.175}$$

As the energy-momentum tensor is diagonal, it simplifies to

$$\begin{aligned}\partial_\mu T^\mu_0+\Gamma^\mu_{\mu\nu}T^\nu_0-\Gamma^\nu_{\mu 0}T^\mu_\nu &= \partial_0 T^0_0\\ &+\Gamma^0_{00}T^0_0+\Gamma^1_{10}T^0_0+\Gamma^2_{20}T^0_0+\Gamma^3_{30}T^0_0\\ &-\Gamma^0_{00}T^0_0-\Gamma^1_{10}T^1_1-\Gamma^2_{20}T^2_2-\Gamma^3_{30}T^3_3\end{aligned} \tag{8.176}$$

As calculated earlier, we have

$$\Gamma^0_{00}=0$$

$$\Gamma^1_{10}=\Gamma^2_{20}=\Gamma^3_{30}=\frac{\dot{a}}{a}$$

and putting

$$T^0_0=\rho$$

$$T^1_1=T^2_2=T^3_3=-p$$

we get

$$\frac{\partial \rho}{\partial t} = -3\frac{\dot{a}}{a}(\rho + p) \tag{8.177}$$

To proceed further, we have to choose an equation of state, a relationship between ρ and p. The perfect fluid relevant to cosmology, obeys the simple equation of state

$$p = \omega\rho \tag{8.178}$$

where ω is a constant independent of time. In view of this result, we re-write the energy conservation equation

$$\frac{\dot{\rho}}{\rho} = -3(1+\omega)\frac{\dot{a}}{a} \tag{8.179}$$

which on integration yields

$$\rho \propto a^{-3(1+\omega)} \tag{8.180}$$

The two most popular examples of cosmological fluids are dust and radiation. Dust is collisionless, non-relativistic matter which obeys $\omega = 0$. As examples, are stars and galaxies for which the pressure is negligible as compared to energy density ρ. The universe where energy density is mostly due to dust, is called matter-dominated. Therefore the energy density in matter-dominated universe, falls off as

$$\rho \propto a^{-3} \tag{8.181}$$

A universe in which most of the energy density is due to radiation is known as radiation-dominated. Radiation is used to describe electromagnetic radiation or massive particles moving at velocities sufficiently close to velocity of light so that these are indistinguishable from photons, atleast as far as their equation of state is concerned.

The radiation has the energy-momentum tensor

$$T^{\mu\nu} = diag(\rho, p, p, p) \tag{8.182}$$

Its trace is

$$T = T^{\mu}_{\mu} = \rho + 3p \tag{8.183}$$

The equation of state for the radiation is

$$p = \rho/3 \tag{8.184}$$

The energy density in radiation falls off as

$$\rho \propto a^{-4} \tag{8.185}$$

The energy density of radiation falls off faster than matter because of the red-shift.

Lastly we consider the vacuum energy density. As the vacuum energy is a constant, therefore it remains the same at all times. The matter and radiation densities decrease with the expanding universe, but the vacuum energy density remains the same and so eventually will dominate the universe.

Example 8.1

Show that the Robertson-Walker metric

$$ds^2 = dt^2 - a^2(t)\left[\frac{dr^2}{1-kr^2} + r^2(d\theta^2 + \sin^2\theta d\phi^2)\right]$$

can be written as

$$ds^2 = a^2(t)\left[d\eta^2 - d\chi^2 - \Sigma^2(\chi)(d\theta^2 + \sin^2\theta d\phi^2)\right]$$

where $\Sigma^2(\chi) = \sin^2\chi$ or χ^2 $\sinh^2\chi$ $(k = 1, 0, -1)$.

Solution

Let
$$r = \begin{cases} \sin\chi & k = +1 \\ \chi & k = 0 \\ \sinh\chi & k = -1 \end{cases}$$

then

$$dr = \begin{cases} \cos\chi d\chi \\ d\chi \\ \cosh^2\chi \end{cases}$$

$$\frac{dr^2}{1-kr^2} + r^2 d\Omega^2 = d\chi^2 + \Sigma^2(\chi)$$

where

$$\Sigma^2(\chi) = \begin{cases} \sin^2\chi & k = +1 \\ \chi^2 & k = 0 \\ \sin h^2\chi & k = -1 \end{cases}$$

Furthermore, let the transformation of variable t to variable η be defined as

$$dt = a(\eta)d\eta$$

Then

$$\begin{aligned} ds^2 &= dt^2 - a^2(t)[d\chi^2 + \Sigma^2 d\Omega^2) \\ &= a^2(\eta)[d\eta^2 - d\chi^2 - \Sigma^2 d\Omega^2) \end{aligned}$$

8.7 NON-STATIC COSMOLOGICAL MODELS. TIME EVOLUTION OF UNIVERSE

We take up the problem of the time evolution of the universe which essentially implies solving for the scale factor $a(t)$. Firstly we take up the case of the early universe, which was radiation dominated. Setting $\rho = 3p$ and $\Lambda = 0$ (for the sake of simplicity), we write the Friedmann equations

$$\frac{3}{a^2}(k+\dot{a}^2)=8\pi\rho$$

$$2\frac{\ddot{a}}{a}+\frac{1}{a^2}(k+\dot{a}^2)=-\frac{8\pi\rho}{3} \tag{8.186}$$

Eliminating ρ from these equations, we have

$$2\frac{\ddot{a}}{a}+\frac{1}{a^2}(k+\dot{a}^2)=-\frac{1}{a^2}(k+\dot{a}^2)$$

or

$$\ddot{a}+\frac{1}{a}(k+\dot{a}^2)=0 \tag{8.187}$$

In the very early universe k/a can be neglected, so that

$$a\ddot{a}+\dot{a}^2=0$$

or

$$\frac{d}{dt}(a\dot{a})=0 \tag{8.188}$$

Integrating, we obtain

$$a\dot{a}=C$$

$$ada=Cdt$$

which gives

Integrating both sides, we have

$$\frac{a^2}{2}=C_1t$$

which shows that

$$a(t)\propto\sqrt{t} \tag{8.189}$$

As will turn out, the expansion of the universe is more rapid in the case of radiation-dominated universe in comparison to the matter-dominated universe. This results from the radiation pressure which is the dominating factor in the early stages of the universe.

In order to illustrate the behaviour of the universe from the start to the end, we take up the simple case, the de Sitter model. This is a flat model devoid of any matter or radiation, *i.e.*, $\rho = 0$ and $p = 0$.

Therefore, setting $k = 0$, we write the line-element as

$$ds^2 = dt^2 - a^2(t)dr^2 - a^2(t)r^2 d\theta^2 - a^2(t)r^2 \sin^2\theta d\phi^2 \tag{8.190}$$

As the universe is expanding, the matter and radiation density will ultimately drop to negligible levels and in its end the universe may be a de-Sitter universe. Under these conditions, Friedmann equations, Eqs. (8.186) and (8.188) become

$$\frac{3}{a^2}\dot{a}^2 - \Lambda = 0$$

$$2\frac{\ddot{a}}{a} + \frac{1}{a^2}\dot{a}^2 - \Lambda = 0 \tag{8.191}$$

From the first equation, we get

$$\frac{\dot{a}^2}{a^2} = \frac{\Lambda}{3}$$

or

$$\frac{\dot{a}}{a} = \sqrt{\frac{\Lambda}{3}} \tag{8.192}$$

Integrating it, we get

$$a(t) = Ce^{\sqrt{\frac{\Lambda}{3}}t} \tag{8.193}$$

where C is a constant of integration. As time elapses, the universe expands, Fig. (8.5).

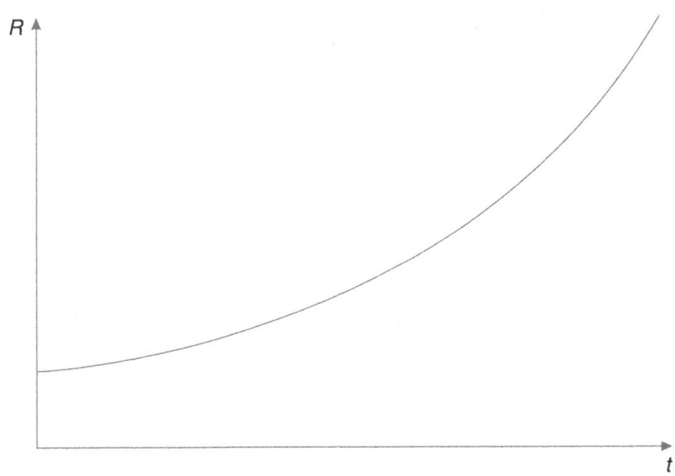

Figure 8.5 *The de-Sitter universe is a universe without matter. It expands exponentially with time.*

Thus, for the de-Sitter universe, ignoring the constant of integration as we are interested to see the expansion of the universe qualitatively, the line-element becomes

$$ds^2 = dt^2 - a^2(t)dr^2 - e^{2\sqrt{\frac{\Lambda}{3}}t} r^2 d\theta^2 - a^2(t)r^2 \sin^2\theta d\phi^2 \tag{8.194}$$

Now we consider the universe that contains matter. Direct solution of Friedmann equations is, in general, too involved and can be carried out numerically. Since recent observations indicate that the universe is flat, we can set $k = 0$. After putting $\rho = 1/a^3$ in Eq. (8.186), we get

$$\dot{a}^2 = \frac{8\pi}{3a}$$

or

$$\sqrt{a}\frac{da}{dt} = \sqrt{\frac{8\pi}{3}}$$

which leads to

$$\sqrt{a}da = \sqrt{\frac{8\pi}{3}}dt \tag{8.195}$$

Integrating both sides, we get

$$\frac{2}{3}a^{\frac{3}{2}} = \sqrt{\frac{8\pi}{3}}t$$

which shows that

$$a(t) \propto t^{2/3} \tag{8.196}$$

A plot of this equation indicates that the universe evolves continuously and a deceleration parameter, $q = 1/2$.

Lastly we consider the case with positive curvature. $k = +1$ and $\Lambda = 0$ (for the sake of simplicity), we get

$$\dot{a}^2 = \frac{8\pi}{3a} - 1$$

$$= \frac{C - a}{a} \tag{8.204}$$

where $\frac{8\pi}{3} = C$, a constant. Solving this equation parametrically, we define

$$a = C\sin^2\tau \tag{8.205}$$

where $\tau = \tau(t)$. Thus, we get

$$\frac{da}{dt} = 2C\sin\tau\cos\tau\frac{d\tau}{dt}$$

or on squaring, one obtains

$$\dot{a}^2 = 4C^2 \sin^2 \tau \cos^2 \tau \left(\frac{d\tau}{dt}\right)^2 \tag{8.206}$$

Comparing it with Eq. (8.204), we get

$$4C^2 \sin^2 \tau \cos^2 \tau \left(\frac{d\tau}{dt}\right)^2 = \frac{C - C\sin^2 \tau}{C\sin^2 \tau} = \frac{\cos^2 \tau}{\sin^2 \tau}$$

or

$$2C\sin^2 \tau d\tau = dt \tag{8.207}$$

Integrating, we get

$$\begin{aligned} t &= 2C\int \frac{1-2\cos 2\tau}{2} d\tau \\ &= C\left(\tau - \frac{1}{2}\sin 2\tau\right) \\ &= \frac{C}{2}(2\tau - \sin 2\tau) \end{aligned} \tag{8.208}$$

The equation for a is

$$\begin{aligned} a &= C\sin^2 \tau \\ &= \frac{C}{2}(1 - \cos 2\tau) \end{aligned} \tag{8.209}$$

These equations enable us to obtain $a(t)$ as a function of t, Fig. (8.6). At $t = \tau = 0$, the size of the universe is $a(0) = 0$; at the Big Bang with zero size. The radius increases with time, reaches a maximum and again contracts to $a(0) = 0$. The maximum radius is the Schwarzschild radius determined by $C/2$.

8.7.1 The Big Bang

The FRW models, provided the matter content is reasonably physical, predict an initial singularity, commonly known as a Big Bang. It can be seen easily as follows.

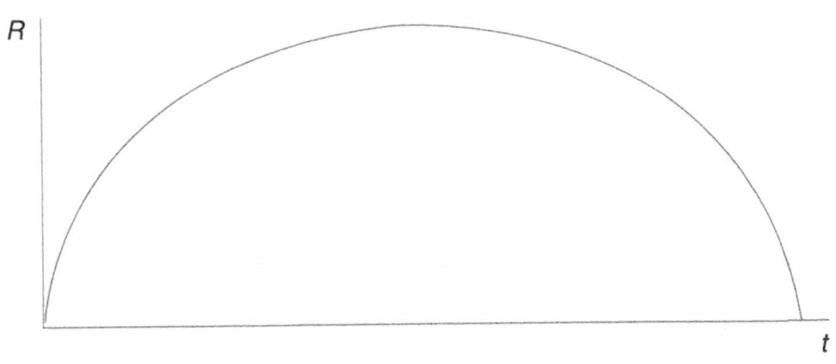

Figure 8.6 *With zero cosmological constant and positive curvature, a dust filled universe, that after maximum expansion, collapses on itself.*

According to the Friedmann Eq. (8.173), with the cosmological constant we have

$$-3\frac{\ddot{a}}{a} = 4\pi(\rho + 3p) - \Lambda \tag{8.210}$$

As long as the right-hand side is positive, one has $\ddot{a} < 0$, so that the universe is decelerating due to gravitational attraction. This is obviously the case for the cosmological constant $\Lambda = 0$, since $\rho + 3p$ is positive for all matter. Furthermore, it equally holds for a negative cosmological constant since its negative energy density is outweighed by 3 times its positive pressure. However, it may not hold in the presence of a positive cosmological constant, which provides an accelerating contribution to the expansion of the universe. We assume for the present *i.e.*, zero or, atleast non-positive.

Since the scale factor by definition, $a > 0$ since we observe a red-shift and $\ddot{a} < 0$, because $\rho + 3p > 0$, it follows that there was no turning point in the past and a must be concave upwards. As such, $a(t)$ must have attained $a = 0$ at some time in the past. We call this point $t = 0$, $a(0) = 0$.

An early universe must be radiation dominated, for which ρa^4 is constant, implying thereby that $\rho \propto 1/a^4$ and as $a \to 0$, this leads to a singularity.

It is natural to wonder at this point, like the case of black holes, whether the singularities as predicted by general relativity in the case of cosmological models, are generic or only an outcome of highly symmetric situations we were dealing. There are singularity theorems applicable to these situations which state that the singularities will occur independently of assumptions about symmetries, provided there is matter content.

8.8 USEFUL TERMINOLOGY

We list below several terms which come across in cosmology.

8.8.1 The Scale Factor

The size of the universe changes with time as it expands. The spatial size of the universe at a given time t, is called the scale factor, denoted by $a(t)$. Cosmological observations show that universe expands as time moves forward, so $a(t) > 0$.

8.8.2 Matter Density

Matter is all the stuff in the universe, stars, planets, comets, asteroids, galaxies, etc. The matter is modelled as dust while deriving the stress-energy tensor for the universe. The energy density of universe is shown as ρ_m, whereas the pressure of matter $p_m = 0$, as the universe expands when time moves forward, ρ_m decreases.

8.8.3 Radiation Density

The energy density of radiation is represented by ρ_R and the pressure by p_R. Radiation is treated as a perfect fluid. As the universe expands, ρ_R decreases like ρ_m, but ρ_R decreases faster than ρ_m, since photons are red-shifted on expansion of universe and hence lose energy.

8.8.4 Vacuum Density

Recent cosmological evidence shows that the expansion of the universe is accelerating and this is consistent with a positive cosmological constant. The cosmological constant represents vacuum energy that is treated as a perfect fluid with the condition that $\rho = -p$.

8.8.5 Matter-Dominated and Radiation-Dominated Universe

The ratio ρ_m/ρ_R, is the deciding factor whether the universe is matter-dominated or radiation-dominated. Currently, $\rho_m/\rho_R \sim 10^3$ which shows that the universe is matter-dominated. However, in the early history of the universe, it was radiation dominated. It is considered that as the universe ages, ultimately it will be dominated by the vacuum energy.

8.8.6 The Hubble Parameter

It indicates the rate of expansion of the universe and is defined as

$$H = \frac{\dot{a}}{a} \tag{8.211}$$

where $\dot{a}$ is the time derivative of the scale factor. Scale factor is the size of the universe. The units of H are $\sec^{-1}$, but for convenience of astronomical applications, the units are stated as km/sec/Mpc, where Mpc are megaparsecs, which is a measure of distance. Furthermore, there is controversy about its value. In reality, there are two schools of thought, one group likes a Hubble constant about 55 km/sec/Mpc and the other the values around 90 km/sec/Mpc.

8.8.7 The Hubble Time

The reciprocal of the Hubble constant is the Hubble time, which is the approximate estimate of the age of the universe. At the Hubble time, all the galaxies in the universe were located at the same point.

8.8.8 Hubble Law

The distance of a galaxy from us is related to its velocity by the Hubble law

$$v = Hd \tag{8.212}$$

8.8.9 The Deceleration Parameter

The rate of change of the expansion of the universe is quantified with the deceleration parameter. It is defined as

$$q = -\frac{a\ddot{a}}{\dot{a}^2} \tag{8.213}$$

8.8.10 Luminosity Distance

Luminosity distance is a term used in astronomy to describe the distance at which an astronomical body would lie based on its observed luminosity, without any attenuation. The luminosity distance does however take into account the fact that the observed luminosity is attenuated by two factors; relativistic red-shift and the Doppler shift of emission, each of which contributes a $(1+z)$ attenuation

$$L_0 = \frac{L_e}{(1+z)^2} \tag{8.214}$$

where z is the red-shift.

The observed flux is given by

$$\Phi_0 = \frac{L_0}{4\pi d_{\text{prop}}^2} = \frac{L_e}{4\pi d_{\text{prop}}^2 (1+z)^2} \tag{8.215}$$

where d_{prop} is the proper distance and L_e is the emitted luminosity.

8.8.11 Critical Density

Whether the universe is open or closed, is dependent on the density of matter in it. If there is enough matter, thus enough gravity, it can exercise enough attraction so as to withhold further expansion and cause contraction leading to a closed universe. The density needed to have a closed universe is called critical density.

In order to derive an expression for critical density, consider a sphere of galaxies of radius R, where R is larger than the distance between cluster of galaxies but smaller than any distance characterizing the universe as a whole. If ρ is the cosmic mass density, the mass of the sphere is

$$M = \frac{4\pi R^3}{3}\rho \tag{8.216}$$

According to Newton's theory of gravitation, the potential energy of any typical galaxy at the surface of the sphere is given by

$$P.E. = -\frac{mMG}{R} = -\frac{4\pi mR^2 \rho G}{3} \tag{8.217}$$

where m is the mass of the galaxy and G, Newton's constant of gravitation

$$G = 6.67 \times 10^{-8}\,\text{cm}^3/\text{gm sec}^2 \tag{8.218}$$

The velocity of the galaxy is given by Hubble law as

$$v = HR \tag{8.219}$$

where H is Hubble constant.

The kinetic energy of the galaxy is

$$K.E. = \frac{1}{2}mv^2$$
$$= \frac{1}{2}mH^2R^2 \qquad (8.220)$$

Thus the total energy of the galaxy, E is

$$E = P.E. + K.E.$$
$$= mR^2\left[\frac{1}{2}H^2 - \frac{4\pi}{3}\rho G\right] \qquad (8.221)$$

The total energy being a constant of motion, must remain constant as the universe expands.

Now if E is negative, the galaxy can never recede to infinity, since at any great distances, the potential energy becomes negligible, and the total energy is just the kinetic energy that is always positive. On the other hand, if E is positive, the galaxy can reach infinity with some remaining kinetic energy. Therefore for the galaxy to have just barely escape velocity, is possible only if E vanishes, *i.e.*, when

$$\frac{1}{2}H^2 = \frac{4\pi}{3}\rho G$$

or

$$\rho_c = \frac{3H^2}{8\pi G} \qquad (8.222)$$

ρ_c is the critical density. Although the above derivation, is based on Newtonian theory of gravitation, but it is valid even when the contents of the universe are highly relativistic, provided ρ is taken as the total energy density divided by c^2.

In order to measure the Hubble constant, all one needs a distance and a red-shift to a galaxy that is distant enough that its peculiar velocity does not matter. Measuring redshift for galaxies is easy, but measuring distances are hard. The Hubble constant is, therefore, not easy to measure and it is not surprising that there is controversy about its value. In reality, there are two schools of thought, one group likes a Hubble constant about 55 km/sec/Mpc and the other the values around 90 km/sec/Mpc.

Taking the value of H as 50 kms per sec per million light years, and recalling there are 9.46×10^{12} kms per light year, we get

$$\rho_c = \frac{3}{8\pi(6.67\times10^{-8}\,\text{cm}^3/\text{gm/sec}^2}\left[\frac{50\text{km/sec}/10^6\,\text{lt yrs}}{9.46\times10^{12}\,\text{km/lt yrs}}\right]^2$$
$$= 5\times10^{-29}\,\text{gm/cm}^3 \qquad (8.223)$$

Avogadro number, 6.02×10^{23} implies that there are 6.02×10^{23} nuclear particles per gm, thus the value of ρ_c corresponds to nearly 3×10^{-5} nuclear particles per c.c. or 0.03 particles per litre.

8.8.12 The Density Parameter

It is the ratio of the observed density to the critical density,

$$\Omega = \frac{8\pi G}{3H^2}\rho = \frac{\rho}{\rho_c} \tag{8.224}$$

The density ρ is obtained by taking into account the contributions from all possible sources such as matter, radiation and vacuum. At present $\Omega \cong 1$, the universe is not closed.

8.9 THE RED-SHIFT

It is proposed to show that the R-W line-element predicts a shift in the wavelength of the radiation emitted by a remote source such as a nebula or a radio galaxy. On account of the isotropy of the model, it follows that any radius vector is a possible trajectory of the light ray, *i.e.*, $\theta, \phi =$ constant, if we use spherical polar coordinates.

The R–W line-element, Eq. (8.143), is

$$ds^2 = dt^2 - \frac{a^2(t)}{1-kr^2}dr^2 - a^2(t)r^2 d\theta^2 - a^2(t)r^2 \sin^2\theta d\phi^2 \tag{8.225}$$

and putting $ds = 0,\ d\theta = 0 = d\phi$, we obtain

$$\left(\frac{dr}{dt}\right)^2 = \frac{(1-kr^2)}{a^2(t)}$$

or

$$\frac{dr}{dt} = \pm\frac{(1-kr^2)^{1/2}}{a(t)} \tag{8.226}$$

The plus sign holds for light travelling away from the origin and minus when towards the origin. We imagine that an observer is located at the origin O of our coordinate system, the two successive light pulses are emitted by a nebula situated at a point $P\,(r, \theta = \text{constant} = \phi)$ at times t_0 and $t_0 + \Delta t_0$. These are received by the observer at the origin at times t and $t + \Delta t$ respectively. As such $t > t_0$ and $t + \Delta t > t_0 + \Delta t_0$. According to Eq. (8.226), we get

$$\int_{t_0}^{t} \frac{dt}{a(t)} = -\int_{r}^{0} \frac{dr}{(1-kr^2)^{1/2}} \tag{8.227}$$

and

$$\int_{t_0+\Delta t_0}^{t+\Delta t} \frac{dt}{a(t)} = -\int_{r}^{0} \frac{dr}{(1-kr^2)^{1/2}} \tag{8.228}$$

Writing Eq. (8.228) in expanded manner, we have

$$\int_{t_0+\Delta t_0}^{t+\Delta t} \frac{dt}{a(t)} = \int_{t_0+\Delta t_0}^{t_0} \frac{dt}{a(t)} + \int_{t_0}^{t} \frac{dt}{a(t)} + \int_{t}^{t+\Delta t} \frac{dt}{a(t)}$$

$$= -\frac{\Delta t_0}{a(t_0)} + \int_{t_0}^{t} \frac{dt}{a(t)} + \frac{\Delta t}{a(t)} \tag{8.229}$$

Substituting Eq. (8.229) in Eq. (8.228), we get

$$\int_{t_0+\Delta t_0}^{t+\Delta t} \frac{dt}{a(t)} = \int_{t_0+\Delta t_0}^{t_0} \frac{dt}{a(t)} + \int_{t_0}^{t} \frac{dt}{a(t)} + \int_{t}^{t+\Delta t} \frac{dt}{a(t)} = -\int_{r}^{0} \frac{dr}{(1-kr^2)^{1/2}}$$

or

$$\frac{\Delta t}{a(t)} = \frac{\Delta t_0}{a(t_0)} \tag{8.230}$$

where use was made of Eq. (8.227). Thus the interval of emission Δt_0 is not equal to the interval of emission Δt.

The proper time of emission $\Delta \tau_0$ is equal to the coordinate time, *i.e.*, $\Delta \tau_0 = \Delta t_0$. Likewise $\Delta \tau = \Delta t$. If the number of waves emitted in proper time interval is n of frequency ν_0 and are received by the observer as n waves with a frequency ν, we get

$$n = \nu_0 \Delta t_0 = \nu \,\Delta t \tag{8.231}$$

Now $\nu\lambda = \nu_0\lambda_0 = c$, we get

$$\frac{\lambda}{\lambda_0} = \frac{a(t)}{a(t_0)} \tag{8.232}$$

Introducing a parameter z defined as

$$z = \frac{\lambda - \lambda_0}{\lambda_0} = \frac{d\lambda}{\lambda} \tag{8.233}$$

one gets

$$1 + z = \frac{a(t)}{a(t_0)} \tag{8.234}$$

The astronomical observations show that there is a red-shift of the stellar radiation, *i.e.*, $\lambda > \lambda_0$. Thus, we infer that

$$a(t) > a(t_0) \tag{8.235}$$

Thus, the radius $a(t)$ (actually $ra(t)$) is increasing with time, *i.e.*, the universe is expanding. If τ is the time of transit of the radiation from P to the origin, then

$$t - t_0 = \tau \tag{8.236}$$

Thus,
$$t_0 = t - \tau; \frac{1}{a(t_0)} = \frac{1}{a(t-\tau)}$$

Expanding $\frac{1}{a(t-\tau)}$ about the time t in a Taylor series, we get

$$\frac{1}{a(t_0)} = \frac{1}{a(t)} + \tau\left(\frac{\dot{a}}{a^2}\right)_t + \frac{\tau^2}{2}\left(-\frac{\ddot{a}}{a^2} + 2\frac{\dot{a}}{a^3}\right)_t + \cdots. \tag{8.237}$$

The suffix t implies that these quantities are evaluated at time of observation t.

Defining

$$H = \frac{\dot{a}(t)}{a(t)} = \text{Hubble Parameter}$$

$$H_1 = \frac{\ddot{a}(t)}{a(t)} = \text{acceleration parameter} \tag{8.238a}$$

$$q = -\frac{\ddot{a}(t)}{a(t)}\left(\frac{a(t)}{\dot{a}(t)}\right)^2 = -\frac{H_1}{H^2} = \text{deceleration parameter} \tag{8.238b}$$

Substituting Eq. (8.238) in Eq. (8.237), and writing Eq. (8.234) as $z = \frac{a(t)}{a(t_0)} - 1$, we get

$$z = H\tau + \tau^2 H^2 (2+q)/2 + \cdots \tag{8.239}$$

or to a first approximation, the above equation becomes

$$z \approx H\tau \tag{8.240}$$

This is a relation of the Hubble type. Attributing the red-shift due to Doppler motion of the nebula as receding from the observer, we have

$$z = \frac{d\lambda}{\lambda} = \frac{\Delta\nu}{\nu} = \frac{v}{c} \tag{8.241}$$

Here v is the velocity of recession of the nebula. If at time t, the source is distant $L = ra(t)$, we get

$$v = \frac{dL}{dt} = r\dot{a}(t) \tag{8.242}$$

Therefore,

$$\frac{d\lambda}{\lambda} = \frac{1}{c} r\dot{a}(t) = \left(\frac{1}{c}\frac{\dot{a}(t)}{a(t)}\right) L \tag{8.243}$$

Since $L/c = \tau$, is the travel time of the radiation, we get

$$\frac{d\lambda}{\lambda} = \tau \frac{\dot{a}(t)}{a(t)} \tag{8.244}$$

According to Hubble velocity–distance relation, the velocity of recession v varies directly with the distance, *i.e.*,

$$v = \frac{dL}{dt} = HL \tag{8.245}$$

where H is the Hubble constant[6]. Putting Eq. (8.242) in the above equation, one gets

$$\dot{a}(t) = Ha \tag{8.246}$$

where we put $L = ra(t)$. Hubble constant has the dimensions of reciprocal time. Usually it is denoted as $H = 1/T$, we obtain

$$\frac{\dot{a}}{a} = H = \frac{1}{T}$$

or

$$z = \frac{\tau}{T} \tag{8.247}$$

Astronomical data confirm the law, Eq. (8.247) but the determination of the value of Hubble constant is hampered by the uncertainty in the establishment of a scale of cosmic distances suitable for distant galaxies.

8.10 PRELIMINARIES FOR EARLY UNIVERSE

It is proposed to consider the background material having a role in the expanding universe, such as black -body radiation, Jeans mass and the energy density of neutrinos, antineutrinos and photons after electron-positron annihilation.

8.10.1 Black-body Radiation in an Expanding Universe

The Planck distribution for the energy du of black-body radiation per unit volume, in a narrow range of wavelengths from λ to $\lambda + d\lambda$, is

$$du = \frac{8\pi hc}{\lambda^5} \frac{d\lambda}{e^{\frac{hc}{kT\lambda}} - 1} \tag{8.248}$$

where T is the temperature; k, Boltzmann's constant; c, the velocity of light; e, the base of Naperian logarithm (2.718) and h, the Planck constant (6.625×10^{-27} erg sec).

In the region of long wavelengths,

$$e^{(hc/kT\lambda)} - 1 \approx \left(\frac{hc}{kT\lambda} \right) \tag{8.249}$$

Therefore, the Planck distribution becomes in the long wavelength approximation

$$du = \frac{8\pi kT}{\lambda^4} d\lambda \tag{8.250}$$

This is Rayleigh–Jeans formula. If this formula is valid down to arbitrarily small wavelengths, then $du/d\lambda \to \infty$ for $\lambda = 0$, and the total energy density in black-body radiation, will become infinite. But the Planck formula for *du* has a maximum at a wavelength

$$\lambda = .2014052\ hc/kT \tag{8.251}$$

and falls steeply for decreasing wavelengths. The total energy density in the black-body radiation is given by

$$u = \int_0^\infty \frac{8\pi hc}{\lambda^5} \frac{d\lambda}{e^{\frac{hc}{kT\lambda}} - 1} \tag{8.252}$$

From tables of definite integrals, we have

$$\int_0^\infty \frac{1}{\lambda^5} \frac{d\lambda}{e^{\frac{hc}{kT\lambda}} - 1} = \frac{1}{15} \left(\frac{\pi kT}{hc} \right)^4 \tag{8.253}$$

Thus Eq. (8.252) becomes

$$u = \frac{8\pi^5 (kT)^4}{15(hc)^3} = 7.56464 \times 10^{-15} [T(^\circ\text{K})]^4\ \text{erg/cm}^3 \tag{8.254}$$

This is Stefan–Boltzmann law.

Let us express the Planck distribution in terms of quanta of light. Each photon has energy *E* as

$$E = hc/\lambda \tag{8.255}$$

Therefore, the number *dN* of photons per unit volume in the black-body radiation in range of wavelengths λ to $\lambda + d\lambda$ is

$$dN = \frac{du}{hc/\lambda} = \frac{8\pi}{\lambda^4} \frac{d\lambda}{e^{\frac{hc}{kT\lambda}} - 1} \tag{8.256}$$

The total number of photons per unit volume is obtained by integrating the above expression, as

$$
\begin{aligned}
N &= \int_0^\infty dN \\
&= 60.42198\left(\frac{kT}{hc}\right)^3 \\
&= 20.28[T(°\text{K})]^3 \text{ photons/cm}^3
\end{aligned}
\tag{8.257}
$$

and the average photon energy is

$$
\begin{aligned}
E_{\text{average}} &= u/N \\
&= 3.73\times10^{-16}[T(°\text{K})]\text{ergs}
\end{aligned}
\tag{8.258}
$$

Let us examine as what happens to the black-body radiation when the universe is expanding. If the size of the universe changes by a factor *f*, the wavelength will change according to the new size as

$$\lambda' = f\lambda \tag{8.259}$$

The energy density du' in the new wavelength range λ' to $\lambda' + d\lambda'$ is less thanin du the old wavelength range λ to $\lambda + d\lambda$, for the following two reasons:

- As the universe size has increased by a factor f^3 and if no photons have been either created or destroyed, the number of photons per unit volume has decreased by a factor $1/f^3$.
- The energy of each photon being inversely proportional to the wavelength, it decreases by a factor $1/f$.

Thus, we infer that the energy density gets decreased by a factor $1/f^3$ times $1/f$ or $1/f^4$ Therefore,

$$
\begin{aligned}
du' &= \frac{1}{f^4}du \\
&= \frac{8\pi hc}{\lambda^5 f^4}\frac{d\lambda}{e^{\frac{hc}{kT\lambda}}-1} \\
&= \frac{8\pi hc}{\lambda'^5}\frac{d\lambda'}{e^{\frac{hcf}{kT\lambda'}}-1}
\end{aligned}
\tag{8.260}
$$

It is identical to the earlier formula for *du* in terms of λ and $\lambda + d\lambda$ except that *T* has been replaced by

$$T' = T/f \tag{8.261}$$

Thus, it is obvious that the black-body radiation continues to be described by Planck distribution but by a temperature that drops inversely to the scale of expansion.

8.10.2 Jeans Mass

For the formation of a gravitationally bound system out of a clump of matter, it is necessary that its gravitational potential energy exceeds its internal thermal energy. Consider a clump of radius r and mass M, its gravitational potential energy is of the order of

$$P.E. \approx -\frac{GM^2}{r} \tag{8.262}$$

The internal thermal energy per unit volume is proportional to the pressure p, so the total internal energy is of the order of

$$I.E. \approx pr^3 \tag{8.263}$$

The gravitational clumping will take place if

$$\frac{GM^2}{r} \gg pr^3 \tag{8.264}$$

But

$$M = \frac{4\pi}{3}\rho r^3 \tag{8.265}$$

where ρ is the density of the bound system.

Writing the gravitational clumping condition in terms of ρ, we get

$$GM^2 \gg p(M/\rho)^{4/3} \tag{8.266}$$

or

$$M \gg M_J \tag{8.267}$$

where $M_J = \frac{p^{3/2}}{G^{3/2}\rho^2}$, known as Jeans mass.

The mass density of photons and neutrinos, just before recombination of hydrogen, at 3,000°K [refer to Eq. (8.282], is

$$\begin{aligned}\rho &= 1.22\times10^{-35}(3\times10^3)^4 \text{ gm/cm}^3\\ &= 9.9\times10^{-22} \text{ gm/cm}^3\end{aligned} \tag{8.268}$$

and pressure

$$p \approx \frac{1}{3}c^2\rho = 0.3 \text{ gm/cm sec}^2 \tag{8.269}$$

Therefore,

$$\begin{aligned}\text{Jeans mass} &= \left(\frac{0.3 \text{ gm/cm sec}^2}{6.67\times10^{-8}\text{ cm}^3/\text{gm sec}^2}\right)^{3/2}\left(\frac{1}{9.9\times10^{-22}\text{ gm/cm}^3}\right)^2\\ &= 9.7\times10^{51}\text{ gm} = 5\times10^{18}\text{ solar mass}\end{aligned} \tag{8.270}$$

After recombination of electrons and positrons, during the whole of radiation dominated era, the pressure dropped by a factor of 10^9, so the Jeans mass became

$$M_J = (10^{-9})^{3/2} \times 5 \times 10^{18}$$
$$= 1 \cdot 6 \times 10^5 \text{ solar mass} \qquad (8.271)$$

This interestingly compares roughly with the mass of large globular clusters within our galaxy.

8.10.3 Neutrino Temperature and Density

Entropy of a system remains fixed as long as the thermal equilibrium continues. The entropy per unit volume S, to a good approximation, at a temperature T is given by

$$S \propto N_T T^3 \qquad (8.272)$$

whereis N_T the effective number of species of particles in thermal equilibrium, whose threshold temperature is below T. In order to keep S constant in an expanding universe, it must be inversely proportional to the cube of the size of universe. Thus if R is the distance between any pair of particles, then

$$SR^3 \propto N_T T^3 R^3 = \text{constant} \qquad (8.273)$$

The neutrinos and antineutrinos had gone out of thermal equilibrium, before the annihilation of electrons and positrons at about 5×10^9 °K, with the result that the only particles in thermal equilibrium were the electron, positron and photon.

The effective number of species is the product of three factors: The first factor is 2 or 1 according to whether the particle has an antiparticle or not; the second factor is the number of possible orientations of the particle's spin; the last factor is 7/8 or 1, according to whether or not the particle is a fermion and obeys Pauli's exclusion principle or not.

Thus, the effective number of species before annihilation

$$N_{\text{before}} = 2 \times 2 \times 7/8 + 1 \times 2 \times 1$$
$$= 7/2 + 2 = 11/2 \qquad (8.274)$$

Now after the annihilation of electrons and positrons at about 5×10^9 °K, the only remaining particle in thermal equilibrium is photon. The effective number of species is

$$N_{\text{after}} = 2 \qquad (8.275)$$

According to the law of conservation of entropy, we have

$$\frac{11}{2}(TR)^3_{\text{before}} = 2(TR)^3_{\text{after}}$$

or

$$\frac{(TR)_{\text{after}}}{(TR)_{\text{before}}} = \left(\frac{11}{4}\right)^{1/3} = 1.401 \qquad (8.276)$$

The heat resulting from the annihilation of electrons and positrons increases the quantity TR by a factor 1.401. Thus after the annihilation process, the photon temperature is higher than the neutrino temperature by a factor

$$\left(\frac{T}{T_\nu}\right)_{\text{after}} = \frac{(TR)_{\text{after}}}{(TR)_{\text{before}}} = \left(\frac{11}{4}\right)^{1/3} = 1.401 \tag{8.277}$$

The neutrino temperature T_ν before the annihilation of electrons and positrons was the same as the photon temperature T. But from then on, T_ν decreased like $1/R$ so subsequently $T_\nu R$ equalled the value TR before the annihilation.

$$(T_\nu R)_{\text{after}} = (T_\nu R)_{\text{before}} = (TR)_{\text{before}} \tag{8.278}$$

Although out of thermal equilibrium, the neutrinos and antineutrinos contribute significantly to the cosmic energy density. The effective number of neutrinos and antineutrinos is 7/2. As the photon has spin 2, the effective number of species of neutrinos and antineutrinos is 7/4 of the effective number of photons. Also the fourth power of neutrinos temperature is lower than the fourth power of photon temperature by a factor $(4/11)^4$.

The ratio of the energy density of neutrinos and antineutrinos to that of photons is

$$\frac{u_\nu}{u_\gamma} = \frac{7}{4}\left(\frac{4}{11}\right)^{4/3} = 0.4542 \tag{8.279}$$

According to Stefan–Boltzmann law, the photon energy density at temperature T is

$$u_\gamma = 7.5641\times10^{-15}\ \text{erg/cm}^3\times[T\ {}^\circ\text{K}]^4 \tag{8.280}$$

The total energy density after electron–positron annihilation is

$$u = u_\gamma + u_\nu = 1.4542u_\gamma = 1.100\times10^{-14}\ \text{ergs/cm}^3[T\ {}^\circ\text{K}]^4 \tag{8.281}$$

The equivalent mass density is obtained by dividing the energy density by the square of the velocity of light. Thus,

$$\rho = u/c^2 = 1.22\times10^{-35}\ \text{ergs/cm}^3[T\ {}^\circ\text{K}]^4 \tag{8.282}$$

8.11 THE STANDARD MODEL OF EARLY UNIVERSE

The standard model of very early universe, is essentially a parts kit for the subatomic world. The standard model consists of six quarks, six leptons, five known bosons and a postulated sixth, the Higgs boson (named after a Scottish physicist, Peter Higgs, also called God particle) and three of the four physical forces, the strong and weak nuclear forces and electromagnetic force.

The basic building blocks of matter are quarks: up, down, strange, charm, top, and bottom (which are referred to as their flavours and these are further divided into the colours—red, green, and blue). These quarks are held together by particles called gluons and together quarks and gluons form protons and neutrons. Leptons are the source of electrons and neutrinos. Quarks and leptons are Fermions. Bosons (named after Indian physicist S.N. Bose) are particles that produce and carry forces, and include photons and gluons. Higgs Boson was postulated as a means of endowing particles with mass. It is believed that scientists at the Large Hadron collider at CERN, have captured the elusive sub-atomic particle, Higgs Boson, called, 'God particle' that gives mass and holds the physical fabric of the universe together. The leaders of the two CERN teams, each presented what was essentially strong evidence of a new particle.

The standard model not only lacks elegance and simplicity but is incomplete as it is silent about the force of gravity. Furthermore, it does not explain mass. The introduction of Higgs Boson is too artificial and its existence or otherwise is yet to be established. As Feynman observed: 'So we are stuck with a theory and we do not know whether it is right or wrong, but we do know that it is a little wrong, or atleast incomplete.'

Physicists proposed the Superstring theory, with a view to draw everything together. It postulates that quarks and leptons that we previously thought as particles, are actually 'strings'—vibrating strands of energy that oscillate in eleven dimensions, comprised of three of space plus time and seven other dimensions that are unknowable to us. The strings are very tiny-tiny enough to pass for point particles. String theory has further spawned 'M theory', which incorporates surfaces known as membranes or simply 'branes'.

The standard model of the universe assumes that the universe began with a hot big bang at sometime in the past, which is considered as the origin of the universe. At the time of big bang, the universe was extremely hot and dense, a singularity in space–time. After the big bang, the universe started expanding and cooled as time progressed and the average velocity of particles decreased. As the strong, electromagnetic and weak reaction rates, were much larger than the expansion rate of the universe initially, it is a good approximation to make that all the elementary particles are in thermal equilibrium. Under these conditions, the universe is dominated by radiation, which implies that the majority of energy is in the form of quarks, gluons and leptons and as the temperature fell below 10^{12} °K, the universe underwent the quark to hadron phase transition and after that the constituents were leptons and hadrons.

Because of the equivalence of mass and energy, in a radiation dominated era, the particles and their antiparticles, are continuously undergoing reactions annihilating each other, and photons can collide and create particle and antiparticle pairs. Under these conditions, the radiation and matter are in thermal equilibrium because they can freely convert back and forth.

Excepting the proton and neutron densities, the densities of leptons are determined by the Fermi–Dirac or Bose–Einstein distributions at the temperature of the universe. As the temperature was much higher than the electron mass energy, the electron and protons were equal in abundance.

The proton and neutron densities relative to densities of other constituents, is an important basis to decide the onset and progress of nucleosynthesis. One can make some observations at this stage. As the universe is charge neutral, the difference between electron and positron densities should equal the proton density. Furthermore, the photon (and neutrino) densities are determined by the temperature of the universe, since the photons are in thermal equilibrium with the other constituents. In reality, the photon density follows a dependence on the age of the universe through its temperature, which falls continuously with the expansion. The photons present in the early universe survive even today as the cosmic background radiation having a temperature of 2.7 K. Thus, the photon density during nucleosynthesis, can be easily determined from the present temperature of the cosmic background radiation. The sum of the neutron and proton densities as a function of the age of the universe, can be estimated from the expansion rate of the universe. From observational cosmology, we find that the ratio of the baryon density and the photon density is extremely small $\sim 10^{-10}$. This ratio has a direct bearing on the amount and species of nuclei even in the very early universe. This ratio was extremely small even in the early universe. This implies that the process of nucleosynthesis can be divided into two independent processes, one of the expansion of the universe and the other of nucleosynthesis. In view of this extremely small ratio of the baryon density to the photon density, the evolution of the universe in the early stages is essentially independent of the baryon density, *i.e.*, how much matter goes into different nuclei and how nucleosynthesis proceeds.

Let us consider the evolution of the early universe and start from the temperature of 10^{12} K or about 100 MeV, when the muons have decayed and photons, electrons, positrons, neutrinos, and antineutrinos constitute the bulk of the universe. Their densities are given by Bose–Einstein or Fermi–Dirac distributions. The chemical potential of all these particles can be taken to be zero. The energy density in the universe, which governs the expansion at this era, is determined by light particles, *i.e.*, photons, electrons and neutrinos alongwith their antiparticles. This is the radiation-dominated phase of the universe because the particles which are responsible for the expansion are those whose rest masses are much smaller than the temperature of the universe. The energy density of the universe is given by $\rho(t) = CT^4$ where the constant C is determined from the number of light particles, their degeneracies and whether these are Bosons or Fermions.

The expansion of the universe is governed by Friedmann equation

$$\dot{a}^2 + k = \frac{8\pi}{3}\rho(t)a^2 \tag{8.283}$$

where $a(t)$ is the length scale, k is the curvature constant, which is zero for flat universe and $1(-1)$ for universe having positive (or negative) curvature and $\rho(t)$ is the energy density.

In addition, the energy conservation equation, is

$$\frac{\partial\rho}{\partial t} = -3\frac{\dot{a}}{a}(\rho + p) \tag{8.284}$$

where p is the pressure. For fluid having relativistic particles, $p(t) = \rho(t)/3$, therefore, the energy conservation equation becomes

$$\frac{\partial\rho}{\partial t} = -4\frac{\dot{a}}{a}\rho \tag{8.285}$$

Rearranging terms, we obtain

$$\frac{\partial\rho}{\rho} = -4\frac{da}{a} \tag{8.286}$$

Integrating and ignoring the integration constant, one gets

$$\ln \rho = \ln a^{-4}$$

or

$$\rho \propto a^{-4} \tag{8.287}$$

As $\rho(t) \propto T^4$, according to Stefan–Boltzmann law, it implies that T varies as a^{-1}.

With Eq. (8.287), Friedmann Eq. (8.283) becomes

$$\dot{a}^2 + k = \frac{8\pi}{3} a^{-2} \tag{8.288}$$

One can show that $\frac{8\pi}{3} a^{-2} \gg 1$, so neglecting the curvature constant on the r.h.s. of Eq. (8.288), we obtain the solution

$$a^2(t) = 2\sqrt{\frac{8\pi}{3}}\, t \tag{8.289}$$

Thus, the length scale of the universe increases as the square root of time. But as the temperature varies like $1/a(t)$, so the temperature of the universe drops as inverse of the square root of time. The time and temperature relation as well as the neutron fraction in the universe is given in Table 8.1.

8.11.1 Evolution of Neutron Fraction

We next consider the neutron fraction or the ratio of the neutron to baryon density as a function of time. As the heavier nuclei are produced through the fusion of neutrons and protons, neutron fraction has central importance in nuclei synthesis. At time ~1/100 second after the expansion of universe begins, the temperature is about 100 billion K and the density is more than a billion times that of water. The universe is expanding rapidly and is very hot, it consists of an undifferentiated soup of matter and radiation in thermal equilibrium. The temperature corresponds to an average energy of particles about 8.6 MeV. The photons are in thermal equilibrium with electrons, positrons, neutrinos, and antineutrinos. The relevant reactions for the neutron fraction to evolve are the weak interactions between baryons, electrons, and neutrinos. These are:

$$n + e^+ \leftrightarrow p + \bar{\nu} \tag{8.290}$$

$$p + e^- \leftrightarrow n + \nu \tag{8.291}$$

$$n \leftrightarrow p + e^- + \bar{\nu} \tag{8.292}$$

$$p \leftrightarrow n + e^+ + \nu \tag{8.293}$$

where $\nu, \bar{\nu}$ stand for the neutrino and antineutrino respectively.

At temperatures above 5×10^9 °K, electrons and protons are abundant and the reaction rates of these weak interactions are larger than the expansion rates. This implies that the baryons are in thermal equilibrium and the ratio of neutron to proton densities is $e^{-\delta m/T}$ where $\delta m = 1.29\,\text{MeV}$, the neutron proton mass difference and T is the temperature. Thus, $T \gg \delta m (= \Delta mc^2)$. The neutron fraction is

$$\frac{e^{-\delta m/T}}{1+e^{-\delta m/T}} = \frac{1}{1+e^{\delta m/T}} \sim 0.5 \tag{8.294}$$

At time $\sim 1/10$ second, the temperature has dropped to several times 10 billion K and the density is a little over 10 million times that of water as the universe continues to expand. As a free neutron is slightly less stable than a free proton, neutron β- decay to proton plus electron plus neutrino with a half-life of approximately 886 seconds. Thus, the initial approximately equal balance between neutrons and protons begins to be tipped in favor of protons. By this time about 62% of the baryons are protons and 38% are neutrons. The decay of neutrons will continue until the simplest nucleus, deutron can form.

Let us follow further sequence of events that take place in the big bang in the beginning of nucleosynthesis.

At time ~1 second, the temperature has dropped to about 10 billion K as the universe continues to expand, and the density is now down to about 400,000 times that of water. About the time the decreasing density and temperature have increased, the mean free time of neutrinos and antineutrinos

Table 8.1 *The time dependence of temperature of the universe and the neutron fraction.*

The time when the temperature is 10^{12} K is taken as zero.

Temperature (°K)	Time (Sec)	Neutron Fraction (X_n)
10^{12}	0.0	0.496
3×10^{11}	0.00113	0.448
10^{11}	0.01078	0.462
3×10^{10}	0.1209	0.380
10^{10}	1.103	0.241
3×10^{9}	13.830	0.170
1.3×10^{9}	98	0.150
1.2×10^{9}	119	0.147
1.1×10^{9}	148	0.143
10^{9}	182	0.137
9×10^{8}	226	0.131
8×10^{8}	290	0.123
7×10^{8}	383	0.122
10^{8}	2080	0.021

increases to an extent that they begin to behave like free particles, no longer in thermal equilibrium with the electrons, positrons or photons. From now on, they cease to play any active role in the big bang, except that their energy will continue to provide part of the source of the gravitational field of the universe. Nothing much changed when the neutrinos go out of thermal equilibrium. The deuterium bottleneck still exists so there are no composite nuclei and the neutrons continue to beta decay to protons. At this temperature, the proton abundance is upto 76% and the neutron abundance has decreased to 24%.

With the decrease of temperature, the neutron fraction decreases as $1/(1+e^{\delta m/T})$. This will continue till the temperature drops to about 5×10^9 K, which corresponds to 0.5 MeV (comparable to the rest mass-energies of electron and positron). At this temperature range, electrons and positrons annihilate and their densities drop to small values comparable to the baryon densities. The reaction rates of Eqs. (8.290) and (8.291) are reduced considerably and the neutron fraction is no longer given by the thermal equilibrium condition. This happens when the universe is about 10 seconds old, Table 8.1.

The neutron fraction will decrease by β-decay and is described by the relation

$$X_n(t) = X_n^0 e^{-\lambda t} \tag{8.295}$$

where $X_n(t)$ is the neutron fraction at time t, λ is the decay constant and X_n^0 is the neutron fraction when the neutron fraction decay is the dominant mechanism for the evolution of the neutron fraction X_n^0 is close to the value when the temperature drops down to 3×10^9 K which happens when the universe is about 10 seconds old. The mean life of neutron is 886 seconds.

A dynamical calculation including weak reaction rates for all the reactions, Eqs. (8.290) and (8.291) for calculation of neutron fraction has to be carried out, taking into account electron positron annihilation and computing the evolution of the temperature of the universe as a function of time. The evolution of temperature and $X_n(t)$ obtained from such a calculation are shown in Table 8.1.

8.11.2 Nucleosynthesis

The formation of nuclei critically depends on the baryon density. As the baryon density is extremely small, nuclei will be formed in two-body process. The first step in nucleosynthesis is the deutron formation as

$$p + n \leftrightarrow d + \gamma \tag{8.296}$$

Deutrons must be formed in enough large quantity for the heavier nuclei to be built up. The rate of deutron formation will critically depend on the deutron binding energy, *i.e.*, 2.2 MeV and the baryon to photon ratio. Normally one is inclined to imagine that when the temperature of the universe falls lower than the deutron binding energy, the neutron capture by protons would dominate over deutron breakup by photons but this does not happen, since the background photon density is high and the deutrons that are formed disintegrate under the influence of the high energy photons present in the background. This is what happens due to small baryon to photon density, even when the temperature drops below the deutron binding energy. Although the number density of high energy photons falls exponentially, still there are quite a large number of high energy photons even at temperatures lower than 2.2 MeV, to cause the photodisintegration of deutrons which are formed.

At time ~13.8 seconds, the temperature has fallen to about 3×10^9 °K and the average energy of the particles in the gas has fallen to about 0.25 MeV. As the energy is too low for photons to produce electron–positron pairs, so they fall out of thermal equilibrium and the free electrons begin to annihilate all the positrons to form photons. The deuterium bottleneck still keeps appreciable deuterium from forming and the neutrons continue to decay, with the result that neutron abundance has fallen to 13% and proton abundance has risen to 87%.

At time ~3 min 45 seconds, the temperature drops sufficiently low (about $10^9\,°\mathrm{K}$) that nuclei can hold together. The deuterium bottleneck is finally broken and a rapid sequence of the following nuclear reactions combines neutrons and protons to form deutron, ^3He, ^3H, ^4He nuclei through the following reactions:

$$d + p \leftrightarrow {}^3He + \gamma \tag{8.297}$$

$$d + n \leftrightarrow {}^3H + \gamma \tag{8.298}$$

$$d + d \leftrightarrow {}^4He + \gamma \tag{8.299}$$

$$H^3 + p \leftrightarrow {}^4He + \gamma \tag{8.300}$$

$$^3He + n \leftrightarrow {}^4He + \gamma \tag{8.301}$$

The buildup of the densities of these nuclei depends on the reaction rates which in turn depend on the binding energies of these nuclei through kinematic factors. The reaction rates are higher if the binding energies are greater and the strongly bound nuclei will build faster. Among these, ^4He with a binding energy of 28 MeV is the most strongly bound nucleus. ^3He, ^3H binding energies are about 7 MeV whereas ^2H as has 2.2 MeV. This means that as soon as deutron density starts building up, neutron and proton capture on ^2H, ^3H, and ^3He starts and as a result ^4He density builds up. A dynamical calculation, which includes the reaction rates of the reactions, Eqs. (8.297)–(8.301), shows that most of the neutrons, which are forming deutrons quickly end up in forming ^4He and very little of deutrons and other nuclei survive. But a lot of protons survive as the neutron fraction is less than 0.5. The amount of ^4He formed depends critically on when the nucleosynthesis starts. The helium fraction formed during nucleosynthesis in early universe can be estimated in almost model-independent manner and the only crucial input is the baryon to photon ratio. If it is assumed that the nucleosynthesis begins at, $10^9\,°\mathrm{K}$ all the neutrons at this temperature will end up in helium. We find from Table 8.1, that X_n at $10^9\,°\mathrm{K}$ is 0.137. Assuming that all these neutrons go into helium, the helium fraction would be $X_n/2$ or 0.0685; which corresponds to $4 \times 0.0685 = 0.274$ by weight as each ^4He nucleus has 4 nucleons. This number is very close to the abundance found in the universe.

The ratio of the number of baryons per photon was one of the contributions of the discovery of 3 K background radiation. These measurements permitted a calculation of the photon energy density in the universe and the range of baryon density gave a baryon/photon ratio of 10^{-9}.

Hydrogen and helium account for nearly all the matter in today's universe and the figures for relative abundance indicate that helium is about 25% by mass and hydrogen about 73% with all other elements constituting less than 2%. The agreement between the observed and estimated percentage of helium strongly favours the big bang model. As there is no known process, which significantly changes the *H/He* ratio, it is assumed to be the ratio, which existed at the time when deutron became stable in the expansion of the universe. This ratio is important as a test of cosmological models, since it will be affected by the time period from the time when the temperature dropped below that necessary to produce protons to neutrons to the time when deutron became stable, stopping the decay of the neutrons.

Basically, the hydrogen–helium abundance helps to model the expansion rate of the early universe. If it were faster, there would be more neutrons and more helium. If on the other hand, it was slower, more of the free neutrons would have decayed before the deutron stability point leading to less of helium.

It seems that heavier elements are not produced in early universe, due to the fact that the nuclei with mass number 5 and mass number 8 are unstable and this is the reason why we do not have ^5He, ^5Li, and ^8Be in nature. The ^4He density being large, one channel to produce heavier elements is through the capture of neutron or proton on helium or through the fusion of two ^4He nuclei, but that does not come to pass. Only other possibility is the production of heavier elements through three body reactions. The possibility of having three helium nuclei within nuclear volume at the same time is extremely small. In reality, ^8Be is a resonant state with somewhat small width and so two helium nuclei may form the resonance and before it decays, the third helium may interact to form ^{12}C. This process is happening in the stars but is very rare in the early universe due to very small baryon density. Thus, the lack of stable nuclei with mass numbers 5 and 8, is the bottleneck for the production of heavy elements. However, some ^7Li formed in early universe through the reaction $^4He + {}^3H \to {}^7Li + \gamma$, but the abundance is nine orders of magnitude smaller than helium.

At time ~35 minutes, the temperature is now about 300 million K and the universe consists of protons, the excess electrons that did not annihilate with the positrons, helium-4, photons, neutrinos and anti-neutrinos. There are no atoms yet because the temperature is still too high for the protons and electrons to bind together.

At time ~700,000 years, the temperature has fallen to several thousand K, which is low enough for the protons and electrons to hold together and begin forming hydrogen atoms. At this stage, the matter and radiation, which have been in thermal equilibrium so far, decouple. The free electrons are held up in atoms, the primary cause leading to scattering of photons is removed and the universe which had been opaque so far, becomes transparent and light can travel large distances before getting absorbed.

Example 8.2

Under which of the following suppositions, the hot big bang would have produced less He^4 than predicted by the standard model? less H^2 (Deuterium).

- The baryon density is greater than the value we assume.
- Weak interaction constant is smaller than the present value.
- There are many more neutrinos than anti-neutrinos or photons in the cosmic background radiation.
- There are many more anti-neutrinos than neutrinos or photons in the cosmic background radiation.

Solution

According to the standard model, helium and deuterium are produced in the following steps:

1. Freezing out a nonequilibrium neutron/proton ratio when the weak interaction for the $n \to p$ reaction, become slower than the time scale for the expansion to drop the temperature.
2. Producing deuterium by $n + p \to d + \gamma$ reaction, while most neutron have not yet decayed to protons.

3. Using up almost all d by the reactions

$$
\begin{aligned}
{}_1^2d + {}_1^2d &\rightarrow {}_1^3H + p \\
{}_1^2d + {}_1^2d &\rightarrow {}_2^3He + n^0 \\
{}_1^2d + {}_2^3He &\rightarrow {}_2^4He + \gamma
\end{aligned}
$$

In view of the discussion above, we take up the cases one by one.

1. If the baryon density in the universe is larger than the actual value, it does not effect the freezing out ratio, but the higher baryon density at any temperature, will lead to increased 'cooking', thus enhance complete transformation to ${}_2^4He$, more ${}_2^4He$, more ${}_1^2d$.
2. Smaller weak interaction constant will make the $n \rightarrow p$ reaction slower, thus more frozen out neutrons, other reaction rates not effected; more ${}_2^4He$, more ${}_1^2d$.
3. The reaction $n + \nu \rightarrow p + e^-$ will be driven to the right by the degeneracy (Fermi) energy of the neutrinos, fewer neutrons will give less ${}_2^4He$, less ${}_1^2d$.
4. This will favour the reaction $p + \bar{\nu} \leftrightarrow n + e^+$, giving more ${}_2^4He$, more ${}_1^2d$.

8.11.3 Difficulties with the Standard Cosmological Model

Standard cosmological model is the 'big bang' and despite enormous evidence in support of it, there are some difficulties with it. These are in the following:

1. The anti-matter problem
2. The monopole problem
3. The galaxy formation problem
4. The horizon problem
5. The flatness problem

Let us discuss these difficulties one by one.

1. **The Anti-matter Problem:** There is predominance of matter over anti-matter in the universe. After the beginning of the particle era, there is no known process, which can alter the net particle number of the universe and by the time the universe is a milli-second old, the matter between the matter and anti-matter is fixed for ever.

   Obviously there is some asymmetry in nature's way of treating matter and anti-matter. One possibility is that of CP symmetry violation in the decay of particles by the weak interaction. The main experimental evidence comes from the decay of neutral kaons, which shows a small violation of CP symmetry. There is a clear distinction between the matter and anti-matter in the decay of kaons to electrons and this could possibly constitute one of the ways to the predominance of matter over anti-matter in the universe.

2. **The Monopole Problem:** Big Bang cosmology predicts that a very large number of heavy, stable 'magnetic monopoles' should have been produced in the early universe. However, magnetic monopoles have never been observed experimentally, so that if they exist at all, they are far more rare than what the Big Bang predicts.

3. **The Galaxy Formation Problem:** Random inhomogeneities in the expanding universe are not adequate to allow the formation of galaxies. The turbulence created by the expansion itself in the rapid expansion, the gravitational attraction is too slow for galaxies to form. The formation of the large-scale structure of the universe is a major unsolved problem in cosmology.
4. **The Horizon Problem:** The cosmic microwave background from opposite directions in the sky is characterized by the same temperature within 0.01%, but the regions of space from which they were emitted at 500,000 years were apart more than the transit time and thus, could not have communicated with each other to establish the thermal equilibrium. Evidently they were beyond each other's horizon.

   The problem is also referred to as the 'isotropy problem', so the cosmic microwave background radiation reaching us from all different directions in space, is so nearly isotropic. How could the temperature of parts of space in opposite directions from us could be exactly the same, when the thermal equilibrium could not be established because of the inability to communicate with each other. If the estimated look back time is 13.72 billion years as obtained from a Hubble constant of 71 km/sec per Megaparsec as suggested on the basis of WMAP (Wilkinson Microwave Anisotropy Probe), then these remote parts of the universe are 27.44 billion years apart; then how can they have the same temperature?
5. **The Flatness Problem:** Observations indicate that the amount of matter in the universe is surely greater than one-tenth and less than ten times the critical amount needed to stop the expansion. The universe is either barely open or barely closed, or 'very nearly flat'.

8.11.4 Inflation Theory

The inflation theory[2] proposed a period of extremely rapid (exponential) expansion of the universe 20 or 30 orders of magnitude following the Planck time of 10^{-43} seconds. It was developed around 1980 by Alan Guth, Andrei Linde, Paul Steinhardt, and Andy Albrecht, to explain several difficulties with the standard Big Bang theory, according to which the universe expands gradually throughout its history. The prospects of an extremely tiny volume which could have been considered to be intimately connected, were spread over the whole of the known universe today, leading to both extreme flatness and the extremely isotropic nature of the cosmic background radiation.

Inflation is now considered an extension of the Big Bang theory since it explains the above difficulties, while retaining the basic paradigm of a homogeneous expanding universe. Moreover, Inflation links modern physics ideas like symmetry breaking and phase transition to cosmology. Inflation theory solves the above difficulties of the standard model as follows:

1. **The Flatness Problem:** Imagine yourself to be on the surface of a soccer ball (a 2-dimensional world). You will realize that you are living in a closed universe and curved space. If the ball expands to the size of earth, it will appear flat but still a sphere on larger scale. Further, assume increasing the size to astronomical scales. Now it will appear flat as far as you can see, although it might have curved to begin with. Inflation stretches very initial curvature of 3-dimensional universe to near flatness.

   Any deviation from 'flatness' should become amplified with time and at this stage of the universe, tiny irregularities should have been much exaggerated. If the density of the present universe appears to be close to critical density, then it must have been even closer to 'flat' in earlier eras. Alan Guth credits a lecture by Robert Dicke as one influence, which prompted him to suggest the inflationary hypothesis. Dicke pointed out that the flatness of today's universe would require that the universe be flat to one part in 10^{14} at one second after the Big Bang. Kaufmann suggests that right after the Big Bang, the density must have been equal to the critical density to 50 decimal points.

2. **The Horizon Problem:** Since inflation supposes a burst of exponential expansion in the early universe, it follows that distant regions were in reality much closer together before the onset of inflation that they would have been with only standard Big Bang expansion. Thus, such regions should have been in causal contact prior to inflation and could have attained a uniform temperature.
3. **The Monopole Problem:** Inflation allows for magnetic monopoles to exist as they were produced prior to the period of inflation. During inflation, the density of monopoles drops exponentially, so that their abundance drops to undetectable levels.
4. **The Structure of the Universe:** Inflation also explains the origin of structure in the universe. Prior to inflation, the portion of the universe was, as we can observe today, was microscopic and the quantum fluctuations in the density of matter on these microscopic scales expanded to astronomical scales during inflation. Over the next several hundred million years, the higher density stars condensed into stars, galaxies and clusters of galaxies.

8.11.5 The Steady State Model

The Big Bang model had a challenger that was called the steady state model. In this theory proposed in the late 1940's by Herman Bondi, Thomas Gold and (in a somewhat different formulation) by Fred Hoyle, the universe is not only homogeneous and isotropic when averaged over a large enough volume, it is the same for all time too.

Since the universe is known to be expanding as inferred from the red shifts of light, it was mandatory to postulate continuous creation of matter in the space between the stars and galaxies, in order to maintain the same density overtime and thereby ensure the perfect cosmological principle of a universe that is unaltered in time on large scales. This assumption is in contradiction of the law of mass-energy conservation, but the required rate of mass creation is too small to be detectable by any experiment so as to be ruled out experimentally. The rate that is required to create about 1 hydrogen atom per cubic centimeter every 10^{15} years.

Initially the steady state theory competed with the Big Bang theory but the eventual cosmological observations lent unambiguous support to the Big Bang theory, while ruling out the steady state model. The two most important cosmological observations in this regard are as follows:

1. The discovery of quasars at great distances made possible with the help of deep space radio telescope observations, showed that there were more quasars in the early universe than the present one. The steady state theory postulates that the universe is unchanging over time on large scales.
2. The discovery of the cosmic microwave background as predicted by the Big Bang theory, could not be explained in any way by the steady state theory.

In view of these facts and the overwhelming success of the Big Bang theory, the steady state theory is not considered viable by most astronomers.

8.12 AGE OF THE UNIVERSE

Astronomers estimate the age of the universe in the following two ways:

1. by looking for the oldest stars, and
2. by measuring the rate of expansion of the universe and extrapolating back to the Big Bang.

1. **The Oldest Stars:** Astronomers figure out the ages of some of the oldest stars in the universe by studying globular clusters. A globular cluster is a dense collection of close to a million stars, all of which formed at roughly the same time. The density of stars near the centre of a globular cluster is enormous.

   The life cycle of a star depends on its mass. High mass stars are much brighter than low mass stars and exhaust their fuel supply of hydrogen rapidly. A star like the sun appears to have been active for 4.6 billion years and has enough fuel in its core to burn at its current brightness for approximately five billion years or so. A star that is twice as massive as the sun will exhaust its supply in only 800 million years. A 10 solar mass star burns nearly a thousand times brighter and has only a 20 million year fuel supply. On the other hand, a star *i.e.*, half massive as the sun, burns slowly enough for its fuel supply to last more than 20 billion years.

   Since all of the stars in a globular cluster are formed at roughly the same time, these clusters can serve as cosmic clocks. If a globular cluster is more than 10 million years old, then all of its hydrogen burning stars will be less massive than 10 solar masses. This implies that no individual hydrogen burning star will be more than 1000 times brighter than the sun. If a globular cluster is more than 2 billion years old, then there will be no hydrogen burning star more massive than 2 solar masses.

   The oldest globular clusters contain only stars less massive than 0.7 solar masses. These low mass stars are much dimmer than the sun. This suggests that the oldest globular clusters are between 11 and 18 billion years old. The uncertainty in this estimate is due to the difficulty in determining the exact distance to a globular cluster (hence, an uncertainty in the brightness and mass) of the stars in the cluster. Another source of uncertainty in this estimate is our ignorance of some of the finer details of stellar evolution.

2. **Extrapolating Back to the Big Bang:** Another way to estimate the age of the universe is to measure the 'Hubble constant'. The Hubble constantis (H_0) a measure of the current expansion rate of the universe. Cosmologists use this measurement to extrapolate back to the Big Bang. The extrapolation depends on the current density of the universe and on the composition of the universe. Because of the uncertainties in distance measurement to other galaxies, there have been considerable uncertainties about the values of Hubble constant.

   Recent measurements of the cosmic background radiation by the COBE (Cosmic Background Explorer Satellite) and WMAP (Wilkinson Microwave Anisotropy Probe) satellites have provided values of Hubble constant which are much more precise.

   Even with a known value of the Hubble constant, the value projected for the age of the universe is model dependent. Much evidence suggests that the universe is very nearly 'flat', *i.e.*, it is expanding at just the right rate to continue to expand indefinitely. Asymptotically, it will approach a zero expansion rate and will not collapse back. A flat universe called an 'Einstein de-Sitter universe' requires a critical density. We presume that we have very close to the critical density, but it is much more than the mass density we see: hence, the presumption of 'dark matter' to make up the critical mass.

Presuming a flat universe, we can estimate an expansion time by making use of the CMB temperature and the critical mass density of the universe. For a Hubble constant of 71 km/s/Mpc, the critical density is $5.33\ \text{GeV/m}^3$.

In the case of a simple expansion, it is a straight-forward matter to calculate the time of origin of the expansion from the measured rate of expansion. But our time-span of observation is so brief compared to the duration of the process as a whole, that we have no other course but to draw the conclusions from an almost instantaneous value.

The Hubble constant has the dimensions of inverse time, so a Hubble time t_H may be obtained by inverting the present value of the Hubble parameter. So

$$H_0 = 71\frac{\text{km/s}}{\text{Mpc}} = 2.3\times 10^{-18}\ \text{s}^{-1}$$

$$t_H = \frac{1}{2.3\times 10^{-18}\ \text{s}^{-1}} = 13.8\times 10^{9}\ \text{years}$$

One has to exercise caution in interpreting the 'Hubble time', since the relationship of the expansion time to the Hubble time is different for the radiation-dominated universe and the mass-dominated universe.

8.13 COSMOLOGICAL CONSTANT IN EINSTEIN'S FIELD EQUATIONS

Einstein introduced the cosmological constant in his field equations for getting around the non-steady state universe. He called it the greatest blunder of his life. The equations with the cosmological constant had the form

$$R_{\mu\nu} - \frac{1}{2}g_{\mu\nu}R + g_{\mu\nu}\Lambda = -8\pi T_{\mu\nu} \tag{8.302}$$

or

$$R^{\mu}_{\nu} - \frac{1}{2}g^{\mu}_{\nu}R + g^{\mu}_{\nu}\Lambda = -8\pi T^{\mu}_{\nu}$$

$$R^{\mu}_{\nu} - \frac{1}{2}g^{\mu}_{\nu}R = -8\pi \tilde{T}^{\mu}_{\nu} \tag{8.303}$$

where $\tilde{T}^{\mu}_{\nu} = T^{\mu}_{\nu} + g^{\mu}_{\nu}\Lambda$

and

$$\tilde{T}^{\mu}_{\nu} = 8\pi\begin{pmatrix} \rho + g^{\mu}_{\nu}\Lambda/8\pi & 0 & 0 & 0 \\ 0 & -p + g^{\mu}_{\nu}\Lambda/8\pi & 0 & 0 \\ 0 & 0 & -p + g^{\mu}_{\nu}\Lambda/8\pi & 0 \\ 0 & 0 & 0 & -p + g^{\mu}_{\nu}\Lambda/8\pi \end{pmatrix} \tag{8.304}$$

The new energy-momentum tensor $\tilde{T}^{\mu}_{\nu}$ shows that the cosmological constant Λ is a source of energy density and the inverse pressure (opposing the pressure of matter). It was on this basis that the name of dark energy was coined.

The density of dark energy does not depend on the scale factor a. The equation of state for the dark energy is $p(\rho)=\rho$. Furthermore, the equation of state for matter $p(\rho)=0$ and for radiation, $p(\rho)=\rho/3$. These all can be written collectively as

$$p(\rho)=\omega\rho \tag{8.305}$$

where the parameter $\omega=1$ for dark energy, $\omega=0$ for matter and $\omega=1/3$ for radiation.

Let us consider a mixture of matter and dark energy

$$\begin{aligned}\rho &= \rho_m+\rho_{de}\\ &= \rho_{m0}\left(\frac{a_0}{a}\right)^3+\rho_{de}\end{aligned} \tag{8.306}$$

Introducing the definitions

$$\Omega_{m0}\equiv\frac{8\pi}{3H_0^2}\rho_{m0}=\frac{\rho_{m0}}{\rho_{cr0}} \tag{8.307}$$

$$\Omega_{de0}\equiv\frac{8\pi}{3H_0^2}\rho_{de0}=\frac{\rho_{de0}}{\rho_{cr0}} \tag{8.308}$$

The first Friedmann equation, Eq. (8.186), is

$$\frac{3}{a^2}(k+\dot{a}^2)=8\pi\rho$$

or

$$\left(\frac{\dot{a}}{a}\right)^2-\frac{8\pi}{3}\rho=-\frac{k}{a^2}$$

or

$$\left(\frac{\dot{a}}{a}\right)^2-H_0^2\Omega_{m0}\left(\frac{a_0}{a}\right)^3-H_0^2\Omega_{de0}=-\frac{k}{a^2} \tag{8.309}$$

From the definition of Hubble rate H

$$\begin{aligned}H &= \frac{\dot{a}}{a}\\ \dot{H} &= -H^2+\frac{\ddot{a}}{a}=-H^2\left(1-\frac{\ddot{a}}{H^2a}\right)\\ &\equiv -H^2(1+q)\end{aligned} \tag{8.310}$$

where the *deceleration parameter* q is defined as

$$q=-\frac{\ddot{a}}{H^2a} \tag{8.311}$$

Non-relativistic matter-dominated universe is modeled by dust approximation, $p = 0$.
Thus, from Friedmann's equation, Eq. (8.186), we get

$$\frac{\ddot{a}}{a} + \frac{4\pi}{3}\rho = 0 \tag{8.312}$$

or in terms of H

$$-H^2 q + \frac{4\pi}{3}\rho = 0 \tag{8.313}$$

Therefore,

$$\rho = \frac{3H^2}{4\pi} q \tag{8.314}$$

The Friedmann Eq. (8.184) becomes

$$3\left(\frac{\dot{a}^2}{a^2} + \frac{k}{a^2}\right) = 8\pi\rho \tag{8.315}$$

or

$$H^2 - 2H^2 q = -\frac{k}{a^2} \tag{8.316}$$

So

$$-k = a^2 H^2 (1 - 2q) \tag{8.317}$$

As both $a \neq 0$ and $H \neq 0$ for the flat universe ($k = 0$), we get

$$q = 1/2 \tag{8.318}$$

Thus,

$$\rho = \frac{3H^2}{8\pi} \tag{8.319}$$

It may be remarked that the quantity q gives the relationship between the density of the universe and the critical density ρ_{cr}, *i.e.*,

$$q = \frac{\rho}{2\rho_{cr}} \tag{8.320}$$

From Eq. (8.317), we have

$$-k = a^2 H^2 (1 - \Omega_T) \tag{8.321}$$

where

$$\Omega_T \equiv 2q = \frac{\rho}{\rho_{cr}} = \Omega_m + \Omega_{de} \tag{8.322}$$

WMAP data show that the universe is nearly flat, so $k = 0$, which gives

$$\Omega_T = \Omega_{T0} = \Omega_{mo} + \Omega_{de0} = 1$$
$$\Rightarrow \Omega_{mo} = 1 - \Omega_{de0} \tag{8.323}$$

Taking $a_0 = 1$, we get from Eq. (8.309)

$$\left(\frac{\dot{a}}{a}\right)^2 = H_0^2\left[(1-\Omega_{de0})\frac{1}{a^3} + \Omega_{de0}\right] \tag{8.324}$$

Solving it, we get

$$\dot{a} = H_0\sqrt{\frac{1-\Omega_{de0}}{a} + \Omega_{de0}a^2} \tag{8.325}$$

and

$$H_0 t_0 = \int_0^1 \frac{da}{\sqrt{\frac{1-\Omega_{de0}}{a} + \Omega_{de0}a^2}}$$
$$= \int_0^1 \frac{a^{1/2}da}{\sqrt{(1-\Omega_{de0}) + \Omega_{de0}a^3}}$$
$$= \frac{2}{3\sqrt{\Omega_{de0}}}\ln\left[2\left(\sqrt{\Omega_{de0}a^3} + \sqrt{\Omega_{de0}(a^3-1)+1}\right)\right]_0^1$$
$$= \frac{2}{3\sqrt{\Omega_{de0}}}\ln\left(\frac{1+\sqrt{\Omega_{de0}}}{\sqrt{1-\Omega_{de0}}}\right) \tag{8.326}$$

Therefore, the age of the universe with dark energy is

$$t_0 = \frac{2}{3H_0\sqrt{\Omega_{de0}}}\ln\left(\frac{1+\sqrt{\Omega_{de0}}}{\sqrt{1-\Omega_{de0}}}\right) \tag{8.327}$$

As $\Omega_{de0} \to 1$, $t_0 \to \infty$, hence, some matter is needed to keep the age of the universe finite.

From WMAP data, we obtain the age of the universe to be 13.7 aeons and from the ω model for the equation of state of matter and dark energy that

$$\Omega_{T0} = \Omega_{m0} + \Omega_{de0} = 1$$
$$\Omega_{de0} = 0.72 \tag{8.328}$$

$$\Rightarrow \Omega_{m0} = 0.28 \tag{8.329}$$

This implies that

$$\frac{\Omega_{m0}}{\Omega_T}\times 100 = 28\% \text{ of the universe is matter} \tag{8.330}$$

$$\frac{\Omega_{de0}}{\Omega_T}\times 100 = 72\% \text{ of the universe is dark energy} \tag{8.331}$$

The WMAP data also indicates that only 4% of the universe is baryonic (normal) matter and the remaining 24% of matter is non-baryonic. It tantamounts to the fact that we are still ignorant of what constitutes the 96% of the universe.

8.14 COSMIC MICROWAVE BACKGROUND RADIATION

In 1948, Ralph Alfer and Robert Herman predicted that intense microwave radiation, greatly diluted and cooled due to the universe's expansion, still fills the universe. In 1965 Arno Penzias and Robert Wilson[3], accidentally and unaware of the prediction, discovered microwave radiation in all directions of the sky. The cosmic microwave background (CMB) radiation provides strong evidence that a Big Bang really occurred. It is a snapshot of the universe at its infancy, when the universe was only a few hundred thousand years old.

The early universe was a hot plasma of photons, electrons and baryons. The electrons were continually scattering from the plasma through Thomson scattering. As the universe expanded, the adiabatic cooling led to the cooling of the plasma, till the temperature was favourable for the electrons to combine with the protons and form hydrogen atoms. This occurred when the temperature was around 3000 K. The photons now scattered from neutral atoms and could travel freely through space. This process is called *recombination* or *decoupling* (referring to the electrons combining with the protons or to the decoupling of radiation from matter).

The photons have been cooling ever since and are now at 2.725 K and their temperature will continue to fall till the expansion of the universe continues, since $T_\gamma \propto a^{-1}$.

The Big Bang theory suggests that the CMB radiation fills all the space in the universe and most of the radiation energy in the universe is in CMB radiation, that makes up a fraction of approximately 5×10^{-5} of the total density of the universe. The discovery of the 3 K microwave background radiation and its detailed anisotropies, was a crucial success of the Big Bang model of the universe.

NASA's Cosmic Background Explorer (COBE) satellite was launched to explore the cosmic background radiation. The data is displayed in Fig. (8.7), the data points are shown superimposed on the theoretical black body curve[4].

The data from COBE discovered fluctuations in that radiation, which are important to Big Bang cosmological calculations. The explorer carried three main instruments: A Differential Microwave Radiometer, a Far-Infrared Absolute Spectrophotometer (cooled to 1.6 K by liquid helium) and the Diffused Infrared Background Experiment, also at 1.6 K. The infrared instrument will measure infrared spectra of the background, which are presumed to be uniform, but any unexpected variations will indicate the presence of energy sources which might have driven turbulence to trigger galaxy formation. The infrared instruments' sensitivity was 100 times greater than that achievable from earth based instruments.

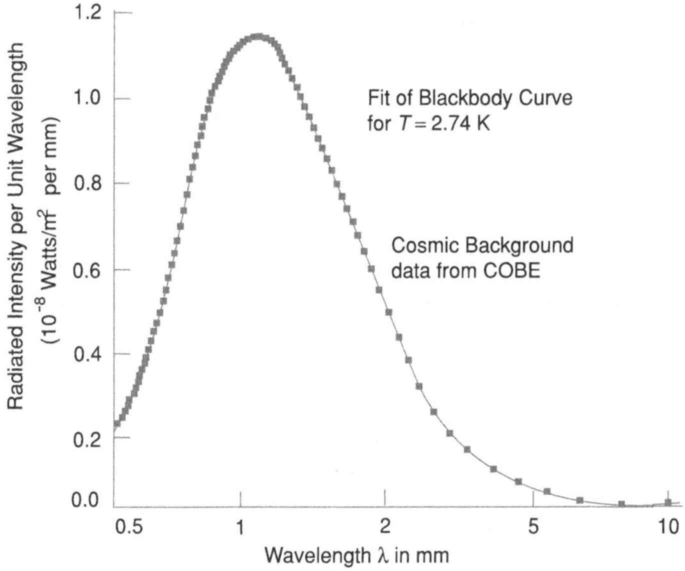

Figure 8.7 *Cosmic background radiation data from COBE.*

The recent Wilkinson Microwave Anisotropy Probe (WMAP) gave a much higher resolution picture of the anisotropies in the cosmic background radiation. These are used to estimate the parameters of the standard adiabatic cold dark matter (ACDM) model of the Big Bang. The information about the shape of the universe can be obtained directly from the CMB radiation.

Let us examine the role of the 3 K in cosmology. The 3 K background implies about 5.5×10^5 photons per litre. The range of estimate for baryon density is from twice critical density at 6×10^{-3}/ litre to the low end estimate of the visible galaxy, 3×10^{-3}/litre. This gives a range of 1×10^8 to 2×10^8 photons/baryon. This estimate of the number of photons per baryon was crucial in calculations of the Big Bang. Further, this estimate of photons per baryon agreed with the modelling of the nucleosynthesis in the Big Bang including the hydrogen/helium ratio.

When the trace quantities of ${}^2_1d$, ${}^3He$, and ${}^7Li$ are examined and made a part of the Big Bang model, it provides a still stringent test of the ratio of photons to baryons. The Particle Data Group gives the baryon/photon ratio as

$$2.6\times10^{-10} < \eta < 6.3\times10^{-10} \text{ baryons/photon}$$

As the conservation of baryon number is a strong conservation principle, it is construed that the ratio is constant throughout the process of expansion of the universe. No known process in nature changes the number of baryons.

8.14.1 Anisotropy of 3 K Background

Anisotropy of nearly 0.1% has been observed in the cosmic microwave background radiation, which is attributed to a Doppler shift caused by the motion of the solar system through the radiation.

If CMB radiation looks like a perfect black body radiation to one observer, it will not look the same to another observer who is moving relatively to the first observer, since the radiation will be

Doppler shifted due to observer's motion. It is necessary to account for the relativistic shift, before one can analyze the CMB spectrum for any anisotropy.

Let ν_1 be the frequency observed by an observer to whom the CMB radiation is perfect black body radiation. The second observer is moving with velocity $\mathbf{v}$ in an arbitrary direction given by a unit vector $\mathbf{n}$ such that $\mathbf{v}\cdot\mathbf{n} = v\cos\theta$. The frequency observed by the second observer ν_2 is given by the relativistic Doppler shift[5], as

$$\nu_2 = \nu_1 \frac{1 - v\cos\theta/c}{\sqrt{1-v^2/c^2}} \tag{8.332}$$

However, we are moving with respect to the reference frame at rest, $\nu_1 \equiv \nu_0$ and observing light in the reference frame 'at rest' has frequency $\nu_2 \equiv \nu_e$, so

$$\frac{\nu_0}{\nu_e} = \frac{\sqrt{1-v^2/c^2}}{1 - v\cos\theta/c} \tag{8.333}$$

This implies that the temperature observed in the directionin θ, $T(\theta)$ terms of the average temperature $\langle T\rangle$ is given as

$$\begin{aligned}
\frac{T(\theta)}{\langle T\rangle} &= \frac{\sqrt{1-v^2/c^2}}{1 - v\cos\theta/c} \\
&= (1-v^2/c^2)^{1/2}(1 - v\cos\theta/c)^{-1} \\
&\approx \left(1 - \frac{1}{2}\frac{v^2}{c^2} + \cdots\right)\left(1 + \frac{v}{c}\cos\theta + \frac{v^2}{c^2}\cos^2\theta + \cdots\right) \\
&\approx 1 + \frac{v}{c}\cos\theta + \frac{v^2}{c^2}\left(\cos^2\theta - \frac{1}{2}\right) + \cdots
\end{aligned} \tag{8.334}$$

The motion of the observer gives rise to a dipole and higher order corrections. The observed dipole anisotropy first detected in 1960, led to the following inference:

$$v_{\text{O}} - v_{\text{CMB}} = 370 \pm 10\,\text{km/sec towards}\,\mathbf{n} \quad \phi = 267.7 \pm 0.8°, \theta = 48.2 \pm 0.5' \tag{8.335}$$

where θ is the colatitude (polar angle) in the range $0 \le \theta \le \pi$ and ϕ is the longitude (azimuth) in the range $0 \le \phi \le 2\pi$. Obvously, $\theta = 0$ at the north pole, $\theta = \pi/2$ at the equator and $\theta = \pi$ at the south pole.

Accounting for the sun's motion in the Galaxy and the motion of the Galaxy within the Local group, it turns out that the Local Group is moving with

$$v'_{\text{LG}} - v'_{\text{CMB}} \approx 600\ \text{km/sec to wards} \quad \phi = 268°, \quad \theta = 27$$

After allowing for the motion from the CMB radiation, the dipolar anisotropy turns out to be about few parts in 10^5. These primordial perturbations, no matter minuscule, served as seeds for the structure of the universe.

8.14.2 Fluctuations in 3 K Background

The COBE satellite has discovered fluctuations in the cosmic microwave background radiation with the use of a differential microwave radiometer. The size of the fluctuations are $\Delta T/T = 6\times10^{-6}$. The 'horizon' is the distance within which there can be causal connections, *i.e.*, within light travel time of each other. The scale of the fluctuations is larger at the time the CMB radiation was emitted, indicating thereby that these fluctuations are primordial, being before the separation of radiation and matter, the transparency point.

Next we discuss the technique for quantification of the small scale fluctuations in the CMB. We define the normalized temperature Θ in direction $\hat{\mathbf{n}}$ on the celestial sphere by the deviation from the average

$$\Theta(\hat{\mathbf{n}}) = \frac{\Delta T}{\langle T \rangle} \tag{8.336}$$

We define the multipole decomposition of $\Theta(\hat{\mathbf{n}})$ in terms of spherical harmonics Y_{lm}

$$\begin{aligned}\Theta(\hat{n}) &= \Theta(\theta,\phi)\\ &= \sum_{l=0}^{\infty}\sum_{m=-l}^{l} \Theta_{lm} Y_{lm}(\theta,\phi)\end{aligned} \tag{8.337}$$

where

$$\Theta_{lm} = \int \Theta(\hat{\mathbf{n}}) Y_{lm}^{*}(\hat{\mathbf{n}})\, d\Omega \tag{8.338}$$

The spherical harmonics $Y_{lm}(\hat{\mathbf{n}})$ are

$$\begin{aligned}Y_{lm}(\hat{\mathbf{n}}) &= Y_{lm}(\theta,\phi)\\ &= \sqrt{\frac{(2l+1)}{4\pi}\frac{(l-m)!}{(l+m)!}}\, P_l^m(\cos\theta) e^{im\phi}\end{aligned} \tag{8.339}$$

$P_l^m(x)$ are associated Legendre functions

$$P_l^m(x) \equiv \frac{(1-x^2)^{m/2}}{2^l\, l!} \frac{d^{m+l}}{dx^{m+l}} (x^2-1)^l \tag{8.340}$$

The spherical harmonics $Y_{lm}(\theta, \phi)$ are orthonormal, *i.e.*,

$$\int_{\theta=0}^{\pi}\int_{\varphi=0}^{2\pi} Y_{lm} Y_{l'm'}^{*}\, d\Omega = \delta_{ll'}\delta_{mm'} \tag{8.341}$$

$\delta_{ll'}$ and $\delta_{mm'}$ are the Kronecker delta functions.

The field of Gaussian random fluctuations is totally characterized by its power spectrum $\Theta_{lm}^{*}\Theta_{l'm'}$. The angular orientation of a fluctuation is described by m whereas the characteristic angular size of the fluctuation is determined from the l value. As the universe is isotropic, the power spectrum is independent of m. Furthermore, the universe being homogeneous, *i.e.*, being same from point to point, one expects the power spectrum to be independent of l.

Finally, the angular power spectrum C_l is defined as

$$C_l = \langle \Theta_{lm}^{*}\Theta_{l'm'} \rangle = \delta_{ll'}\delta_{mm'} C_l \tag{8.342}$$

The brackets denote the average over the skies with the same cosmology. The estimate of C_l is then the average over m.

The power spectrum is displayed as (the power per logarithmic interval in l)

$$\Delta T^2 = \frac{l(l+1)}{2\pi} C_l T_{\text{CMB}}^2 \tag{8.343}$$

where T_{CMB} is the blackbody temperature of the CMB radiation. The power spectrum of the CMB radiation temperature anisotropy in terms of the angular scale is shown in Fig. (8.8).

The anisotropy of the CMB radiation is classified into two types, primary anisotropy which is due to interactions occurring at the last scattering surface or before, and secondary anisotropy which is the outcome of the interactions with hot gas or gravitational potential after the last scattering surface till the time of observation.

The structure of CMB radiation is primarily determined by two effects: acoustic oscillations and diffuse damping (also called collisionless damping).

8.14.3 Acoustic Oscillations[6]

In the early universe before decoupling, rapid scattering couples photons and baryons into a plasma which behaves as perfect fluid. It is the competition in the photon-baryon plasma that gives rise to the acoustic oscillations. The pressure of the photons tends to erase the anisotropies whereas the

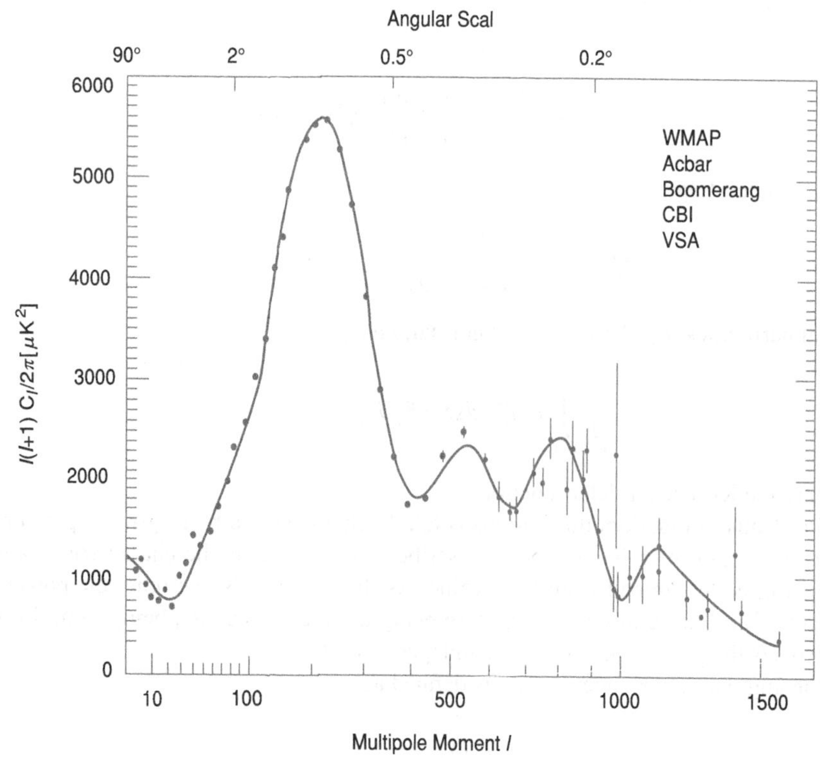

Figure 8.8 *The power spectrum of CMB radiation. The solid line shows a theoretical model.*

gravitational attraction of the baryons tends to collapse the fluid to form dense haloes. The competition between these two opposing effects creates acoustic oscillations. These acoustic oscillations give rise to the characteristic peak structure of the power spectrum. The peaks correspond to resonances in which the photons decouple when a particular mode is at its maximum amplitude.

The angular scale of the peaks contain interesting physical signatures. The angular scale of the first peak determines the curvature of the universe. The ratio of the odd peaks to the even ones, determines the reduced baryon density. The third peak can be used to extract the dark matter density.

Furthermore, the important information about the primordial density perturbations can be derived from the location of the peaks. There are two fundamental types of density perturbations, called the adiabatic and isocurvature. Both these types of perturbation contribute to any general density perturbation. Different theories used to explain the primordial density perturbation spectrum predict different mixtures of these two fundamental types.

According to adiabatic density perturbation, the fractional over density in each matter component, such as baryons, photons, neutrinos is the same. To illustrate it, if there is 1% more energy in baryons than average in one spot, then there is also 1% more energy than average in other components, such as photons and neutrinos. Cosmic inflation predicts that the primordial perturbations are adiabatic.

However, according to isocurvature density perturbations, the sum of the fractional overdensities is zero, *i.e.*, it is a zero sum perturbation. Cosmic strings predict that the primordial density perturbations are mostly isocurvature perturbations.

These two types of perturbations produce different peak locations and thus the CMB spectrum is able to distinguish these two. Adiabatic density perturbations[4] produce peaks whose angular scales (l-values of the peaks) are in the ratio 1:2:3, whereas the isostructure density perturbations predict the l-values of the peaks in the ratio 1:3:5. Observations support the primordial density perturbations to be entirely adiabatic, thus, lending crucial support for inflation and ruling out other models of structure formation, involving cosmic strings.

Collisionless damping comes into play when the treatment of the primordial plasma as a fluid starts breaking down. This is brought about by the two effects: the increasing mean free path of the photons as the primordial plasma becomes progressively rarefied in an expanding universe and the other being the Compton scattering which occurs due to the finite thickness of the last scattering surface (LSS), which is the cause of the rapid increase of the mean free path during decoupling. The thickness of LSS implies that the decoupling of the photons and baryons does not happen instantaneously, but instead is spread over an appreciable fraction of the age of the universe upto that era. The WMAP team finds that decoupling took place roughly 115,000 years and when it was complete, the universe was 487,000 years old.

References

1. Robertson, H.P., (1933) 'Relativistic Cosmology' in *Reviews of Modern Physics*, 5, pp. 62–90.
2. Alan H. Guth, 'The Inflationary Universe: The Quest for a New Theory of Cosmic Origin,' (1998).
3. Penzias, A.A. and Wilson, R.W. (1965). 'A Measurement of Excess Antenna Temperature at 4080 mc/s', *Astrophysical Journal*, 142, pp. 419–421.
4. Mather, J.C., *et al.*, *Astrophysical Journal*, 354, L37, 1990.
5. Puri, S.P., (2012). Special Theory of Relativity, Pearson Education.
6. Hu, W. and White, M. Cocconi (1996). 'Acoustic Signatures in the Cosmic Microwave Background,' *Asrophysical Journal*, 471, pp. 30–51.

Problems

8.1 The hubble constant has dimensions of inverse seconds, but is usually expressed in kilometers per second per megaparsec (1 Mpc $\sim 3\times10^{19}$ km). The value of H_0 is a subject of observational controversy, but as an estimate we may take $H_0 = 50\ \mathrm{km\,s^{-1}Mpc^{-1}}$. Under these assumptions show that $\rho_c = 5\times10^{-30}\,\mathrm{g\,cm^{-3}}$.

8.2 For a radiation filled universe show that as we approach the big bang $(t=0)$ we have $a(t)\propto t^{1/2}$, irrespective of k. Now suppose the radiation to be thermal with temperature T, so that $\rho\propto T^4$. Show that shortly after the big bang $T\propto t^{-1/2}$.

8.3 Einstein modified the gravitational field equations by introducing a term proportional to the metric

$$R_{\mu\nu}-\frac{1}{2}Rg_{\mu\nu}-\lambda g_{\mu\nu}=-8\pi GT_{\mu\nu}$$

where λ is a constant, the cosmological constant. Show that in the Friedmann–Robertson–Walker models, this corresponds to redefining the density and pressure as

$$\bar{\rho}=\rho+\frac{\lambda}{8\pi G};\quad \bar{p}=p-\frac{\lambda}{8\pi G}$$

Show that, for dust (*i.e.*, $p=0$, not $\bar{p}=0$) the Einstein equations do have a static solution and that k must be unity for ρ to be positive.

8.4 Show that in $k=1$ universes light leaving the origin at $t=0$ just has time to go around and return to the origin in the case of dust, and in the case of radiation only has time to go half way.

8.5 Show that if a space is empty and isotropic, it must be Minkowski space.

8.6 Show that the Ricci tensor for the metric

$$ds^2=d\chi^2+f^2(\chi)(d\theta^2+\sin^2\theta d\varphi^2)$$

is

$$R=-\left(2\frac{f''}{f}-\frac{1}{f^2}+\frac{f'^2}{f^2}\right)$$

8.7 Consider Einstein's field equations for $k=0,\pm1$, homogeneous, isotropic and empty universe (*i.e.*, $\rho=p=0$) and with a positive cosmological constant Λ.

(a) What are the resulting spacetime metrics?

(b) Calculate the spacetime curvature tensors $R_{\mu\nu\sigma\rho}$ and

(c) The Ricci scalars for these metrics.

(d) Show that these metrics are spacetime homogeneous and isotropic.

(Since all have the same constant Ricci scalar, they are equivalent–they are the different forms of the de-Sitter universe.)

8.8 Show the $\mu\nu=22$ and $\mu\nu=33$ components of the Einstein equations for the FRW metric yield the same equation $\frac{\ddot{a}}{a}=\frac{4\pi G}{3}(\rho+3p)+\frac{\Lambda}{3}$ as the $\mu\nu=11$ component.

9

Astrophysics

9.1 INTRODUCTION

As a brief introduction to astrophysics, we consider the model of a star in hydrostatic equilibrium. The polytropes are a family of equations of state for a gas in hydrostatic equilibrium and in these equations, the pressure is given as a power law in density. The polytropes serve as simple model of a star, such as the Eddington Standard model. The lane–Emden equation, which describes the polytropes, is derived and solved analytically. Furthermore, we derive the Chandrashekhar mass limit for the formation of a black hole.

Lastly, we discuss how the Pauli Exclusion and Heisenberg Uncertainty principles govern the formation of white dwarfs, neutron stars, and black holes. It is believed that black holes form at the end of a star's life–time when a massive star has exhausted its fuel and collapses.

9.2 TOLMAN—OPPENHEIMER–VOLKOFF EQUATION

The earlier astrophysical theories were based on Newtonian equations of gravity and mechanics. This approach seemed plausible in many situations, as the crucial factor $GM/(c^2R)$ was really small as compared to unity. However, when this factor is of the order of unity, the use of general relativity becomes inevitable. The general relativistic hydrostatic equilibrium equation for a spherically symmetric body of isotropic fluid, is referred to as the Tolman-Oppenheimer-Volkoff equation (TOV). The equation is

$$\frac{dp(r)}{dr} = -\frac{G}{r^2}\left[\rho(r) + \frac{p(r)}{c^2}\right]\left[M(r) + 4\pi r^3 \frac{p(r)}{c^2}\right]\left[1 - \frac{2GM(r)}{c^2 r}\right]^{-1} \tag{9.1}$$

where r is a radial coordinate, $\rho(r_0)$ and $p(r_0)$ are the density and pressure respectively of the fluid at $r = r_0$.

The derivation of the TOV equation is based on the following assumptions:

- Matter is described by a perfect fluid, represented by the energy-momentum tensor

$$T_\mu^\nu = (\rho c^2 + p)u_\mu u^\nu - p\delta_\mu^\nu \tag{9.2}$$

  u^μ is the four-velocity of the fluid.
- It is assumed that the star is spherically symmetric and the metric inside the star is

$$ds^2 = e^\nu c^2 dt^2 - e^\lambda dr^2 - r^2(d\theta^2 + \sin^2\theta d\phi^2) \tag{9.3}$$

  where (r, θ, ϕ) are spherical polar coordinates.

- The system is in equilibrium. This assumption leads to two constraints. First, the metric coefficients are independent of time, so that ν and λ are functions of r only, $\nu = \nu(r)$, $\lambda = \lambda(r)$.Second, the velocity has no space-like components in the metric, so that

$$u^0 = e^{\nu/2},\ u^1 = u^2 = u^3 = 0 \tag{9.4}$$

With these values of components of four-velocity vector, the energy-momentum tensor has only diagonal components non-vanishing

$$T_\mu^\nu = \begin{pmatrix} \rho c^2 & 0 & 0 & 0 \\ 0 & -p & 0 & 0 \\ 0 & 0 & -p & 0 \\ 0 & 0 & 0 & -p \end{pmatrix} \tag{9.5}$$

- The metric inside the star is determined by the Einstein field equations

$$R_\mu^\nu - \frac{1}{2} R\delta_\mu^\nu = -\frac{8\pi G}{c^4} T_\mu^\nu \tag{9.6}$$

From metric Eq. (9.3), metric elements are

$$g_{00} = e^{\nu(r)} c^2 \tag{9.7}$$

$$g_{11} = -e^{\lambda(r)} \tag{9.8}$$

Hydrostatic density is given by

$$T_{00} = \rho(r) e^{\nu(r)} c^4 \tag{9.9}$$

and hydrostatic pressure is

$$T_{11} = p(r) g_{11} = -p(r) e^{\lambda(r)} \tag{9.10}$$

Einstein's tensor element derived from the metric is

$$G_{11} = R_{11} - \frac{1}{2} R g_{11} = \frac{-r\nu'(r) + e^{\lambda(r)} - 1}{r^2} \tag{9.11}$$

Writing Eq. (9.11) in the light of Eq. (9.6), we get

$$-\frac{8\pi G}{c^4} p(r) e^{\lambda(r)} = \frac{-r\nu'(r) + e^{\lambda(r)} - 1}{r^2} \tag{9.12}$$

A relation $T^{\nu}_{\mu;\nu} = 0$ holds identically for all systems and this gives

$$\frac{dp(r)}{dr} = -\left(\frac{\rho c^2 + p}{2}\right)\frac{d\nu}{dr} \tag{9.13}$$

Differential equation of state for hydrostatic equilibrium becomes

$$\frac{dp(r)}{dr} = -\left(\frac{T_{00}g^{00} + T_{11}g^{11}}{2}\right)\frac{d\nu}{dr} = -\left(\frac{\rho(r)c^2 + p(r)}{2}\right)\frac{d\nu}{dr} \tag{9.14}$$

g^{00} and g^{11} are contravariant Einstein tensor elements.
Solving Eq. (9.14) for $\frac{d\nu(r)}{dr}$, we obtain

$$\begin{aligned}\frac{d\nu(r)}{dr} &= -\frac{dp(r)}{dr}\left(\frac{2}{\rho(r)c^2 + p(r)}\right)\\ &= \frac{1}{r}\left(\frac{8\pi G r^2 e^{\lambda(r)} p(r)}{c^4} + e^{\lambda(r)} - 1\right)\end{aligned} \tag{9.15}$$

Solving the above equation for $p(r)$, we get

$$\frac{dp(r)}{dr} = -\frac{(\rho(r)c^2 + p(r))(e^{\lambda(r)} - 1) + 8\pi G r^2 e^{\lambda(r)} p(r))}{2c^4 r} \tag{9.16}$$

From the solution of Schwarzschild metric, we know that

$$e^{\lambda(r)} - 1 = \frac{r_s}{r - r_s} \tag{9.17}$$

where r_s, the Schwarzschild radius $= \frac{2GM(r)}{c^2}$.
From Eq. (9.17), we obtain

$$e^{\lambda(r)} = \left(1 - \frac{r_s}{r}\right)^{-1} \tag{9.18}$$

Equation of state for hydrostatic pressure, Eq. (9.16), in view of Eq. (9.18), becomes

$$\frac{dp(r)}{dr} = -\frac{(\rho(r)c^2 + p(r))(c^4 r_s + 8\pi G r^3 p(r))}{2c^4 r(r - r_s)} \tag{9.19}$$

Substituting the value of r_s, we get

$$\frac{dp(r)}{dr} = -\frac{G(\rho(r)c^2 + p(r))(M(r) + 4\pi r^3 p(r)/c^2)}{r(c^2 r - 2GM(r))} \tag{9.20}$$

Factoring out a factor from the numerator, the Eq. (9.20) becomes

$$\frac{dp(r)}{dr} = -\frac{G}{r^2}\frac{(\rho(r)+p(r)/c^2)(M(r)+4\pi r^3 p(r)/c^2)}{(1-GM(r)/(c^2 r))} \tag{9.21}$$

This is the Tolman-Oppenheimer-Volkoff equation. It was derived by J. Robert Oppenheimer and George Volkoff[1], based on spherically symmetric metrics, which were analysed by Richard C. Tolman[2–3]. Making use of the equation of state for a degenerate Fermi gas of neutrons, Oppenheimer and Volkoff estimated an upper limit of ~0.7 solar mass for the gravitational mass of a neutron star. Since TOV is not realistic for a neutron star, the estimated limited mass is incorrect. The modern estimates for this mass range from 1.5 to 3.0 solar masses[4].

9.3 DEGENERACY OF MATTER

The matter in white dwarfs and neutron stars is degenerate, so first we consider the degenerate matter, as that will lay the basis for our discussion of these bodies. In addition, we will introduce polytropes as simple equilibrium models.

A fermion is a particle with a half integral spin and obeys Fermi–Dirac statistics. According to Pauli Exclusion Principle, no two fermions can occupy the same quantum state. This is equivalent to the requirement that the volume per fermion is proportional to $\lambda_c^3 \sim (\hbar/mc)^3$, where m is the mass of the fermion; $\hbar$, the rationalized Planck constant; c, the velocity of light and λ_c, the Compton wavelength. The average number density of the fermions is then $n_f \sim \lambda_c^{-3}$. In white dwarfs, it is the electrons, which play a part in their equilibrium; so the mass density is n_f times the mass per electron. However, in neutron stars, the mass density is n_f times the mass of the neutron.

The relative densities of white dwarfs and neutron stars, with the approximation that mass per electron is of the order of mass of the neutron, are

$$\frac{\rho_{ns}}{\rho_{wd}} = \frac{\lambda_{c,n}^{-3}}{\lambda_{c,e}^{-3}} = \frac{m_n^3}{m_e^3} \approx (2000)^3 = 8\times10^9 \tag{9.22}$$

If the fermion density is very high in a gas, the lower momentum states are filled, so the fermions must then occupy states of higher momentum. Under such conditions, the gas is said to be (partially) degenerate and the high momentum fermions make a large contribution to the pressure of the gas.

9.3.1 Complete Degeneracy

ε_f, Fermi energy is defined as the energy of the highest occupied quantum state in a system of fermions at absolute zero temperature. The distribution function of F-D distribution is

$$f(p) = \frac{1}{e^{(E(p)-\mu)/T}+1} \tag{9.23}$$

where μ is the chemical potential. For $T = 0,\ \mu = \varepsilon_f$, so as the temperature is lowered, the distribution function tends to unity for particle energies $E \le \mu$, and zero for $E \ge \mu$, thus, $f(p)$ becomes a step function

$$f(p) = \begin{cases} 1 & if\ E \le \mu \\ 0 & if\ E \ge \mu \end{cases} \tag{9.24}$$

The number density of fermions is given by

$$\begin{aligned} n &= g\int_0^\infty f(p)\frac{d^3p}{(2\pi\hbar)^3} \\ &= 2\int_0^{p_f} \frac{4\pi p^2 dp}{(2\pi\hbar)^3} \\ &= \frac{8\pi}{h^3}\int_0^{p_f} p^2 dp \\ &= \frac{8\pi}{h^3}\frac{1}{3}p_f^3 = \frac{8\pi}{3h^3}p_f^3 \end{aligned}$$

or

$$p_f = \left(\frac{3h^3}{8\pi}n\right)^{1/3} \tag{9.25}$$

where $g = (2S + 1)$, is the statistical degeneracy factor for a particle with spin S. p_f is the Fermi momentum corresponding to the Fermi energy

$$\varepsilon_f = \frac{p_f^2}{2m} \Rightarrow p_f = \sqrt{2m\varepsilon_f} \tag{9.26}$$

The energy density is given by

$$\begin{aligned} \rho &= g\int_0^\infty E(p)f(p)\frac{d^3p}{h^3} \\ &= \frac{8\pi}{h^3}\int_0^{p_f} E(p)p^2 dp \end{aligned} \tag{9.27}$$

where $E(p)$ is the kinetic energy per fermion.

For non-relativistic fermions

$$\mathbf{p} = m\mathbf{v}$$

and
$$E(p)=\frac{p^2}{2m}$$

The energy density for non-relativistic fermions is

$$\rho=\frac{8\pi}{h^3}\int_0^{p_f}\frac{p^2}{2m}p^2dp$$
$$=\frac{8\pi}{2mh^3}\int_0^{p_f}p^4dp$$
$$=\frac{8\pi}{2mh^3}\frac{1}{5}p_f^5$$
$$=\frac{8\pi}{5h^3}\left(\frac{3h^3}{8\pi}n\right)\varepsilon_f$$
$$=\frac{3}{5}n\varepsilon_f \tag{9.28}$$

For non-relativistic particles

$$p=\frac{2}{3}\rho$$
$$=\frac{2}{3}\frac{3}{5}n\varepsilon_f$$
$$=\frac{2}{5}n\frac{p_f^2}{2m}$$
$$=\frac{n}{5m}\left(\frac{3h^3}{8\pi}n\right)^{2/3}$$
$$=\frac{h^2}{20m}\left(\frac{3}{\pi}\right)^{2/3}n^{5/3} \tag{9.29}$$

This is the equation of state for a non-relativistic completely degenerate fermion gas.

Next we take up the case of complete degeneracy of relativistic fermions. In this case

$$p\gg mc \text{ and } E(p)=\sqrt{p^2c^2+m^2c^4}\approx cp \tag{9.30}$$

The energy density for relativistic fermions is

$$\rho = \frac{8\pi}{h^3}\int_0^{p_f} cpp^2 dp$$

$$= \frac{8\pi c}{h^3}\int_0^{p_f} p^3 dp$$

$$= \frac{8\pi c}{h^3}\frac{1}{4}p_f^4$$

$$= \frac{2\pi}{h^3}p_f^3(cp_f)$$

$$= \frac{2\pi}{h^3}\left(\frac{3h^3}{8\pi}n\right)\varepsilon_f$$

$$= \frac{3}{4}n\varepsilon_f \tag{9.31}$$

Therefore, for relativistic particles

$$p = \frac{1}{3}\rho$$

$$= \frac{1}{3}\frac{3}{4}n\varepsilon_f$$

$$= \frac{1}{4}ncp_f$$

$$= \frac{1}{4}nc\left(\frac{3h^3}{8\pi}n\right)^{1/3}$$

$$= \frac{hc}{8}\left(\frac{3}{\pi}\right)^{1/3} n^{4/3} \tag{9.32}$$

This is the equation of state for extreme relativistic, completely degenerate fermion gas. For complete or nearly complete degeneracy, p is independent of temperature.

Let us estimate the threshold density for complete non-relativistic degeneracy. There is non-degeneracy if the pressure of the star is entirely contributed due to the thermal equilibrium of the particles. However, the complete non-relativistic degeneracy will result when the pressure is balanced by the pressure due to non-relativistic degeneracy of electrons.

The pressure of the ideal gas in thermal equilibrium is given by its equation of state

$$p = \frac{\rho kT}{\bar{\mu} m_H} \tag{9.33}$$

where $\bar{\mu}$ is the mean molecular weight defined as

$$\frac{1}{\bar{\mu}} = \sum_i \frac{\bar{n}_i}{m_i} m_H \tag{9.34}$$

m_H is the mass of the hydrogen atom and $\bar{n}_i = \rho_i/\rho$ is the abundance of species by weight. Combining Eq. (9.29) with Eq. (9.34), we obtain

$$\frac{\rho kT}{\bar{\mu}\, m_H} = \frac{h^2}{20m_e}\left(\frac{3}{\pi}\right)^{2/3} n_e^{5/3} \tag{9.35}$$

The number density n_e of electrons in terms of density ρ, is

$$n_e = \frac{\rho}{m_H \bar{\mu}_e} \tag{9.36}$$

Taking $\bar{\mu} = \bar{\mu}_e \approx 1$, Eq. (9.35) becomes

$$\frac{\rho kT}{m_H} \approx \frac{h^2}{20m_e}\left(\frac{\rho}{m_H}\right)^{5/3}$$

or

$$\rho = m_H \left(\frac{20m_e k}{h^2}\right)^{3/2} T^{3/2}$$

$$= (1\cdot 67\times 10^{-24}\,\text{g})\left(\frac{20(9.11\times 10^{-28}\,\text{g})(1\cdot 38\times 10^{-16}\,\text{erg/K}}{(6.63\times 10^{-27}\,\text{erg/s})^2}\right)^{3/2} T^{3/2}$$

or

$$\approx 10^{-8}\, T^{3/2} \tag{9.37}$$

Therefore, $\rho \approx 10^{-8} T^{3/2}$ is the prerequisite condition for the electron gas to be completely degenerate. Let us consider the requirement from non-relativistic degeneracy to extreme relativistic degenerate. For relativistic particles $p_f \gg m_e c$ but the "transition occurs", say at $p_f = 2m_e c$.
For relativistic particles p_f is

$$p_f = \left(\frac{3h^3}{8\pi} n\right)^{1/3} = \left(\frac{3h^3}{8\pi}\frac{\rho}{m_H \bar{\mu}_e}\right)^{1/3} = 2m_e c \tag{9.38}$$

Taking $\bar{\mu}_e = 1$, we get

$$\rho \approx \frac{64\pi m_H (m_e c)^3}{3h^3}$$

$$= \frac{64\pi}{3}\,\frac{(1\cdot 67\times 10^{-24}\,\text{g})[(9\cdot 11\times 10^{-28}\,\text{g})(3\times 10^{10}\,\text{cm/s})]^3}{(6\cdot 63\times 10^{-27}\,\text{ergs})^3}$$

$$\approx 10^7\,\text{g/cm}^3 \tag{9.39}$$

Thus, the requirement for the gas to be completely, relativistically degenerate, is

$$\rho > 10^7 \, \text{g/cm}^3 \tag{9.40}$$

These degenerate forms of matter characterize brown dwarfs, white dwarfs (electron degeneracy) and neutron stars (neutron degeneracy).

9.4 MODEL OF A STAR IN HYDROSTATIC EQUILIBRIUM

Consider a thin shell of thickness *dr*, Fig. (9.1). There is an inward force acting on the shell due to the gravitating mass and the outward force due to gas pressure. These forces are

$$F_g = -G\frac{M(r)[\rho(r)4\pi r^2 dr]}{r^2} \tag{9.41}$$

$$F_p = 4\pi r^2[p(r+dr) - p(r)] = 4\pi r^2 dp \tag{9.42}$$

$M(r)$ is the mass interior to the shell

$$M(r) = 4\pi\int_0^r \rho(r)r^2 dr$$

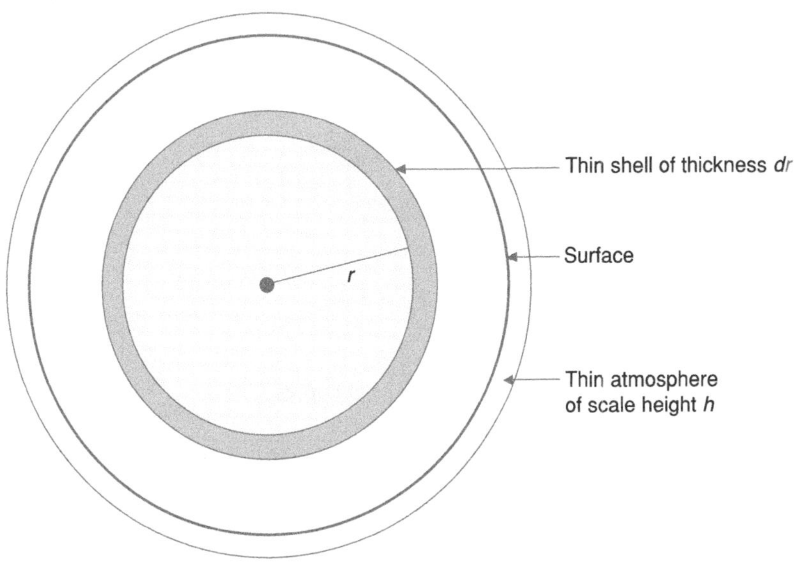

Figure 9.1 *A sphere of gas in hydrostatic equilibrium.*

In hydrostatic equilibrium, these forces balance each other. Thus,

$$F_p = F_g$$

$$4\pi r^2 dp = -G\frac{M(r)[\rho(r)4\pi r^2 dr]}{r^2} \tag{9.43}$$

which leads to the equation of hydrostatic equilibrium

$$\frac{dp}{dr} = -\rho(r)\frac{GM(r)}{r^2} \tag{9.44}$$

In comparison to stellar radius, the stellar atmospheres are usually thin so it is possible to approximate the force of gravity as a constant through the atmosphere. Therefore,

$$g \equiv \frac{GM}{R^2} \approx \text{const} \tag{9.45}$$

Let h be the height of the atmosphere, then equation of hydrostatic equilibrium, Eq. (9.44), becomes

$$\frac{dp}{dh} = -\rho g \tag{9.46}$$

We replaced the r-derivative by a h-derivative. Substituting the value of ρ from the equation of state for ideal gas

$$p = \frac{\rho kT}{\bar{\mu} m_H} \Rightarrow \rho = \frac{\bar{\mu} m_H}{kT}\rho \tag{9.47}$$

into Eq. (9.46), we get

$$\frac{dp}{dh} = -\frac{\bar{\mu} m_H g}{kT} \tag{9.48}$$

Defining the 'scale height' of the atmosphere as

$$H = \frac{kT}{\bar{\mu} m_H g} \tag{9.49}$$

and defining the initial condition $p(0) = p_0$, Eq. (9.48) becomes

$$\frac{dp}{dh} = -\frac{p}{H}$$

or

$$\frac{dp}{p} = -\frac{dh}{H} \tag{9.50}$$

Integrating, we get

$$\log p = -\frac{h}{H} + C$$

or
$$p(h) = Ce^{-h/H}$$

or
$$p(h) = p_0 e^{-h/H} \tag{9.51}$$

The equation of hydrostatic equilibrium is to be accompanied by an equation of state.

9.5 POLYTROPIC STARS

In the nineteenth century, before astronomers learned the physics of radiative (and convective) heat transfer and of nuclear energy generation, they used an instructive, but ad hoc, bypass of the problem. This bypass assumes that, apart from the equation of state that relates p, ρ and T, there exists another relation between pressure and density of a polytropic form

$$p = \kappa \rho^{(n+1)/n} \tag{9.52}$$

where κ is called the polytropic constant and n, not necessarily an integer, is called the polytropic index. To-day, we know that there are a number of ways in which such an assumption can be physically justified.

A gas governed by a polytropic process has the equation of state

$$pV^{\gamma} = \text{const} \tag{9.53}$$

Since $\rho = M/V$, we have

$$p \propto V^{-\gamma} \propto \left(\frac{M}{\rho}\right)^{-\gamma}$$

or
$$p = \kappa \rho^{\gamma} \tag{9.54}$$

Gas obeying an equation of state of this form is called a polytrope. By way of examples of polytropic gases, one can mention: non-relativistic completely degenerate gas $(\gamma = 5/3)$, relativistic completely degenerate gas $(\gamma = 4/3)$, isothermal gas $(\gamma = 1)$ and gas and radiation pressure $(\gamma = 4/3)$.

Let us consider the Eddington standard model. It is a polytrope with $\gamma = 4/3$ and the model of a star supported both by radiation pressure

$$p_r = \frac{1}{3}\rho_r = \frac{1}{3}\frac{\pi^2}{15}T^4 = \frac{1}{3}aT^4 \tag{9.55}$$

and the ideal gas pressure

$$p_g = \frac{\rho k T}{\bar{\mu} m_H} \tag{9.56}$$

Let a constant β denote the relative contribution of gassy pressure to the total pressure, $p = p_r + p_g$. Thus,

$$p_g = \beta p \Rightarrow \beta = \frac{p_g}{p}$$

$$\Rightarrow p_r = (1-\beta)p \tag{9.57}$$

Therefore,

$$p_r = (1-\beta)p = \frac{1}{3}aT^4$$

$$\Rightarrow T^4 = \frac{3(1-\beta)}{a}p \tag{9.58}$$

Further, we eliminate the temperature T from the equation of state

$$\beta^4 p^4 = p_g^4 = \left(\frac{\rho k}{\bar{\mu} m_H}\right)^4 T^4$$

$$= \left(\frac{\rho k}{\bar{\mu} m_H}\right)^4 \frac{3(1-\beta)}{a} p$$

$$\Rightarrow p^3 = \left(\frac{k}{\bar{\mu} m_H}\right)^4 \frac{3(1-\beta)}{a\beta^4} \rho^4$$

$$\Rightarrow p = \left(\frac{k}{\bar{\mu} m_H}\right)^{4/3} \left(\frac{3(1-\beta)}{a\beta^4}\right)^{1/3} \rho^{4/3} \tag{9.59}$$

In the above equation, the term multiplying $\rho^{4/3}$ is constant, provided β is constant and $\bar{\mu}$ is constant if the composition of the gas does not change. If these conditions are fulfilled, then we have the Eddington standard model

$$p = \kappa \rho^{4/3} \tag{9.60}$$

where $\kappa = (\frac{k}{\bar{\mu} m_H})^{4/3} (\frac{3(1-\beta)}{a\beta^4})^{1/3}$.
This model is a special case of the Lane–Emden equation which describes the polytropes in hydrostatic equilibrium.

The pressure support in a star can originate from several sources. These are in the following:

- Thermal pressure of the hot gas, $p = \frac{\rho k T}{\bar{\mu} m_H}$. Stars supported by thermal pressure are the multitude of luminous, shining stars. Our galaxy contains about 10^{11} such stars.
- Degeneracy pressure of electrons, $p \propto \rho^{5/3}$ (non-relativistic) and $p \propto \rho^{4/3}$ (relativistic). Stars supported primarily by pressure due to the degenerate electron gas, are called white dwarfs.

- Degeneracy pressure of neutrons and repulsive strong interaction above nuclear density. Such stars are called neutron stars.
- A giant bag of up, down and strange stars, the pressure arising from their degeneracy pressure and a confinement pressure (called bag pressure). Such stars are called quark stars or storage stars.

Normal stars and white dwarfs are in the regime where non-relativistic version of the hydrostatic equilibrium equation can be used. The energy per particle in these stars does not exceed the rest mass of the main mass providers (protons and neutrons) and the radii of these stars are much larger than the Schwarzschild radii.

We shall treat the case of white dwarfs since their equation of state has pressure as a function of density alone (such equations are barotropic) and finding the solution of the hydrostatic equilibrium is a straight forward affair.

9.6 THE LANE–EMDEN EQUATION

The polytropes are a family of equations of state for gas in hydrostatic equilibrium and in these equations, the pressure is given as a power-law in density

$$p = \kappa \rho^{\gamma} \tag{9.61}$$

where κ and γ are constants. The Lane–Emden equation combines the above equation of state for polytropes and the equation of hydrostatic equilibrium

$$\frac{dp}{dr} = -\rho(r)\frac{GM(r)}{r^2} \tag{9.62}$$

Solving Eq. (9.62) for $M(r)$ we obtain

$$\Rightarrow \frac{dM}{dr} = -\frac{1}{G}\frac{d}{dr}\left(\frac{r^2}{\rho}\frac{dp}{dr}\right) \tag{9.63}$$

and comparing it with the result we obtain from considering the spherical shell in hydrostatic equilibrium

$$dM = 4\pi r^2 \rho dr$$

$$\Rightarrow \frac{dM}{dr} = 4\pi r^2 \rho \tag{9.64}$$

we obtain

$$\frac{dM}{dr} = -\frac{1}{G}\frac{d}{dr}\left(\frac{r^2}{\rho}\frac{dp}{dr}\right) = 4\pi r^2 \rho$$

$$\Rightarrow \frac{1}{r^2}\frac{d}{dr}\left(\frac{r^2}{\rho}\frac{dp}{dr}\right) = -4\pi G\rho \tag{9.65}$$

Putting the value of dp/dr, from Eq. (9.61) into the above equation, we obtain

$$\frac{1}{r^2}\frac{d}{dr}\left(\frac{r^2}{\rho}\kappa\gamma\rho^{\gamma-1}\frac{d\rho}{dr}\right)=-4\pi G\rho \tag{9.66}$$

Defining

$$\rho=\lambda\theta^n$$

$$\gamma=\frac{n+1}{n} \tag{9.67}$$

the Eq. (9.66) becomes

$$\frac{1}{r^2}\frac{d}{dr}\left[\frac{\kappa r^2}{\lambda\theta^n}\frac{n+1}{n}(\lambda\theta^n)^{1/n}\frac{d(\lambda\theta^n)}{dr}\right]=-4\pi G\lambda\theta^n$$

$$\Rightarrow\left[\frac{n+1}{4\pi G}\kappa\lambda^{\frac{1-n}{n}}\right]\frac{1}{r^2}\frac{d}{dr}\left(r^2\frac{d\theta}{dr}\right)=-\theta^n \tag{9.68}$$

Next we introduce a radial variable ξ with a view to making the equation dimensionless

$$\xi\equiv\frac{r}{\alpha}$$

$$\alpha\equiv\sqrt{\frac{n+1}{4\pi G}\kappa\lambda^{\frac{1-n}{n}}} \tag{9.69}$$

Thus, Eq. (9.68), becomes

$$\alpha^2\frac{1}{(\alpha\xi)^2}\frac{d}{d(\alpha\xi)}\left[(\alpha\xi)^2\frac{d\theta}{d(\alpha\xi)}\right]=-\theta^n$$

$$\Rightarrow\frac{1}{\xi^2}\frac{d}{d\xi}\left(\xi^2\frac{d\theta}{d\xi}\right)=-\theta^n \tag{9.70}$$

which is the Lane–Emden equation for polytropes in hydrostatic equilibrium. It is a second order differential equation and requires two boundary conditions in order to seek its precise solution.

These boundary conditions are in the following:

- If the central density $\rho_c=\lambda$, then

$$\rho=\lambda\theta^n\Rightarrow\theta(0)=1 \tag{9.71}$$

- At $r=0$, $\frac{dp}{dr}=-\rho g=-\rho_c g_c=0$, because $g_c=0$, there being no mass inside zero radius.

Thus,

$$\frac{dp}{dr} = \kappa\gamma\rho^{\gamma-1}\frac{d\rho}{dr} \propto \frac{d\theta}{d\xi} \Rightarrow \left.\frac{d\theta}{d\xi}\right|_{\xi=0} = 0 \tag{9.72}$$

In general, the Lane–Emden equation needs to be solved numerically. Only for $n = 0$, 1, and 5 do analytic solutions for θ exist[5].

Let us derive these analytic solutions as it is advisable from both the pedagogical and intuitive standpoint.

9.6.1 Analytic Solution for *n*=0

Substituting $n = 0$ in the Lane–Emden Eq. (9.70), we get

$$\frac{1}{\xi^2}\frac{d}{d\xi}\left(\xi^2\frac{d\theta}{d\xi}\right) = -1$$

$$\Rightarrow \int \frac{d}{d\xi}\left(\xi^2\frac{d\theta}{d\xi}\right)d\xi = -\int \xi^2 d\xi$$

$$\Rightarrow \xi^2\frac{d\theta}{d\xi} = -\frac{1}{3}\xi^3 + c$$

$$\Rightarrow \frac{d\theta}{d\xi} = -\frac{1}{3}\xi + \frac{c_1}{\xi^2} \tag{9.73}$$

Using the boundary conditions, we obtain

$$\left.\frac{d\theta}{d\xi}\right|_{\xi=0} = 0 \Rightarrow c_1 = 0$$

$$\Rightarrow \frac{d\theta}{d\xi} = -\frac{1}{3}\xi \Rightarrow \theta = -\frac{1}{6}\xi^2 + c_2$$

$$\Rightarrow \theta(0) = 1 \Rightarrow c_2 = 1 \Rightarrow \theta_0 = 1 - \frac{1}{6}\xi^2 \tag{9.74}$$

We see that this configuration has a boundary at $\xi = \sqrt{6}$ where $\theta_0 \to 0$.

9.6.2 Analytic Solution for *n*=1

Putting $n = 1.$ in Eq. (9.70), we obtain

$$\frac{1}{\xi^2}\frac{d}{d\xi}\left(\xi^2\frac{d\theta}{d\xi}\right) = -\theta \Rightarrow \frac{d}{d\xi}\left(\xi^2\frac{d\theta}{d\xi}\right) = -\xi^2\theta \tag{9.75}$$

Defining the variable χ

$$\chi(\xi) \equiv \xi\theta(\xi) \Rightarrow \theta = \frac{\chi}{\xi} \tag{9.76}$$

Now
$$\frac{d\theta}{d\xi} = \frac{d}{d\xi}\left(\frac{\chi}{\xi}\right) = \frac{\xi\chi' - \chi}{\xi^2} \tag{9.77}$$

Thus, the Lane–Emden Eq. (9.75), becomes

$$\frac{d}{d\xi}\left(\xi^2 \frac{d\theta}{d\xi}\right) = \frac{d}{d\xi}(\xi\chi' - \chi) = \chi' + \xi\chi'' - \chi' = \xi\chi''$$

$$\Rightarrow \frac{\xi\chi''}{\xi^2} = -\frac{\chi}{\xi} \Rightarrow \chi'' = -\chi \Rightarrow \chi'' + \chi = 0 \tag{9.78}$$

This is the equation of a harmonic oscillator with trial solutions

$$\chi(\xi) = A\sin\xi + B\cos\xi \tag{9.79}$$

or in terms of $\theta = \frac{\chi}{\xi}$, it becomes

$$\theta(\xi) = A\frac{\sin\xi}{\xi} + B\frac{\cos\xi}{\xi} \tag{9.80}$$

Applying the first boundary condition

$$\theta(0) = 1 \Rightarrow B = 0 \text{ since } \lim_{\xi \to 0} \frac{\cos\xi}{\xi} = \infty$$

$$A = 1 \quad \text{since} \quad \lim_{\xi \to 0} \frac{\sin\xi}{\xi} = 1$$

$$\Rightarrow \quad \theta_1(\xi) = \frac{\sin\xi}{\xi} \tag{9.81}$$

The second boundary condition

$$\left.\frac{d\theta}{d\xi}\right|_{\xi=0} = 0$$

is obviously satisfied as shown below. Applying L'Hospital's rule

$$\lim_{\xi\to 0}\frac{\xi\cos\xi - \sin\xi}{\xi^2} = \lim_{\xi\to 0}\frac{-\xi\sin\xi + \cos\xi - \cos\xi}{2\xi} = -\frac{1}{2\xi}\lim_{\xi\to 0}\sin\xi = 0 \tag{9.82}$$

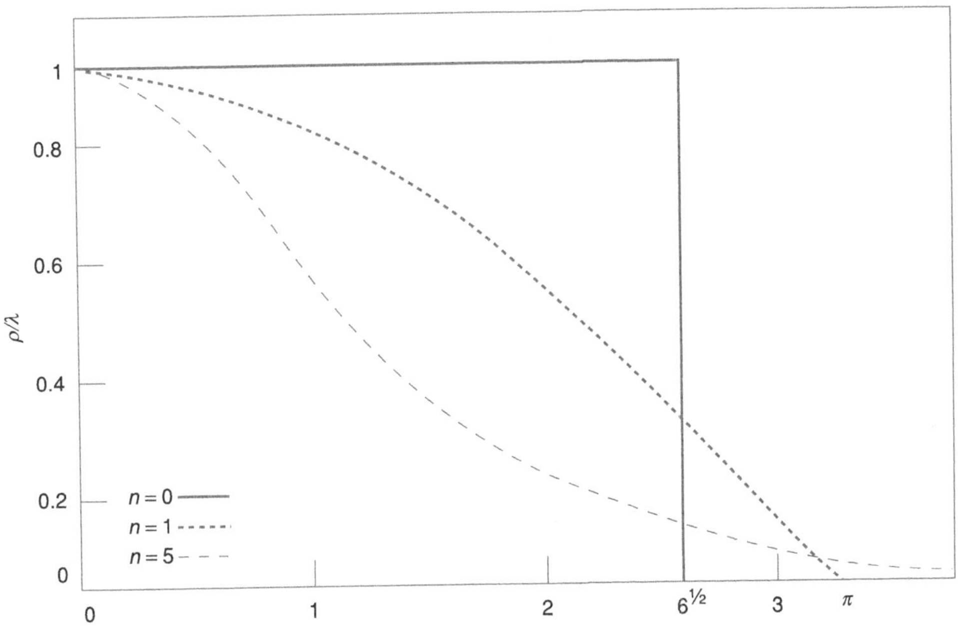

Figure 9.2 *Analytic solutions for the Lane–Emden equation for n=0,1,5.*

This configuration has a boundary at $\xi = \pi$, where $\theta_1 = 0$.

9.6.3 Analytic Solution for *n*=5

The solution of Lane–Emden equation for $n = 5$ is tractable, but quite complicated to integrate. The solution[5] is

$$\theta_5(\xi) = \frac{1}{\sqrt{1+\frac{1}{3}\xi^2}} \tag{9.83}$$

This configuration is unbounded: $\xi \in (0,\infty)$ and $\lim_{\xi\to\infty} \theta_5 = 0$.

The solutions for $n = 0, 1$ and 5 are plotted in Fig. (9.2) below.

9.7 THE CHANDRASHEKHAR MASS LIMIT[5]

Consider a star that as a result of gravitational contraction, has become so dense that it is supported by a completely degenerate, extreme relativistic gas., *i.e.*, $\rho > 10^7\ \text{gcm}^{-3}$. The pressure in terms of density is obtained by combining the equations

$$p = \frac{hc}{8}\left(\frac{3}{\pi}\right)^{1/3} n^{4/3} \tag{9.84}$$

and
$$n = \frac{\rho}{m_H \bar{\mu}} \tag{9.85}$$

Thus, we obtain

$$p = \frac{hc}{8}\left(\frac{3}{\pi}\right)^{1/3}\left(\frac{\rho}{m_H \bar{\mu}}\right)^{4/3}$$
$$= \frac{(6.63\times10^{-27}\,\text{ergs})(3\times10^{10}\,\text{cm/s})}{8}\left(\frac{3}{\pi}\right)^{1/3}\frac{1}{(1.67\times10^{-24}\,\text{g})^{4/3}}\left(\frac{\rho}{\bar{\mu}}\right)^{4/3}$$

$$\Rightarrow \quad p = 1.24\times10^{15}\left(\frac{\rho}{\bar{\mu}}\right)^{4/3} \tag{9.86}$$

This is the equation for a polytrope with $\gamma = 4/3$ and $\kappa = \frac{1.24\times10^{15}}{\bar{\mu}^{4/3}}$. The value of $n = \frac{1}{\gamma-1} = 3$. The mass for this configuration is obtained as follows:

$$M_3 = \int_0^{r_{max}} \rho(r)d^3r$$
$$= 4\pi \int_0^{r_{max}} \rho(r)r^2 dr$$
$$= 4\pi \int_0^{r_{max}} \lambda\theta^3(\alpha\xi)^2 d(\alpha\xi)$$
$$= 4\pi\lambda\alpha^3 \int_0^{\xi_{max}} \left[-\frac{d}{d\xi}\left(\xi^2\frac{d\theta}{d\xi}\right)\right]d\xi$$
$$= 4\pi\lambda\alpha^3 \left[-\xi^2\frac{d\theta}{d\xi}\right]_{\xi_{max}} \tag{9.87}$$

where we made use of Lane–Emden Eq. (9.70). The constant λ, Eq. (9.69), for $n = 3$ becomes

$$\alpha = \sqrt{\frac{n+1}{4\pi G}\kappa\lambda^{\frac{1-n}{n}}} \Rightarrow \alpha = \sqrt{\frac{1}{\pi G}\kappa\lambda^{-2/3}}$$

$$\Rightarrow \quad \lambda\alpha^3 = \lambda\left[\frac{\kappa}{\pi G}\lambda^{-2/3}\right]^{3/2} = \left[\frac{\kappa}{\pi G}\right]^{3/2} \tag{9.88}$$

$[-\xi^2 \frac{d\theta}{d\xi}]_{\xi_{max}}$ term in Eq. (9.87) has been evaluated numerically[6] and is given by

$$\left[-\xi^2\frac{d\theta}{d\xi}\right]_{\xi_{max}} = 2.02 \tag{9.89}$$

Thus, the total mass is

$$M_3 = 4\pi \left[\frac{\dfrac{1\cdot 24\times 10^{15}}{\bar{\mu}_e^{4/3}}}{\pi(6\cdot 6\times 10^{-8})} \right]^{3/2} \times 2.02$$

$$= 4\pi \left[\frac{1.24\times 10^{15}}{\pi(6.6\times 10^{-8})} \right]^{3/2} \times \frac{2.02}{\bar{\mu}_e^{2}}$$

$$= \frac{1\cdot 16\times 10^{34}}{\bar{\mu}_e^{2}} = \frac{1\cdot 16\times 10^{34}}{\bar{\mu}_e^{2}} \frac{\text{Solar Mass}}{1\cdot 99\times 10^{33}}$$

$$\Rightarrow \quad M_3 = \frac{5\cdot 81}{\bar{\mu}_e^{2}} \text{Solar Mass} \tag{9.90}$$

Next we compute $\bar{\mu}_e$ for a star with relativistic matter degeneracy. In such a star, we define the matter density, due essentially to the ions, as $\rho = m_H \mu_e n_e$. In addition, we consider contribution from hydrogen (H), helium (He) and elements with atomic weight greater than 4(Z). Thus, from the definition in Eq. (9.34), we get

$$\frac{1}{\bar{\mu}_e} = \sum \frac{m_H}{m_e} \bar{n}_i^e$$

$$= \frac{m_H}{m_e} \left[\frac{\rho_H^e}{\rho} + \frac{\rho_{He}^e}{\rho} + \frac{\rho_Z^e}{\rho} \right]$$

$$= \frac{m_H n_H}{\rho} + \frac{2\dfrac{m_H}{m_{He}} m_{He} n_{He}}{\rho} + \frac{\dfrac{A}{2}\dfrac{m_H}{m_Z} m_Z n_Z}{\rho}$$

$$= \frac{\rho_H}{\rho} + \frac{2}{4}\frac{\rho_{He}}{\rho} + \frac{A}{2A}\frac{\rho_Z}{\rho} \tag{9.91}$$

According to conservation of mass condition, we have

$$X + Y + Z = 1 \quad \Rightarrow \quad Z = 1 - X - Y \tag{9.92}$$

Thus,

$$\frac{1}{\bar{\mu}_e} = X + \frac{1}{2}Y + \frac{1}{2}(1 - X - Y) = \frac{1}{2}X + \frac{1}{2} = \frac{1+X}{2}$$

$$\Rightarrow \quad \frac{1}{\bar{\mu}_e} = \frac{1+X}{2} \quad \Rightarrow \quad \bar{\mu}_e = \frac{2}{1+X} \tag{9.93}$$

It is plausible to assume that the stars undergoing extreme relativistic degeneracy are highly evolved and are near the end of their life cycle. Thus, these stars have burned up most of their hydrogen fuel, so

$$X \approx 0 \quad \Rightarrow \quad \bar{\mu}_e = 2 \tag{9.94}$$

Lastly combining Eq. (9.94) with Eq. (9.90), we obtain the Chandrasekhar mass limit

$$M_{Ch} = \frac{5.81}{\bar{\mu}_e^{\,2}} \text{ Solar Mass} = \frac{5.81}{2^2} \text{ Solar Mass}$$

$$\Rightarrow \quad M_{Ch} = 1.45 \text{ Solar Mass} \tag{9.95}$$

Masses larger than this cannot be supported by electron degeneracy pressure against self gravity. Such configurations if devoid of thermal support, must then collapse to higher density into a neutron star or a black hole[6].

9.8 FORMATION OF WHITE DWARFS, NEUTRON STARS AND BLACK HOLES $\rho > 10^7\,\text{gcm}^{-3}$

As a result of supernova explosions, stars are created by the coalescence of hydrogen in gaseous nebulae, which get highly compressed and heated through the gravitational interaction. At a temperature of 10^7K, nucleosynthesis of hydrogen into helium starts and a lot of electromagnetic radiation is released. Helium accumulates at the centre of the star and gets compressed and heated enough so as to start the nuclear fusion of helium into heavier elements. Till this stage, the star is held in 'near equilibrium' by the opposing force of gravity, which compresses the star and the outward pressure exerted by the electromagnetic radiation produced in the nuclear reaction. As the star burns hotter and synthesizes heavier elements, which go on accumulating in the core, the outward pressure of electromagnetic radiation becomes progressively less effective in counterbalancing the force of gravity. The nuclear fusion reactions are exothermic till iron, so after iron, the temperature is insufficient to produce heavier elements and the star has consumed its nuclear fuel. Without the countervailing outward pressure of electromagnetic radiation, one would expect that the core will become unstable and start contracting, but another mechanism sets in to counteract the gravitational collapse. It is this force that retains the balance against the force of gravity and as a consequence leads to the various stellar compact objects: white dwarfs, neutron stars and ultimately the black holes.

The bulk properties of the metal, such as heat capacity, compressibility and thermal conductivity are the properties of the electron gas and not of the crystalline lattice of positive ions of the metal. Let us calculate the bulk modulus of a metal and see if such a simple model of a metal (or a star) explains the experimental results. We invoke two principles from quantum mechanics. These are as follows:

- Pauli Exclusion Principle
- Heisenberg Uncertainty Principle

According to Pauli Exclusion Principle, no two electrons have the same set of quantum numbers or be in the same quantum state. For our purpose, it implies that two electrons cannot be at the same point in space.

According to Heisenberg uncertainty principle, the uncertainties in the simultaneous measurement of two canonically conjugate quantities like position (Δx) are related to each other as

$$\Delta x \Delta p \geq \hbar / 2 \tag{9.96}$$

where $\hbar = 1.05 \times 10^{-34}$ Joule-sec, is the rationalized Planck constant.

We will examine how these principles act together to give the order of magnitude of the bulk modulus of the metal correctly. Let us imagine that there are N electrons in a box of volume R^3, Fig. (9.3). Thus, each electron will have R^3/N space and the interparticle spacing is $a_0 = R / N^{1/3}$ (neglecting the e-e interactions and temperature). As the electrons are spatially confined within a region of linear size, a_0, the uncertainty in momentum $\Delta p \approx \hbar / a_0$. Δp^2 is the variance of a large set of measurements of momentum, so

$$\begin{aligned} \Delta p^2 &= \langle (p - \langle p \rangle)^2 \rangle \\ &= \langle p^2 \rangle - \langle p \rangle^2 \end{aligned} \tag{9.97}$$

Obviously, the average value of p^2 must be greater than or equal to Δp^2 .

Next we calculate the energy of an electron gas. The kinetic energy of a particle of mass m and speed v is

$$\varepsilon = \frac{1}{2} m v^2 = p^2 / 2m \tag{9.98}$$

Taking the minimum value of momentum $p^2 = \frac{\hbar^2}{a_0^2}$, we get for the energy of a single electron

$$\varepsilon = \frac{\hbar^2}{m_e a_0^2} \tag{9.99}$$

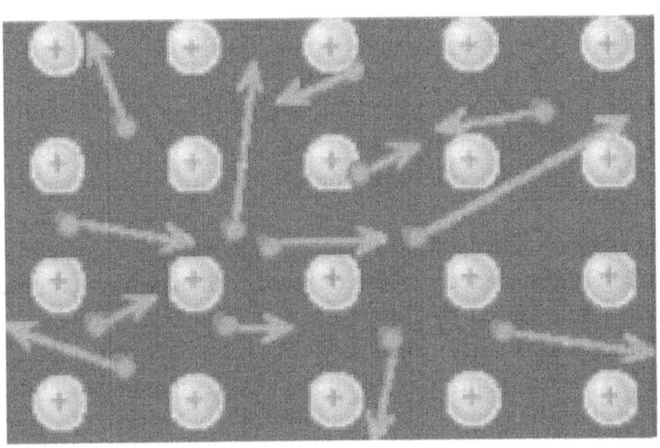

Figure 9.3 *A simple model of a metal.*

Thus the total kinetic energy of N electrons is

$$E_e = N\varepsilon$$

$$= \frac{\hbar^2}{m_e a_0^2} N = \frac{\hbar^2}{m_e R^2} N^{5/3} \tag{9.100}$$

As R decreases, the energy E_e increases implying that there is an effective repulsive force resisting compression, although we have neglected the *e-e* interactions. The cause of the repulsive force is the uncertainty principle.

Further, we calculate the compressibility of a metal. Let us imagine that an applied pressure P causes a decrease ΔV in volume V of the material.

The bulk modulus B is given by

$$B = \frac{\text{Stress}}{\text{Strain}}$$

$$= -V \frac{\partial P}{\partial V} \tag{9.101}$$

The outward pressure resisted by the electron gas is

$$P = \frac{F}{A}$$

$$= -\frac{1}{A}\frac{\partial E_e}{\partial R} = 2\frac{\hbar^2}{m_e}\left(\frac{N}{V}\right)^{5/3} \tag{9.102}$$

Thus,

$$B = -V \frac{\partial P}{\partial V}$$

$$= \frac{10}{3}\frac{\hbar^2}{m_e}\left(\frac{N}{V}\right)^{5/3} \approx 10^{-10} - 10^{-11}\,\text{N/m}^2 \tag{9.103}$$

The volume per electron has been taken to be $1\,\text{nm}^3$. The experimental value for steel and aluminium are $B_{\text{steel}} \approx 6\times10^{-10}\,\text{N/m}^2$ and obviously, we infer that Heisenberg principle is the cause for the rigidity of metals. Next we will show that the same principle is responsible for the prevention of the collapse of stars under their gravity.

Let us now consider the case of compact objects. For a stable equilibrium in a star, it is imperative that there exists some force that counterbalances the attractive force of gravity. In massive stars, such a force is provided by the thermally excited atoms emitting electromagnetic radiation. However, in the case of a white dwarf, the countervailing force originates on account of uncertainty principle. The elements constituting the star, mostly iron, exist in an ionized state due to high temperatures. Thus, a star can be considered as a gas of positively charged atomic nuclei and negatively charged electrons,

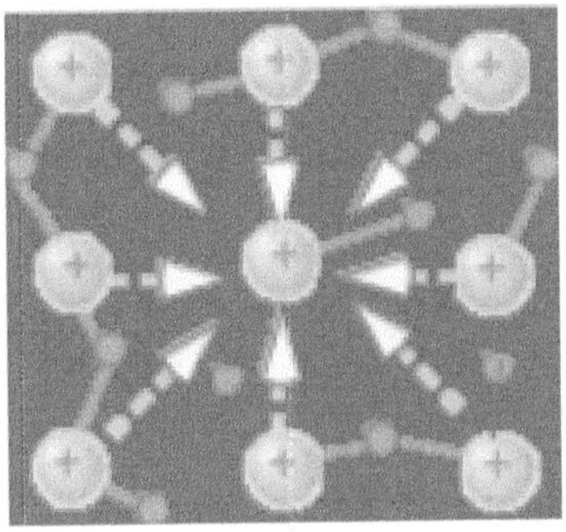

Figure 9.4 *A star consisting of positively charged atomic nuclei and electrons.*

Fig. (9.4). As an atomic nucleus is a few thousand times heavier than the electrons, the compressive force of gravity is provided by the nuclei. However, the electrons are strongly electrostatically attached to the positive nuclei and occupy the same space as nuclei. The electrons are pushed apart due to the uncertainty principle, thus countervailing the compressive force of gravity.

Let us evaluate this force of repulsion provided by the electronic cloud. A sphere of mass M and radius R has potential energy

$$E_g = -\frac{GM^2}{R} \tag{9.104}$$

where G is the gravitational constant $= 7\times10^{-11}\,\mathrm{Nm^2/kg^2}$. The negative sign is there as the force of gravity is attractive. With decreasing R, the energy decreases and becomes more negative.

The kinetic energy of electrons as calculated earlier is

$$E_e = \frac{\hbar^2 N^{5/3}}{m_e R^2} \tag{9.105}$$

We express the gravitational energy in terms of N. The mass of the star is approximately the collective mass of the nucleons. For light elements, the number of nucleons is roughly double the number of electrons. Assuming that for the heavier elements constituting the star, the average number of nucleons per electron is μ, we get the expression for mass of the star as

$$M = \mu m_n N \tag{9.106}$$

where m_n is the mass of the nucleon.

The total energy E is given by

$$\begin{aligned} E &= E_e + E_g \\ &= \frac{\hbar^2 N^{5/3}}{m_e R^2} - \frac{G\mu^2 m_n^2 N^2}{R} \end{aligned} \tag{9.107}$$

Plotting E as a function of R, Fig. (9.5), we see that the energy goes through a minimum for a value R_0, where the force $F = -\partial E/\partial R$ is zero implying that the star is in a state of thermal equilibrium. The value of R_0 is given by

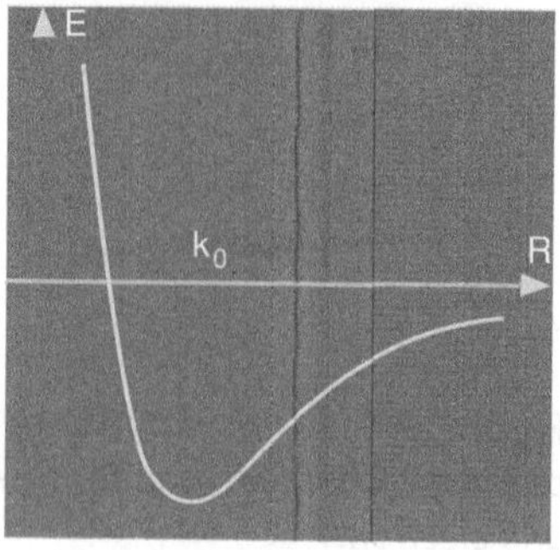

Figure 9.5 *Energy versus R.*

$$R_0 = \frac{\hbar^2 N^{-1/3}}{G m_e \mu^2 m_n^2} \approx 10^7 \text{m} \tag{9.108}$$

where we used, $N = 10^{57}$, which is a reasonable value for a star like our sun. R_0 relates to a star that is a little bigger than earth and is a reasonable estimate for a white star. One can calculate the mean density ρ, assuming the radius R and it turns out that $\rho = 10^9 \text{ kg/m}^3 = 10^5 \times$ densityof steel. Due to this high density, the electrons are much closer to the nucleus than the ordinary matter. However, in the case of extremely massive stars, the huge compressive force of gravity, will force the electrons to have relativistic kinetic energy and enter the extreme relativistic region.

The energy of a relativistic particle is

$$\varepsilon^2 = c^2 p^2 + m^2 c^4$$

or

$$\varepsilon = \sqrt{c^2 p^2 + m^2 c^4}$$

$$= mc^2 + \frac{p^2}{2m} + \text{terms} \geq \left(\frac{p}{mc}\right)^4 \tag{9.109}$$

Assuming an extreme relativistic case, *i.e.*, $p = mc$ or equivalently $p^2 / m = mc^2$, the higher terms cannot be neglected. The different form for the energy of the electron (linear rather than quadratic in p) will have a special consequence for the stability equation for R_0.

According to uncertainty principle, the momentum of an electron within the star is

$$p = \frac{\hbar}{a_0} = \frac{\hbar N^{1/3}}{R} \tag{9.110}$$

The total energy for the electron due to uncertainty principle, is

$$E = N\varepsilon = Npc \approx \frac{\hbar c N^{4/3}}{R} \tag{9.111}$$

The total energy due to both the processes, becomes

$$E = E_e + E_g$$

$$\approx \frac{\hbar c N^{4/3}}{R} - \frac{G \mu^2 m_n^2 N^2}{R} \tag{9.112}$$

The energy $E(R)$ has a completely different behaviour than the relativistic case. The force $F = -\partial E / \partial R$ is just equal to E/R. Plotting $F = -\partial E / \partial R$ as a function of R, Fig. (9.6), we see that if $N > N_c = 1/\mu^3(\hbar c/(Gm_n^2))^{3/2}$, the total energy is positive and the force will cause expansion. On the other hand, if $N < N_c$, the total energy is negative and the force will lead to contraction, unless some other force intervenes.

The critical value of N for which the energy becomes negative is

$$N_c = \frac{1}{\mu^3}\left(\frac{\hbar c}{Gm_n^2}\right)^{3/2} \tag{9.113}$$

Expressing the expression for N_c in terms of critical mass for the star, M_c we get

$$\begin{aligned} M_c &= \mu N_c m_n \\ &= \frac{1}{\mu^2 m_n^2}\left(\frac{\hbar c}{G}\right)^{3/2} \end{aligned} \tag{9.115}$$

If $M > M_c$, the star will continue to collapse and the electrons will be pushed closer and closer to the nucleus, Fig. (9.6). At some stage, a nuclear reaction begins to take place and an electron combines with the proton giving a neutron and a neutrino. A fairly massive star is unstable against such an interaction till all the electrons and protons are converted into neutrons and ultimately leaving behind a neutron star, which is chargeless and non-luminous.

The nuclear force between the neutrons and protons is attractive and one wonders what prevents further collapse of the neutron star. It is again the Pauli Exclusion Principle as in the case of electrons. When neutrons are confined, they avoid each other and acquire large kinetic energy due to uncertainty principle. If the neutrons are non-relativistic, the same calculation for the radius R_0 will hold with the only difference that m_e is replaced by m_n. This will reduce the radius R_0 of the neutron star by a factor of ~2000, the ratio of m_n to m_e.

However, if the neutron star is massive so as to make neutrons relativistic, the further collapse of the neutron star to the end point stellar black hole, is possible if the total energy is negative like the

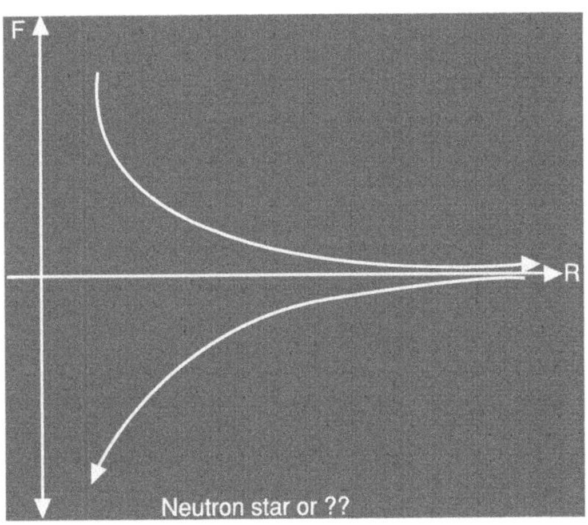

Figure 9.6 *The collapse of the star.*

case of white dwarf. The expression for the critical mass M_c can be adapted to neutrons by putting $\mu = 1$. Since $\mu = 2$ for a white dwarf, one can infer that a star four times more massive than a neutron star is liable to continued collapse.

References

1. Oppenheimer, J.R. and Volkoff, G.M. (1939). *Phys. Rev.*, 55(4), 374–381.
2. Tolman, R.C. (1939). Proceedings of the National Academy of Sciences, 20(3), 169–176.
3. Tolman, R.C. (1939). *Phys. Rev.*, 55(4), 364–373.
4. Bombaci, I. (1996). *Astronomy and Astrophysics,* 305, 871–877.
5. Chandrasekhar, S. (1939). An Introduction to the Study of Stellar Structure, University of Chicago Press, Chicago, pp. 93–94.
6. Bowers, R. and Deeming, T (1984). Table 4.2 of *Astrophysics I: Stars*, Jones and Bartlett Publishers, Potola Valley.

Problems

9.1 Solve the stellar structure equations for a star having uniform density $(\rho = \text{constant})$ and show that the mass is limited by the inequality

$$M < \frac{4}{3}(3\pi\rho)^{-1/2}$$

9.2 Consider a star of uniform density. Show that the equilibrium is only stable if

$$\Gamma = -\frac{V}{P}\frac{\partial P}{\partial V} > \frac{4}{3}$$

where V is the volume of the star and P, the pressure.

Epilogue

In 1955, at the time of Einstein's death, scientists still had almost no evidence of general relativity in action. However, that changed dramatically in the 1960s, when astronomers began to discover neutron stars, white dwarfs and black holes—that due to their massive bodies could curve severely the shape of space-time. Furthermore, currently general relativity is so well understood that it is used to weigh galaxies and locate distant planets.

An attempt is made to make a brief mention of all those high-flying experiments that are either currently on-going or are at the planning stage, and whose results finally may either confirm (or contradict) Einstein's most important predictions in special and general relativity.

Project: **Fermi Gamma-Ray Space Telescope**
Prediction: Value of c is a universal constant.
Start date: 2008 energy

The speed of light is in vacuum and its constancy is implied by a postulate of the special theory of relativity. The Fermi Gamma-Ray telescope will be used to observe light from gamma-ray bursts, which are distant, short-lived explosions. As all the light starts at the same time, it should reach the telescope at the same time. Some theories predict that light may move slightly faster or more slowly depending on its energy. Thus, the journey, which started billions of years ago may or may not, end in a difference of a few milliseconds.

Project: **Pierre Auger Observatory**
Prediction: Special relativity saps the energy of the cosmic rays reaching earth.
Start date: Ongoing

Cosmic rays are ultrarelativistic particles coming from distant galaxies, originated millions of light-years away. According to special relativity, they should slow significantly by the time they reach earth. Pierre Auger Observatory, in Argentina, a network of 1600 fluorescence detectors, tracks the energy of these particles once these enter the earth's atmosphere. They hope to pinpoint where these cosmic rays come from and why some of the particles have so much energy. A recent study suggested they could originate in giant jets shooting out from supermassive black holes.

Project: **Gravity Probe B (GP-B)**
Prediction: Objects warp space-time.
Start date: 2004

It was a joint project of NASA and Stanford University, launched in 2004, to measure the following two key predictions of Einstein's theory of general relativity, by monitoring the orientations of ultrasensitive near-perfect niobium-coated quartz gyroscopes relative to a distant star.

1. Geodesic effect that earth's mass warps its local space-time. This has been confirmed successfully.
2. Frame dragging effect in which earth drags space-time with it slightly as our planet rotates. The latest data indicates clear observations of frame dragging.

Project: **Laser interferometer Gravitational-wave Observatory (LIGO)**
Prediction: The changing curvature of space-time produces gravitational waves.
Start date: Ongoing

General theory of relativity holds that any massive object curves the space-time around it and as a consequence creates the force of gravity. When large objects move suddenly, as in a supernova explosion, the curvature should change drastically producing ripples in space-time, called gravitational waves. LIGO is looking for evidence of these waves by searching for small variations in the path of laser light as it bounces back and forth inside two and a half mile-long arm of the observatory's two detectors. LIGO at present is the most sensitive experiment designed to detect gravitational waves. So far (as of December 2007), no detections have been reported, but a second observatory, called 'Advanced LIGO' will have an event rate at 100 times that of LIGO, hopefully recording several events per year. Its objectives are as follows:

1. Prove by direct observation that gravitational waves exist.
2. Ascertain that gravitational waves propagate at the velocity of light as postulated.
3. Show that gravitational waves cause the predicted displacement in the matter they travel through.
4. Establish the presence of black holes and study their dynamics.
5. Study the reverberations of cataclysmic cosmic events.

Project: **Large Hadron Collider (LHC)**
(Lack of) Prediction: Creating a theory of everything.
Start date: 2009

Einstein was obsessed till his end with the creation of a unified theory, encompassing all the four forces of nature, but he could not succeed in formulating it. Finding Higgs particle, the particle, which is supposed to lend mass to different particles, will reinforce the standard model of particle physics. LHC, the world's most powerful accelerator, is expected to find quark-gluon plasma, an energetic state of matter. It may find particles not known before and reveal extra dimensions as postulated by some variants of string theories.

Project: **Satellite Test of Equivalence Principle (STEP)**
Projection: Gravity affects all objects the same way.
Start date: 2012

Some variants of string theory predict the existence of a very weak force that would make gravity slightly different depending on object's composition.

One test mission, called the Satellite Test of the Equivalence Principle (STEP) is being developed by Stanford. University and an international team of collaborators. Its design uses four pairs of test masses instead of just one pair. It is to ensure that any deviation seen in how the test masses fall, is truly caused by a violation of the equivalence principle, not by any other disturbance or imperfections in the hardware. STEP will be able to detect a deviation as small as one part in a million trillion (10^{18}) and that is 10^5 times more sensitive than any current measurement.

Another satellite-based experiment, the French developed Micro-Satellite ă trainee Compensee pour l'observation du Principle d' Equivalence (MICROSCOPE), is scheduled to launch in 2010. MICROSCOPE will have two cylinders, one made of titanium and the other of platinum in perfect orbit in a satellite. It is capable of detecting a violation of the equivalence principle as small as one part in a million billion (10^{15}).

The third experiment is the Italian satellite Galileo Galilei (GG), which operates in much the same way as the. STEP and MICROSCOPE, except that it uses one pair of test masses. It will have a sensitivity of one part in a hundred million billion (10^{17}).

Project: **Joint Dark Energy Mission (JDEM)**
Prediction: The cosmological constant explains the accelerating expansion of the universe.
Start date: 2016

Dark energy was discovered in 1998 by Department of Energy (DOE) and NASA funded scientists working at the Lawrence Berkeley National Laboratory and other institutions. Dark energy is the name given to the cause of the accelerating expansion of the universe and comprises nearly 74% of the total mass and energy of the universe, yet what it is and how it behaves is a complete scientific mystery.

The objectives of the Joint Dark Energy Mission (JDEM) are to explore the properties of dark energy and measure how cosmic expansion has changed over time. The JDEM Observatory is comprised of a large optical/near infrared wide field telescope operating in space, allowing the study of the cosmic acceleration using multiple techniques. Dark energy was discovered by observing explosions called Type 1a supernova. By observing Type 1a supernova, one can map out their distance and the cosmic expansion.

JDEM will run another Einstein-inspired test: a study of gravitational lensing, which measures the bending of light to determine the distribution of mass in the universe.

The Wide-Field Infrared Survey Telescope (WFIRST) is a NASA observatory designed to settle essential questions in both exoplanet and dark energy research and will advance topics ranging from galaxy formation to the study of objects within our own galaxy.

Project: **International X-Ray Observatory (IXO)**
Prediction: Deformation of space-time by massive objects.
Start date: 2020

The International X-ray Observatory (ISO) is a new X-ray telescope with joint participation from NASA, the Europian Space Agency (ESA) and Japan's Aerospace Exploration Agency (JAXA). It is far more sensitive than the current Chandra X-ray Observatory. IXO will examine the most turbulent areas of the universe to pin down the effects of general relativity. IXO scientists expect to see a drastic warping of space-time and a confirmation of general relativity where its influence over cosmic events is most powerful.

Focus areas for IXO

1. Black Holes
2. Galaxies
3. Dark Matter, Dark energy

Big questions for IXO

1. How do planets, stars, galaxies and cosmic structure come into being?
2. What are the origin, the evolution and fate of the universe?
3. When and how did the elements of life in the universe arise?

Project: **Laser Interferometer Space Antenna (LISA)**
Projection: Detection of gravitational waves.
Start Date: 2020

LISA will consist of three identical spacecraft whose positions mark the vertices of an equilateral triangle, five million km on a side, in orbit around the sun. LISA can be thought of as a giant Michelson interferometer in space. The spacecraft separation sets the range of gravitational wave frequencies, LISA can observe (from 0.03 millihertz to above 0.1 hertz).

The natural free-fall orbits of the three spacecrafts around the sun maintain the triangular formation with the triangle appearing to rotate about the centre once per year. LISA will be able to detect the influence of a gravitational wave that causes a change in distance between the spacecraft on the scale of picometers (10^{-12} metre). LISA will be free of interference from earth's motion, so it should be able to detect the subtle ripples of space-time.

Bibliography

This is a list of useful works. References to the original papers on topics of specialized interest are given in references at the end of each chapter.

Introductory Texts on General Relativity

1. R.A. D'Inverno, *Introducing Einstein's Relativity*, Oxford University Press, Oxford, 1992.
2. J.B. Hartle, *Gravity: An Introduction to Einstein's General Relativity*, Addison-Wesley, New York, 2002.
3. L. Hughston, and K.P. Tod, *Introduction to General Relativity,* Cambridge University Press, Cambridge, 1991.
4. H. Stephani, *General Relativity: An Introduction to the Theory of Gravitational Field,* Cambridge University Press, 1990.
5. Paul A. Dirac, *General Theory of Relativity,* Princeton University Press, 1996.
6. Richard C. Tolman, *Relativity, Thermodynamics and Cosmology,* Dover, 1987 (Reprint).
7. Richard P. Feynman, *Lectures on Gravitation,* Addison-Wesley, Longman, 1995.
8. Bernard F. Schutz, *Afirst Course in General Relativity,* Cambridge University Press, Cambridge, 1985.
9. A. Einstein, *Meaning of Relativity*, 5th ed. Princeton, 1953.
10. A. Einstein. *Relativity: The Special and General Theory.* Methuen, London, 1940.
11. J. V. Narlikar, *Lectures on General Relativity and Cosmology,* Macmillan, India, 1978.
12. S. K. Srivastava, *General Relativity and Cosmology,* Prentice-Hall of India, 2008.

More Comprehensive Texts on General Relativity

1. C.W. Misner, K.S. Thorne and J.A. Wheeler, *Gravitation,* W.H. Freeman, New York, 1977.
2. R.M. Wald, *General Relativity,* University of Chicago Press, Chicago, 1984.
3. Stephan W. Hawking and G. F. Ellis, *The Large Scale Structure of Space-Time,* Cambridge University Press, 1975.
4. Stuart L. Shapiro and Saul A. Teukolsky, *Black Holes, White Dwarfs and Neutron Stars: The Physics of Compact Objects,* John Wiley & Sons, 1983.
5. Steven Weinberg, *Gravitation and Cosmology: Principles and Applications of the General Theory of Relativity,* John Wiley & Sons, 1972.
6. S. Chandrasekhar, *Selected Paperson Relativistic Astrophysics*, University of Chicago Press, 1990.
7. A. Lightman and R.H. Price, *Problem Book on Relativity & Gravitation,* Princeton University Press, Princeton, 1975.
8. L. Landau and E. Lifshitz, *The Classical Theory of Fields,* Addison-Wesley, Cambridge, Mass., 1975.

Philosophy and Relativity Theory

1. Lawrence Sklar, *Space, Time and Spacetime,* University of California Press, 1974. This book won a prize for the exceptionally clear quality of its exposition.
2. Hans Reichenback, *Philosophy of Space and Time,* Dover, 1998. A classic book clearly written and highly influential.
3. Michael Friedman, *Foundations of Space-Time Theories. Relativistic Physics and Philosophy of Science,* Princeton University Press, 1983.

Semi-Popular Books

1. Kip S. Thorne, *Blackholes & Time Warps: Einstein's Outrageous Legacy,* W.W. Norton, 1995.
2. Steven Weinberg, *The First Three Minutes: A Modern View of the Origin of the Universe*, Basic Books, 1988.
3. Robert M. Wald, *Space, Time and Gravity: The Theory of the Big Bang and Black holes,* University of Chicago Press, 1992.
4. Clifford M. Will, *Was Einstein Right? Putting General Relativity to the Test*, 2nd ed, Basic Books, 1993.

Index

NOTES

NOTES

NOTES

NOTES